Climates

Climates: Architecture and the Planetary Imaginary

James Graham
Editor

Caitlin Blanchfield
Managing Editor

Alissa Anderson
Jordan H. Carver
Jacob Moore
Contributing Editors

Columbia Books on
Architecture and the City

Lars Müller Publishers

The Avery Review
www.averyreview.com

Political Ecologies

Corporealities

Enclosures

Climatic Imaginaries

THE EDITORS

Consider the footprint—a metaphor that has long described the interface between architecture and the earth on which it sits. If the poetics of the footprint began with the idea of humans leaving their marks on the "sands of history," for architecture it took on the literal meaning of a foundation's outline or a demarcation of property. Only far more recently has the metaphor begun to refer to architecture's relationship to the *Earth* on which it sits—the way an individual or a building or a city occupies a share of finite planetary resources. The 1960s and '70s ushered in a broader sense of environmental awareness for the discipline, though it took until the 1990s for "ecological footprint" to become a commonly used term (aligning, not coincidentally, with the rise of "sustainability" as an architectural imperative).[1] In this understanding of its footprint, the outline of a building is redrawn to include economies of material extraction and the vast amounts of energy—embodied and operational—that give it physical form and allow it to be occupied in comfort. The now ubiquitous "carbon footprint" is an almost exclusively twenty-first-century concept, aligning with a general public acceptance of the realities of global warming and an accompanying acknowledgment of our individual participation (as citizens of the planet and as designers or builders) in the expenditure of the carbon-based energy so central to our lives and patterns of habitation.[2]

But as Jason Moore's recent *Capitalism in the Web of Life* asks—is this image the best we can conjure? The footprint metaphor relies on an "image of nature as passive mud and dirt" upon which we step; it upholds a dualism of "nature" and "society" as discrete spheres of action.[3] Moore asks us instead to consider what he calls the *co-production* of nature and society, to replace the footprint with a web of mutual imbrication.

The expiration of this division of the human from the natural (originating in antiquity and continually affirmed in Romantic and Modernist thought) speaks to the geological epoch of the Anthropocene, in which humankind has assumed an ecological agency of planetary scale through our transformation of the Earth's environment. Distributing thin but persistent layers of carbon and plastic across

1

The term "ecological footprint" was coined in the early 1990s and made its official debut in Mathis Wackernagel and William E. Rees, *Our Ecological Footprint: Reducing Human Impact on the Earth* (Philadelphia: New Society Publishers, 1996). Interestingly, the metaphor stemmed in part from considering another footprint metaphor—the "footprint" of their new desktop computers on their desks. The footprint concept would soon percolate into architecture, perhaps most famously in the writing of William McDonough, who proposed that we might "create ecological footprints to delight in, not lament." See William McDonough, *Cradle to Cradle: Remaking the Way We Make Things* (New York: North Point Press, 2002).

2

On this, see Timothy Mitchell, *Carbon Democracy: Political Power in the Age of Oil* (New York: Verso Books, 2011).

3

Jason W. Moore, *Capitalism in the Web of Life: Ecology and the Accumulation of Capital* (New York: Verso Books, 2015), 6.

atmospheres, oceans, and landscapes on which we have only seldom, if ever, left literal footprints, the notion that humankind's ecological effects are confined to our immediate environs is now acknowledged as untenable. Moreover, Moore's line of thought reminds us that "nature" is, in a very real sense, what we see in it. It is historical, epistemic. Our *conception* of things like air, ocean, rock, ice, and weather condition our engagement with them. This could mean many things for the field of architecture; perhaps most central, for this book, is the recognition that our environment is not just a resource to be managed or an externality to which we must adapt but one of the chief figurations of shared or contested cultural values.

This book arises from a conviction that there is more to say about climate than the discourses of green or clean energy, eco-friendliness, resilience, or adaptation allow us to consider. These are vital fields of research and should be fundamental to the contemporary practice of architecture and urbanism (not to mention politics, government, and daily life), but climate registers on many levels within our lived realities. The essays contained here expand an inherited view of the encounter between humans, buildings, and the planet that has become so naturalized—so *environmental*—that it takes some effort to dislodge it from the (nonetheless highly urgent) realms of governmental negotiation, dire calculation, and pragmatic problem solving. Loosening up, undermining, and otherwise challenging any singular or inherent notion of climate, these authors begin to show how climate change discourse can actually begin to rearticulate our definitions of environment, and, moreover, how it can open up our willingness to see and shape the historical and cultural frames that set those definitions in the first place. In other words, by recognizing the plural "climates" that humans have constructed and instrumentalized for various ends, they show us that the relationships between people and the built and natural environments are limited only by our imaginations.

Talk of "climatic imaginaries" might seem problematic to some readers— global warming is very real, nothing imaginary about it. As Bruno Latour wrote in his canonical essay "Has Critique Run Out of Steam?" the language of social construction in science—meant to help uncover the implicit biases of the apparatuses of scientific knowledge, to help reveal ideological assumptions—has come to be harnessed by "dangerous extremists … to destroy hard-won evidence that could save our lives."[4] Seen in this light, the notion of climate as cultural construct might undermine the more urgent work of galvanizing action, right now, and positing solutions. Furthermore, as one earth scientist who declined to contribute to this volume put it, such talk can simply seem *impractical*, an academic exercise that evades more elemental responsibilities for design.

Our view is that these seemingly separate modes of operation in fact rely on one another to produce the most meaningful results. Design, and architecture in particular, always functions well beyond its most traditionally and narrowly construed role as a social or corrective technology. In many cases, architecture redefines the problems themselves. It educates. It calls to action and gathers publics. It does so messily, always burdened by its complicity in the expenditure of energy and material. But it asks questions of those with whom it engages that might not otherwise be asked. And in doing so, architecture fills a critical void in times of crisis. Architecture, seen from a certain vantage point, has always addressed the imaginary, the abstract, and attempted to concretize social, cultural, and historical aspirations into solid forms. In this sense, the discipline and its material traces are

4

Bruno Latour, "Why Has Critique Run Out of Steam? From Matters of Fact to Matters of Concern," *Critical Inquiry*, vol. 30, no. 2 (Winter 2004): 227.

uniquely positioned to propose new and novel means for asking and answering questions pertaining to the imaginary and the real, where our climate is concerned.

The past decades of architectural historiography have been marked by a turn toward "the global," whether as a means of questioning the globalizing tendencies of multinational corporatism or questioning the forms of violence that undergird familiar hegemonic narratives of European and American modernism. This much needed decentering of our field's self-understanding is now being joined by a turn toward "the planetary"—the notion that architecture needs to think about the Earth not only as a host to cultural diversity but as a host to life itself. One could think of Martin Heidegger's notion that the "fundamental event of modernity is the conquest of the world as picture"—that is, that humankind had become a *subject* of study (among others), with the globe itself enframed by the calculating nature of modern science.[5] This view encompasses literal world pictures, the desire for the representation of some kind of whole, as well as conceptual ones, in that they attempt to freeze the world and its population at a certain moment in time and within certain mediating forms of knowledge and understanding. The globe, like climate, is historical.

Inescapable in all of this is the question of capitalism—or, rather, how architecture participates in multivalent and transnational systems of labor, resource extraction, and wealth accumulation. Moore argues that "Capitalism is not an economic system; it is not a social system; it is *a way of organizing nature.*" It is a "world-ecology, joining the accumulation of capital, the pursuit of power, and the co-production of nature in dialectical unity."[6] Architecture has long been implicated as a vehicle for capitalist expansion and the ills it produces. Yet Moore's definition could easily be used to describe the discipline, and even purpose, of architecture as both a profession and a means of ordering the material world, and this helps us perceive our entwinement in this "co-production" in subtler and more potentially political ways. Working from this notion allows for refocusing the agency of architecture to make more direct interventions into capitalism's more negative effects.

After all, architecture's planetary imaginary has long been political. The nineteenth century saw the flourishing of Saint-Simonian thought, which envisioned a new form of political economy that redefined the globe itself as an integrated and *organized* entity. The rise of glass architecture, most notably the Crystal Palace of the 1851 Great Exhibition in London—since diagnosed by Peter Sloterdijk as a harbinger of the "world interior of capital"—was, importantly, a climatic architecture, drawing on Joseph Paxton's knowledge of horticulture and positing something of a planet-in-miniature. In the early decades of the twentieth century, "the planetary" was a central concern of radical social movements, notably the Russian avant-gardes. Kazimir Malevich's "planets" combined the overthrow of global capitalism with the conquest of gravity itself, imagining floating Suprematist compositions—occupied by a new socialist society—that revealed the world-making of capitalism by contradistinction. This outlook on the planetary was frequently climatic. In Nikolai Fedorov's posthumously printed *Philosophy of the Common Task* of 1906, the newly collectivized utopia he envisioned grew literally out of attempts to control the weather by seeding clouds with magnesium as well as surrounding the planet with a network of "electric rings," strategically altering its milieu rather than containing it: "Regulation, the control of the blind force of nature, can and must become the great task common to us

5

Martin Heidegger, "The Age of the World Picture" [1938], in *Off the Beaten Track*, ed. and trans. Julian Young and Kenneth Haynes (Cambridge: Cambridge University Press, 2002), 71.

6

Moore, *Capitalism in the Web of Life*, 2–3.

Plate from *Pro dva kvadrata [About Two Squares]*, El Lissitzky, published 1922.

all"—revolution as the redefinition of climate.[7] When Konstantin Melnikov proposed in his entry for the Green City competition of 1929 to "rationalize the sun," he was alluding to just this mix of managerial utopia and planetary politics, not (as has commonly been assumed) making a glibly Futurist gesture toward the science-fictive.[8]

As we blithely engineer, and urgently attempt to reverse-engineer, our own global climate—to "rationalize the atmosphere" on the terms of global capitalism—are these imaginaries so preposterous? Might this longer history of planetary imaginaries help us to see and critique the managerialisms latent in our own ideas of sustainability? Without trafficking in nostalgia for failed political experiments and outmoded avant-gardes, can we continue to think about planetary commoning, if not communality, within the multiple registers (aesthetic, technical, social) that architecture has at its disposal? Within today's cultural milieu, the utopian world picture of something like El Lissitzky's *About Two Squares*—which shows a perfectly red, perfectly round Earth approached by the titular squares, representing the dueling world-producing systems of capitalism and communism—has long since been replaced by the technoscientific world picture of the Blue Marble, taken by the astronauts of Apollo 17.[9] But to divorce the scientific facticity of climate change from its aesthetic, cultural, and political bearings is also a reduction of the ground on which architecture can operate.

7

N. F. Fedorov, *What Was Man Created For? The Philosophy of the Common Task: Selected Works*, trans. Elisabeth Koutiassov and Marilyn Minto (Lausanne: Honeyglen/L'Age d'Homme, 1990), 33–37. For an English-language summary of Fedorov's thought, see Stephen Lukashevich, *N. F. Fedorov (1828–1903): A Study in Russian Eupsychian and Utopian Thought* (Cranbury, NJ: Associated University Presses, 1977), esp. 267–272.

8

S. Frederick Starr, *Melnikov: Solo Architect in a Mass Society* (Princeton, NJ: Princeton University Press, 1978), 179.

9

On the material production of this and other "blue marbles," see Laura Kurgan, *Close Up at a Distance: Mapping, Technology, and Politics* (New York: Zone Books, 2013), esp. 9–14.

The "Blue Marble" photograph of the Earth, taken during the Apollo 17 mission on December 7, 1972. Courtesy of NASA Johnson Space Center.

—

In the spring of 2015, we issued an open call for essays that explored the stakes of climatic thinking within architecture and its allied fields. In keeping with the central idea of the *Avery Review*—to engage the work of others using the genre of the critical essay—we asked for reviews of "climates in architecture." The specific questions we posed included: How does climate inflect our understanding of things like human settlement, global migration, spatial violence, and resource extraction? How does climate figure, historically and at present, in our conception of what architecture is and does? What are the material and conceptual infrastructures that render climate legible, knowable, and actionable, and what are the spatial implications of these infrastructures? How do these interrelated questions offer new vantage points on the architectural ramifications of climate change, extending and amplifying our understanding of ideas like resiliency, sustainability, and ecotechnology?

The following December, Columbia University's Graduate School of Architecture, Planning and Preservation (GSAPP) hosted a conference titled "Scales of Environment," which drew together historians, scientists, architects, designers, and scholars on the occasion of the "COP21" United Nations Paris Climate Conference, to collectively investigate how a changing climate is reframing and redefining architecture and urbanism, particularly in the scales it addresses. The conference and this publication are motivated by shared concerns and a shared desire to enrich the conversation at the intersections of architecture and climate change—a desire that is palpable in the essays of the conference participants included in this volume. Their essays, and those we received through our open call, are joined by two conversations (page 21 and page 163), one short story (page 212), and one experimental history (page 372), as well as a dossier of some thirty precedents for thinking about climate change (pages 261–291), which were nominated and described by a group of invited architects. Together they constitute a wide-ranging and often wildly creative dialogue operating in the real, the imaginary, and that messy hybrid of the two we call thinking.

The contents that follow have been loosely grouped according to four preoccupations that we saw emerge across this project—"loosely" in the sense that most pieces included here weave together all four. The section titled "Earths" illuminates the multiple worlds that coexist on our planet (and beyond) as well as the scales at which they are constructed. "Political Ecologies" questions the regimes of knowledge implicit in ecological thought and its attendant subjectivities. "Corporealities," meanwhile, takes the body as a site of investigation, marking its many interfaces with issues of environment. And concluding the book, "Enclosures" addresses what we might traditionally call "architecture," bound up as it is in social and scientific systems.

So, what do we talk about when we talk about climate? A great deal, it turns out. Architecture is, of course, implicated in the enormous expenditure of resources exacerbating climate change, just as it props up the power structures that distribute the planet's precarity so unequally; and yet, design can respond with ingenuity, creativity, and even, dare we say, a little levity toward the situation at hand. It is architectural thinking, though, that this volume celebrates—the many ways of spatially, historically, and speculatively understanding the worlds we find and make.

—James Graham, Caitlin Blanchfield, Alissa Anderson,
   Jordan H. Carver, and Jacob Moore

Ea

ths

According to NASA, there are—roughly speaking—a billion Earths in this galaxy. Radically dislodging our anthropocentric (and geocentric) moorings, perhaps no other estimation flows as naturally from considering a planetary imaginary. But redefining our galactic vision isn't all that this overwhelming number of "other Earths" demands. Seeing our Earth as one in a billion simultaneously challenges us to redefine—to speak roughly about—the particular version we call home. In doing so, we discover that "the planetary" must be thought of as more than a signifier of size, and that conditions resisting easy scalability must not only be addressed but somehow considered together. Employing methods of description that transgress historical periods, political borders, and received materialities allows us to locate rare earths we might otherwise overlook in the great beyond of our own backyard.

In an interview with the *Avery Review*'s editor, James Graham, Dipesh Chakrabarty insists that thinking about climate science through the humanities—as "comparative planetary studies"—might open architectural history toward a more critical examination of its own shortsightedly anthropocentric tendencies. Building on this call to rethink the writing of history in light of climate change, Deborah Coen traverses the landscapes of the Hapsburg Empire, suggesting that we have something to learn from the multiscalar approach of imperial geography. Turning our attention from representations of the past to those of the future, Jacob Boswell outlines two distinct visions of the apocalypse with startling implications for the architectures of the present, while Rania Ghosn and El Hadi Jazairy review the aesthetic and political ramifications of Bruno Latour's *Gaïa Global Circus*, illuminating performance as a tool for understanding climate change. Reflecting on a performance of a whole other kind, Catherine Seavitt Nordenson considers paradigms of conservation and biodiversity in Pope Francis's *Laudato Si'*. Meredith Miller unearths the biodiversity of new material epochs, asking how plastiglomerates and "post-rock" architectures might help narrate the heretofore untold. Charting territories submerged and unseen, Jeanne Gang makes a case for architectural and representational intervention on an earth that is paradoxically closer to home and yet often more difficult to grasp than those light-years away. Finally, leaving the planet altogether, Felicity Scott traces how visions of space colonization in the 1970s also revealed the assumptions of neoliberalism that were then colonizing our own Earth.

The Universals and Particulars of Climate

Dipesh Chakrabarty
in conversation with James Graham

JAMES GRAHAM

You've argued that one role of the humanities in a time of climate crisis is to help think about the cultural values attached to climate. You mention that the oft-cited 2 degrees Celsius—the maximum global warming target of international climate talks for the past few years—isn't scientifically dangerous so much as *culturally* dangerous, in the sense that it's a politically constructed threshold at which we've agreed that what we value as "human civilization" is threatened. "Danger" is a statement of value rather than a statement of science. So I'd be interested to hear more about how the humanities or creative fields are implicated in the ways we talk about climate change.

DIPESH CHAKRABARTY

There are two aspects to this question. One is the point I've made on other occasions—climate change as such is a geophysical problem defined by climate and earth system scientists. But the word "dangerous" cannot be a scientific word. As historian Julia Adeney Thomas has pointed out, "dangerous" here assumes questions of value (priorities) and scale. These are typically issues addressed by scholars in the humanities. So "dangerous climate change" is indeed a humanistic concept. But climate scientists, by ascribing to humans a geophysical agency of enormous scale (an agency capable of putting off the next ice age, say), pose challenging problems to humanities scholars who are used to thinking of "agency" in particular ways, as figurations of autonomy and sovereignty. So some people take issue with one of my propositions in the "Climate of History" essay, which is that humans have no ontological access to being certain kinds of entities—say, a species or a geophysical force.[1] These scholars often argue in favor of art as a form of auto-access or self-knowledge—that things like films or paintings or installation art can bring about a kind of consciousness of humanity's planet-shaping capacities. There's a lot of "Anthropocene art" that attempts similar things, and much of it is very creative and suggestive. But that does not refute my point about trying to access, or trying even to bring within a sensible grasp, something that is of a scale that is strictly incomprehensible. This is not to dispute the role that art can play in giving us some way of imagining things that defy ordinary human experience. When I was growing up in India, an American physicist named George Gamow was very popular among high school or undergraduate students. He had many books on relativity and other wonders of physics, and he would try to represent them pictorially—to represent curved space, for instance. To think that these representations bring things that defy human experience within sensible (as distinct from conceptual) grasp is, I think, problematic.

JG

Could we list architecture as a marker of "dangerous climate change" too, not just as a technical object but also as a carrier of values? Its ability to represent the scope of climate change is similarly compromised. The scale at which it can address the climate crisis, as an architectural object, is limited to a somewhat localized part of a far more complicated set of interactions between buildings and environment.

DC

Architecture is a nice blend of technical things and humanistic impulses. I'm fascinated by Kate Orff's work on the New York City coastline, which was shown at the Columbia conference. There's a very constructive side to that kind of work: "Okay, a lot of the coastline may go underwater, which means we'll be left with a new coastline. How do we build that new coastline, having learned some of the lessons of previous mistakes?" It's an incredibly interesting proposition. And that kind of work also points to how the realities of climate change are seeping into people's consciousness—that we share this planet with other creatures and other things, that we are connected, that our institutions (whether capitalism, housing, the state) are all embedded in processes that support life.

This is why I'm so interested in that division that begins in Aristotle and runs through Agamben, between *bios*, the life of a citizen, and *zoe*, life in general. (Some Aristotle scholars do not agree with this proposition but I am following Arendt and Agamben in holding on to it, provisionally, to make a point.) To some degree it de-centers the human, by which I mean that it stops us from thinking a thought that many religions have helped us to think—that the world was made for us. The real lesson to learn from the humanities is not that we can be non-anthropocentric, but that we can at least see the mistakes of imagining ourselves at the center of things. We often tend to assume that this planet and everything on it is ours, and that those things that aren't for us have to be kept at a distance or gotten rid of. Unlike the ecomodernists, I don't think the world can be our garden.[2] You can't put a human order on it.

> JG
>
> It seems to me that you're asking for a sort of generosity or even empathy in how we think about climate change. This shows up in how you talk about fossil fuels, which you've framed as a problematic form of dependency but not a moral failing, exactly. But it's also an interspecies kind of empathy, in that you ask us to consider the question of habitability, not just for humans, but for animals. In architecture at least, we tend to focus on the urban problems of sea-level rise—and an idea of resiliency framed by water—that leaves out some of these questions of habitability and habitat.

DC

The real problem is that a lot of stuff in the world is inimical to human life, or at least puts us in danger. The primary duty of the state and public institutions is generally thought to be to secure human lives, which in turn becomes a problem of population—our population size has to be such that in securing the lives of all humans we don't endanger the forms of habitability that other life forms have, or at least not to the degree to which we will if we are securing the lives of nine or ten billion or even more people by the end of this century, because securing the lives of these vast global populations means that those people should enjoy the fruits of development. Without development, you can't secure their lives. That's where we come to a real dilemma. Any positive proposal of reduction in population—which has to be a part of how we conceive of moderating our effects on the planet—in effect will be an anti-poor proposal and therefore morally unacceptable. We have to find some natural way of first allowing populations to peak and then finding harmless ways, or at least democratically acceptable ways, for reducing our numbers.

And at the same time, even if it happens over generations, we have to give up what we're calling "fossil fuel attachment." The attachment is not to fossil fuels as such, but to what fossil fuels have made possible. I often emphasize to people that you can't get too moralistic about fossil fuels because a lot of our moral possibilities and universes have been enhanced by fossil fuels.

> JG
> I'm really interested in how population has been central to your thinking about climate crisis, in that you're echoing—and, at the same time, recasting—certain terminology that began with a set of 1970s ecologisms. There's obviously been quite a bit of criticism over the past decades of population alarmists like Paul Ehrlich. But I'm thinking especially of your use of the term "ecological overshoot," which is an incredibly useful and poignant phrase that also derives from that early-'70s sense of crisis. That was William R. Catton's term. So I'm curious about how we should be reading those figures, or not, for the present moment.

DC

I've tried to distinguish between that position and mine by making a distinction between *evolutionary* growth and the pace of it, and the growth that we've seen at a much faster rate than the pace of evolution. This is why I was using the work of the Israeli historian Yuval Noah Harari, who argues that the rise of humans to the top of the food chain wasn't an evolutionary rise. We didn't individually become majestic animals. We used our brains and our symbolic systems and our capacity to create technology, and our ability to band together in larger numbers, to make it possible for us to be at the top of the food chain. If other animals are growing at the evolutionary pace, then our fast-paced growth of the last, let's say, seventy to eighty thousand years doesn't give other animals and their ecosystems the chance to adjust to our growth. There are certain moments along our journey to the top of the food chain when things get ratcheted up. The last of those moments was the post–Second World War world, after which population and consumption and many other indices of human growth and footprint really shot up. The question is not whether we can feed this population—though that is a complex question, given how climate change will affect agriculture. The real question is what kind of implications would there be for other forms of life, if ten billion people on this planet ate well. It's clear that if other forms of life are destroyed or put into crisis, then we suffer because the food chain suffers. You have to think about the general distribution of life. Even in this moment when most people don't get enough to eat, humans and the animals that we keep and eat claim 95 percent of what the biosphere produces.

And there can of course be genuinely alarmist scenarios. Some scientists are claiming that if the oceans get hotter by 6 degrees Celsius, then the phytoplankton—which currently supply about 60 percent of the oxygen in the atmosphere—will die off, meaning there will be radically less oxygen to go around. These types of scenarios depend on the actual increase in temperature under consideration—they're not produced by alarmism or rhetoric; they're simply produced by the logic of the argument.

> J G
> Thus the discussion around the idea of "pre-traumatic stress disorder" among climate scientists, which Emily Scott mentioned at the conference. The science doesn't need alarmism to be deeply alarming.

D C

Though when you de-center the conversation from the human again, and approach it from the perspective of planetary life, the problem becomes a different one. If there's a massive extinction, we may go extinct entirely, or only a minuscule number of humans may survive—like the avian dinosaurs survived even when the dinosaurs went extinct. The most dominant species never continues its dominance after a major extinction. When dinosaurs and reptiles were ruling the planet, mammals were basically rodents. It's only when the reptiles ceased to be dominant that mammals took over, and of the mammals, we eventually became the dominant ones. Since earlier mammals were all forms of rodents, our ancestors basically all had night vision—we and the monkeys, to some degree, lost it.

> J G
> That's amazing.

D C

It had to do with monkeys needing to find fruit in the forest. As the forests became denser, trees couldn't depend on wind to disseminate seeds, so they had to find other ways. So fruits came along, and the trees needed animals that would actually eat the fruit. They used color to attract animals to eat the fruit, so there was a kind of reverse engineering that brought color vision to the eyes of apes and monkeys, and eventually us. It's important that it happened as a kind of collaboration with the larger ecological systems that were maintaining varied forms of life at the same time.

    The point of all this is really to ask: What would our theories of capitalism, our theories of governance, our theories of economy, and our theories of politics look like if we saw all of that history, biological and otherwise, embedded in the processes that support life—that we were and still are a part of the history of life on this planet?

> J G
> One of the tropes that's found in most discussions of climate change—and your work in subaltern studies seems incredibly useful for thinking through it—is this universal, globalized "we" that's spoken of as the subject of climate. It's a language of solidarity and commonality, which is certainly important in many ways. That's the kind of world that was being represented in Paris. But we're also reminded—and I'm thinking here of Adrian Lahoud's presentation at the climate conference—that we have to temper our use of that "we" with a recognition that

climate change is going to have radically particular effects in different places, that there's an uneven distribution of both the production of climate change and its effects on the ground. The "we" has a way of breaking down.

DC

Not only that, but this question of unevenness also applies to the making of the knowledge of climate change. It is defined largely by big American Cold War science. It came out of the studies of how one could colonize and weaponize space and atmosphere. That led to NASA research on life or its absence on other planets. It's not insignificant that people like James Lovelock and Jim Hansen come out of NASA. Lovelock used to study Mars before he developed his Gaia hypothesis. Hansen used to study Venus, which he saw as a case of runaway planetary warming. He wondered if similar processes were taking place on Earth. As he tells the story, he took a few months' leave to work on it and discovered that global warming was indeed happening here. So he left Venus and came back to this planet. *[Laughter]* I often think of climate change science as a problem of comparative planetary studies.

JG

Comparative planetary studies—that's great.

DC

But to return to your question about unevenness, it has been interesting for me to observe the conversations around climate in India. I haven't looked at the situation in China so much because of the language question, but I imagine it might be the same. Compared to the Western discourse on climate change, the Indian discussion in the public sphere has a much smaller bandwidth. The argument is mainly anti-Western: "The West did it; they should pay for it. We need to develop." There are some other voices that are marginalized, but this is the main position in India, whereas if I look at the literature coming out of England, France, Germany, the United States, to some degree Italy, there are many more voices at play—Marxists debating with non-Marxists, energy specialists and policy specialists and economists, all writing different kinds of things.

What we saw in Paris was a meeting of two hundred nations, but nations that aren't equally resourced to create a real public sphere of climate discussion. The discussion on globalization in India is far wider, richer, and more nuanced than the discussion on climate change. So I thought, why is it that climate change, which is a planetary issue, is a much less "global" issue than globalization? Why do people come to Paris from India and other places with such a monochromatic understanding of what's going on? In America, there is the possibility of talking about much broader stakes and longer scales—there are people like David Archer, a geophysicist who wants to talk about climate over the span of a hundred thousand years. Those voices, and voices like mine, are more able to exist.

JG

I was recently revisiting your *Provincializing Europe,* first published some fifteen years ago, and I was struck by your opening lines from Gadamer. The quotation begins, "Europe since 1914 has become provincialized," which is a sensible

> epigraph, given your title. But you continue quoting Gadamer
> as he writes that "only the natural sciences are able to call forth
> a quick international echo."[3] This seemed, in some sense, like
> a bit of an uncanny glimpse of your work to come on the topic
> of climate change.

DC

That's so interesting that you say it. Finish your thought, and I have something to
add to it.

> JG
> Well, on the one hand, the passage contains that classic mod-
> ernist division of natural history from human history that
> you're out to trouble. But it also seems like it might point to
> a possible relationship between the project of subaltern stud-
> ies and climate change. You argued in "Climate of History"
> that climate change was in some way challenging post-colonial
> thought. But it seems to me that for you, in *Provincializing
> Europe*, there's always a kind of oscillation between univer-
> sals—which allow us to think certain forms of freedom—and
> a very particular resistance to those universals, and we can find
> that same oscillation in the debate around climate.

DC

I try to sit on a fence between the universal and the particular, sure. In the case
of climate science, the West is still seen as the master bearer of technology, and
therefore of the science that supports technology and also the economies that support technology. These are societies that can produce a wider-ranging discourse
about this problem, because it's really the science and technology domains of these
societies that are *defining* the problem. When Indian scientists write, they write
about whether particular glaciers are expanding or staying the same or receding.
There's a fundamental difference in the scope of the discourse. Coming back to
your question, when I was writing *Provincializing Europe*, which was rejoining the
debate on globalization, it was much easier to make a place-specific argument—to
say, "Look, capital itself can be place-specific." Whereas with climate change, the
formulations of the problem literally relate to an interplanetary space. The scales
are as big as the scales on which you study black holes or other planets. And this
is made possible by the discussion of science—although of course the value judgment of "danger" always brings the humanities back into the question.

Where the Gadamer quotation becomes relevant to me—and I hadn't thought
about it this way until you mentioned it—is that countries like China and India
are now investing a lot in science, technology, and R&D. In India we talk about
a crisis of the humanities because the country isn't investing enough in them. But
when I look at the climate debate, it's clear that it's a debate that is both provincial
and universal at the same time. It's provincial in the sense that the richest debate is
located in only a part of the world—what we would broadly call the public sphere
of the West. That "quick international echo" isn't fully there. Even people who are
critical of what they see as the very particular interests of the West in the debate,
like Marxists, are still operating within a public sphere that is basically Western.

What would a *Provincializing Europe* approach be for climate crisis? That's a hard question to answer until the public spheres in other countries are actually resourced to widen their voices and to create a more plural discourse.

> JG
> It seems to me that one of the geographical rifts occurs because the scientific discourse is being primarily framed through these Western channels, while the question of something like climate justice is generally thought to be sited in a different kind of territory.

DC

Yes—climate justice discourse really comes out of a pamphlet called "Global Warming in an Unequal World: A Case of Environmental Colonialism," which was published in 1991 by two respected Indian environmental activists. Its rhetoric is absolutely anchored in a Third World–ist, anti-colonial understanding of the global economy. But keep in mind that back in 1991, China was way, way down the list in terms of total emissions; India was barely in the picture. In the last twenty-five years, China has become the largest total emitter; its per capita emissions are higher than the EU's. India has become third or fourth in total emissions, depending on how you calculate the EU's emissions. The climate justice argument has largely kept its picture of the world as it was in 1991. That's why I'm saying that there's a lack of nuance and a small bandwidth of discussion. Indian intellectuals are incredibly up-to-date with respect to globalization, and I kind of wonder why that doesn't happen with climate—with climate science or the idea of climate justice. One reason might be that Indian scientists have not written any books discussing or explaining to Indians the nature of the problem.

> JG
> And you mentioned the problem of research infrastructure, which creates certain self-perpetuating patterns within discourse.

DC

If the problem of climate change hadn't arisen, I think I wouldn't have noticed the poverty of this public sphere, which really has to do with the historically evolved role of scientists. This wasn't true of the period of nationalist struggle, when scientists were part of the larger public sphere. But they are becoming extremely technical, and they write technical papers. There are good geologists and good oceanographers who write on particular problems. But there is no Jim Hansen or David Archer speaking to or from India.

> JG
> So in that way, another role of the humanities in the discussion of climate change includes, as always, to assist in the work of constructing that public sphere.

DC

This also returns to your question about the universal "we." The climate scientists' way of producing a sense of a crisis is to frame it as a crisis for everybody.

When Jim Hansen went to speak to the Senate in 1988, his sense of urgency was not altruistic. He wasn't saying, "I'm concerned that Indians will suffer." He was concerned that everybody would suffer, including Americans. So his book is called *Storms for My Grandchildren*—not *Storms for the Grandchildren of My Friends in India*. Even though the science is provincial in the sense that it's located in particular places, dependent on particular technologies and institutions, it defines its crisis, at its extreme, as a crisis of life on this planet. That's why I go back to interplanetary studies—the original question of climate science was "Why is Mars so cold and Venus so hot, and both so incapable of supporting life?"

> J G
> One of the central observations of your "Climate of History"—and this gets to the question of the public sphere in certain ways—is that the period of 1750 to the present, known broadly to historians as the Enlightenment, with all of its attendant ideas about rationality and human freedoms, is also the period that scientists identify as the Anthropocene. These philosophical and scientific periodizations are bound up in each other. And we know from so many scholars, like Timothy Mitchell, that the institutions of democratic governance as we know them are bound up in carbon-based energy infrastructures.[4] Could we talk more about the new kind of politics that might be demanded of this moment, which in certain ways might run counter to those freedoms promised by the Enlightenment? The formation of a genuine politics around climate change seems to be a major challenge, right?

D C

Technically speaking, the origins of the Anthropocene period are still being debated. But the Paris Conference on climate change is a good example of the contradictions we inhabit. Recently, at a conference on art and the environment in Florence, I was listening to a very interesting philosopher. His argument was that if you have a genuine crisis, which we do, then to simply say, "we have a crisis," while keeping on doing the same things, leaves you with a merely managerial approach. You acknowledge the crisis but you go on as before. A *critique* can emerge out of a crisis only if you arrest the things that you're actually doing.

My response to that was, "Look, because you're a philosopher and thinking philosophically, you probably think in more absolute terms than a historian would." Because as a historian, I think the reality for all the negotiators at the climate conference in Paris was that they were not negotiating climate change in a world in which climate change was the only global problem. There are other problems operating alongside it. Some people say, "We have a lot of poor people to bring out of poverty." Others say, "If I shrink my economy and de-scale my growth and become economically weaker, and therefore militarily weaker, my enemies will take advantage of that." To evenly contract economies at the same time, in all parts of the world, is an impossible condition to achieve in the real world.

And then you can look at the politics of climate change as the process of trying to produce this global humanity. But then, as I was saying to this friend in Florence, you are trying to produce it in a world in which moral life is already riven by other questions, where humans don't see climate change as the only problem. That's why it's a predicament that our collective animal life has produced—a moral problem that comes in the shape of an odd question. We are asked, "What should we do?" But the moment we ask, "What should we do?" we discover that the "we" needs to be constructed. By the time that we are able to produce that "we," even if we are successful in that nearly impossible project, it may be too late in terms of the IPCC [Intergovernmental Panel on Climate Change] timetable.

> JG
> With a few weeks of distance from the Paris Conference, how
> do you think we did at constructing that "we" thus far?

DC
I think the basic thing about the Paris Conference is that the "we" was realized to the degree that all the signatory countries now have at least deliberately made themselves available for peer-group pressure and review. But in terms of commitments, just to give you a single example, they agreed that from 2020 onward they'll create an annual $100 billion fund for climate-related adaptation and mitigation. But $100 billion is a very small sum. I was reading somewhere that the total sales figure for cell phones currently is $147 billion. Energy production is subsidized to a much, much higher degree (one figure I have come across is $5.3 trillion globally). So if we get serious, we'll need to reorient those numbers. In India, they've put the fossil fuel guys in charge of the transition to solar.

> JG
> Ah! Of course. It's a global phenomenon, I'm sure. It's like the
> rebranding of BP to connote green energy production.

DC
When I was more of a Marxist, I used to read Paul Baran and all these people on American monopoly practices. I remember it being discussed that the American automobile companies would take out patents on electric cars only to defer their production, so that they could make maximum profits on petroleum-based cars— and then one day move on to electric or whatever. With the energy industry in control of the transition, they can obviously decide to some degree how to amortize their investments in fossil fuels as the transition happens. You can imagine all the financial processes involved.

What happens, philosophically, is that I think we're living in a world where the climate crisis is producing one timetable for action. Global capitalism then produces another timetable, or multiple timetables, for actions. The companies do it internally, in terms of scenario planning. They do it sector-wide. Then there are national timetables. You can see that there are many negotiations going on around different regimes of temporality and different regimes of politics and legality. I often say that we will enter an age of mismatched calendars. To imagine that the IPCC timetable could be met seems to me too optimistic, but what produces some optimism is the fact that countries have all agreed to make themselves more accountable, though maybe not in a legally binding way, to their peers.

Another ground for optimism is that as the crisis deepens, we will be forced to act—it's not a static situation. Last December, India had huge floods in Chennai;

it recently snowed in Rajasthan, which is one of the hottest parts of the country. Weird things are happening, and that—coupled with the large costs of these kind of weather events—may galvanize more action than we can imagine at this point. We may have to go down the melancholy path of climate change. But I don't see a scenario in which action issues from sheer collective wisdom, or an operative political agency that is being driven simply by wisdom.

> J G
>
> I wanted to posit one more connection between your work on climate change and the disciplinary history of architecture, maybe a slightly spurious one. But I was very interested of your readings of Benedetto Croce and R. G. Collingwood, following Vico, in "Climate of History." They're also figures within the history of art and architecture, in that they've both theorized aesthetics: in particular, how aesthetic fields behave linguistically. You quoted Collingwood: "So far as man's conduct is determined by what may be called his animal nature, his impulses and appetites, it is non-historical; the process of those activities is a natural process." So for Collingwood, the distinction of *bios* and *zoe* also maps onto a kind of historical versus non-historical binary.
>
> The reason I've latched on to this passage is that at roughly the same moment as Croce is writing, we have Banister Fletcher writing his *History of Architecture on the Comparative Method*. It includes his much discussed "tree of architecture," where the categories of "historical" and "non-historical" styles are used to distinguish the main trunk of that tree—the Western classical architectural tradition, which for Fletcher ran from Greece to America—from the dead-end branches, which carried the "non-historical" styles, by which Fletcher bracketed out the "non-Western" architectures of India, China, Peru, Egypt, etc. These were architectures that sat outside of his developmental narrative. So it seemed like there was something in this question of *zoe* and *bios* that could relate to the historiography of architecture, and how we've framed thatproblematic "Enlightenment" arc of architectural tradition relative to other architectures on the planet.

D C

It seems to me that, in effect, in this empirical view of architecture, he's relegating the Aztec architecture to the realm of nature, and therefore the *zoe*. From that point of view, you could think of the massive Aztec structures as kinds of human replications of an insect mound—constructing at a human scale what a social insect might create. That returns us to what I was saying I liked about Kate Orff's work. By looking at the marine creatures in the ocean, by looking at the wind, her project was really looking at habitation as something that exists in the *zoe*, and therefore in the question of life in general. It was giving up on an idea that I've also often critiqued, which is the ecomodernist notion of stewardship.

I see this distinction in Collingwood going back to Kant's "Speculative Beginning of Human History" of 1786, where he basically says that the humanists should study the moral life of the species and not its animal life, which will be studied by scientists. But the problem today is that our animal life is so expanded because of our population and technology—the amount of stuff we eat, the energy we process, all the things we do—that we're facing a crisis for the general distribution of life, such that even scientists are forced to ask a moral question: "So what should we do?" And there, of course, you encounter all the problems of moral life.

So in a sense, it seems to me that architecture may have maintained the *zoe/bios* division through the writing of its history, which has allowed the main Western trajectory to think of itself as belonging exclusively to *bios*. Whereas in fact, the challenge today is for Western architecture to take itself out of *bios* and into *zoe*.

JG
Returning architecture to the *zoe*—that seems like a wonderful place to end. Thank you so much, Dipesh.

1

Dipesh Chakrabarty, "The Climate of History: Four Theses," *Critical Inquiry*, vol. 35, no. 2 (Winter 2009): 197–222. See also Dipesh Chakrabarty, "Human Agency in the Anthropocene," *Perspectives on History* (December 2012): 35–36.

2

For a description of the ecomodernist movement, see Michelle Nijhuis, "Is the 'Ecomodernist Manifesto' the Future of Environmentalism?" *The New Yorker*, June 2, 2015, http://www.newyorker.com/tech/elements/is-the-ecomodernist-manifesto-the-future-of-environmentalism.

3

Dipesh Chakrabarty, *Provincializing Europe: Postcolonial Thought and Historical Difference* (Princeton, NJ: Princeton University Press, 2000), 3.

4

See, for example, Timothy Mitchell, *Carbon Democracy: Political Power in the Age of Oil* (New York: Verso, 2011).

Dipesh Chakrabarty is the Lawrence A. Kimpton Distinguished Service Professor in History, South Asian Languages, and Civilizations at the University of Chicago. He is also a faculty fellow of the Chicago Center for Contemporary Theory and an associate faculty member of the Department of English and School of Law. He is a founding member of the editorial collective of *Subaltern Studies*, a consulting editor of *Critical Inquiry*, and a founding editor of *Postcolonial Studies*. His books include *Provincializing Europe: Postcolonial Thought and Historical Difference* (Princeton, 2000; second edition, 2008); *Habitations of Modernity: Essays in the Wake of Subaltern Studies* (Chicago, 2002); and the forthcoming *Climate of History* (Chicago).

James Graham is the founding editor of the *Avery Review*, and the Director of Publications at Columbia University's Graduate School of Architecture, Planning, and Preservation.

Seeing Planetary Change, Down to the Smallest Wildflower

DEBORAH R. COEN

Anthropogenic climate change is a process unfolding on multiple scales at once, posing a challenge that is aesthetic as well as cognitive. Scientists have known for more than a century that changes in climate at local and regional scales can magnify global-average changes. For instance, the melting of polar ice reduces the reflectivity of the earth's surface and allows it to absorb more heat, which accelerates global warming and increases local melting. Moreover, to describe the climate problem simply as "global" is to ignore the fact that responsibility for the buildup of greenhouse gas emissions is not evenly distributed across humanity, nor is vulnerability to the consequences of a warming planet. How then might we conceive and represent these multiple, interacting dimensions of change?

To pursue this question is to become aware of the ways in which political structures constrain the scalar imagination. As the legal theorist Michael Gerrard observes, there is a very narrow range of spatial scales on which legislation has addressed the climate problem.[1] To put the issue in historical perspective: we live in the age of the nation-state, in which a single political model is so dominant that it has become difficult even to imagine governance on other scales, despite the fact that most of the world entered this epoch only recently, in the aftermath of the First and Second World Wars. The nation-state model constrains policymaking and even the study of climate change, since ecosystems do not respect political boundaries and science is often a national enterprise.[2]

By contrast, the term *ecology* was coined in central Europe in the 1860s, a historical moment marked by a diversity of state forms in Europe and throughout the world—the result, in part, of the reconfiguration of Europe and its empires that took place during and after the French Revolutionary and Napoleonic Wars, and that lasted until the consolidation of nation-states in the late nineteenth and early twentieth centuries.[3] Even within individual states and empires in the mid-nineteenth century, there existed varying degrees of local autonomy.

Why does this matter for thinking about climate change? Because of the historical relationship between science and the state. As I have found in my current research, key elements of our present-day understanding of climate change—including our conception of the multiple scales involved—depended for their formulation on the structure of nineteenth-century states and the research they sponsored. It was principally in Europe's continental empires before the First World War (in imperial Russia as well as imperial Austria) that climate came to be studied as a multi-scale system, from the dimensions of the planet to those of agriculture and human health.[4] Until then, scientists had studied climate either as a planetary physics or as a local descriptive geography. The science that brought these disparate approaches together was

1

Michael Gerrard, presentation for *Climate Change and Scales of the Environment*, Columbia University Graduate School of Architecture, Planning and Preservation, December 2014.

2

On scientific nationalism, see Carol E. Harrison and Ann Johnson, eds., *Osiris* 24, *National Identity: The Role of Science and Technology* (2009).

3

Jürgen Osterhammel, *The Transformation of the World: A Global History of the Nineteenth Century*, trans. Patrick Camiller (Princeton, NJ: Princeton University Press, 2014), 573.

4

On Russian agricultural climatology, see David Moon, *The Plow that Broke the Steppes* (Oxford: Oxford University Press, 2013).

5

For an overview of the history of climate science since antiquity, see Matthias Heymann, "The Evolution of Climate Ideas and Knowledge," *Wiley Interdisciplinary Reviews: Climate Change* 1 (2010), 581–597.

6

William Ashworth, "The Habsburg Circle," in *Patronage and Institutions: Science, Technology, and Medicine at the European Court, 1500–1700*, ed. Bruce T. Moran (Woodbridge: Boydell Press, 1991), 137–167.

7

Josef Chavanne, "Höhenschichtenkarte," in *Physikalisch-Statistischer Hand-Atlas von Oesterreich-Ungarn* (Vienna: Ed. Hölzel, 1887), ix.

8

Max Eckert, *Die Kartenwissenschaft: Forschungen und Grundlagen zu einer Kartographie als Wissenschaft* (Berlin: W. De Gruyter, 1921), vol. 1, part 6, esp. 469.

9

By contrast, elevation maps in Wilhelmine Germany tended to use a simple two-color scheme (green for lowlands, brown for mountains) and to neglect spot heights. Ingrid Kretschmer, "The First and Second Austrian School of Layered Relief Maps in the Nineteenth and Early Twentieth Centuries," *Imago Mundi* 40 (1988): 9–14, 12.

dynamic climatology, the analysis of the transformation of typical air masses as they flow across the surface of the earth. The critical ingredients were nineteenth-century innovations in thermodynamics and fluid mechanics. In the place of the Enlightenment's axially symmetric model of the global circulation, dynamic climatology mapped flows that became more complex as one zoomed in either spatially or temporally.[5]

My research focuses on the development of dynamic climatology in the Habsburg Monarchy, which covered nearly 700,000 square kilometers of central, eastern, and southern Europe in the nineteenth century, an area nearly twice as large as the reunified state of Germany today. With the decline of the Ottoman threat in the eighteenth century and the rise of cultural nationalism, the Habsburg dynasty was in search of a modern ideology of supranationalism. In the wake of the liberal-nationalist revolution of 1848, the dynasty increasingly legitimated its rule in central Europe on the basis of the emerging discipline of economic geography. Ministers and scholars in the service of the empire portrayed its territory as an organic system in which flows of air, water, capital, goods, workers, and information bound together its diverse regions. In keeping with the agricultural basis of much of the economy, the empire's regions were defined, to a first approximation, by climatic difference. A metaphor borrowed from atmospheric science envisioned economic and cultural exchange as a circulation driven by the "evening out of neighboring contrasts." This ideology set the agenda for many different fields of natural and social science in imperial Austria in the second half of the nineteenth century, including climatology, botany, geology, and ethnography. All were used as tools to flesh out this new "Austrian Idea" by studying the concrete conditions of (economic) unity in (environmental) diversity.

If we recall that the mathematical challenge of squaring the circle was known in Leibniz's day as the *Problema austriacum*, then the conundrum of visualizing "unity in diversity" became the "Austrian Problem" for the nineteenth century.[6] The challenge of representing local detail within an empire-wide overview propelled novel techniques in cartography. Consider the *Physical and Statistical Hand-Atlas of the Austro-Hungarian Monarchy* (1882–87), the first publication of its kind, which applied innovative methods for visualizing physical geographic and demographic ("statistical") data. The challenge is easiest to grasp on the map of topographic elevation. As its creator put it, "Austria-Hungary in particular is a country in which one can clearly follow the interactions between the physical conditions of the land and the physical-cultural conditions of the populace; in both ways, the complex structure of the land offers a diversity of conditions of development and formation."[7] From this perspective, the ability to represent fine gradations of height might allow for a finer appreciation of cultural variation. But how was one to represent elevation on a map that encompassed the towering heights of the Alps and the monotonous expanse of the Hungarian plain? By the early nineteenth century, there were several methods in use for representing elevation, including hatching, contour lines, and profiles.[8] What Habsburg cartographers pioneered was the use of color for the purposes of contouring, where it had previously been used to demarcate territorial holdings. Colors could be used to give an immediate visual impression of elevation, particularly on small-scale maps such as overviews of the Monarchy as a whole.[9] It was not simply the application of color that distinguished the Austrian experiments in mapping elevation; it was the depth of theoretical engagement with the problem of representing vertical scale, resulting in the competing systems

Map of the distribution of heat in July in Austria-Hungary, Josef Chavanne, 1887.

of Karl Hauslab and Karl Peucker, the latter based on studies of the physics, psychology, and physiology of color perception (a lively field of research
in Vienna at the time).[10] In the 1887 *Atlas*, the map of elevation employed
Hauslab's scheme, deviating only in the choice of pale blue for the highest
peaks, the realm of eternal frost. In fact, color was used throughout the atlas
to make legible neighboring contrasts of terrain, climate, economy, or culture
on an exceedingly small scale.

The Austrian Problem cut clear across the natural and human sciences. In
anthropology, linguistics, architecture, art history, and beyond, researchers
were likewise setting out to document the Monarchy's cultural multiplicity.
This brings us to Vienna's *fin-de-siècle* contributions to architectural theory.
Architectural historians might be familiar with the conceptual legacy of the
mid-century enterprise to preserve Habsburg art monuments, namely Alois
Riegl's concepts of *Stimmung* (mood or ambience) and "age-value." They
are unlikely to realize, however, that Riegl was articulating an aesthetic ideal
shared by climatologists.[11]

Commonalities among these scientific and humanistic projects were
overdetermined by their institutional histories, shared personnel, and mutual
methodological influence. For instance, the director of the Vienna Administrative Statistical Office, Karl von Czoernig, oversaw research on topics
ranging from ethnography to art history to climatology and public health.
The new minister of trade, Carl Ludwig Bruck, whose vision of Austria's
economic geography would prove so influential, also directed the commission
charged with preserving Austria's art historical monuments. These projects
shared a common grounding in the "positive" methods of natural history: art
historians insisted on the direct observation of original works of art, not cop-
ies, and they valued these objects not merely as things of beauty but as clues
to an evolutionary process.[12] More importantly, these projects shared a politics. They rested, first, on the principle that no cultural or linguistic tradition

10

See my *Vienna
in the Age of Uncer-
tainty: Science,
Liberalism, and
Private Life* (Chi-
cago: University
of Chicago Press,
2007), ch. 5.

11

Diana Reynolds
Cordileone has
demonstrated
the influence of
natural science on
the Vienna School
of Art History;
here, I draw on
her analysis but go
beyond it to show a
more fundamental
common ground
in the conditions
for spatial research
that derived from
the structure of the
supranational state.
Diana Reynolds
Cordileone, *Alois
Riegl in Vienna
1875–1905: An
Institutional Biog-
raphy* (Burlington:
Ashgate, 2014).

12

Cordileone, *Alois
Riegl*, ch. 2.

was too minor to merit scholarly attention. As Crown Prince Rudolf insisted, "none of the crown lands of Austria-Hungary should be regarded as unworthy of a loving, detailed depiction."[13] Second, each of these sciences sought to study the cultural effects of vigorous mixing or exchange. In ethnography, for instance, the seminal three-volume survey by Karl von Czoernig charted a history of migrations across the Habsburg lands in order to demonstrate that the ethnic diversity of the monarchy reached down to such minute dimensions that no division of the territory along national lines was conceivable.[14] Analogously, the Vienna School of Art History challenged the naiveté of Romantic-nationalist celebrations of folk art by revealing the hybrid histories of these traditions.[15] As Alois Riegl later put it, using language that echoed the "dynamic" thinking of Habsburg climatology: "On the occasion when the unfamiliar meets the unfamiliar in a close and sustained relationship the process of development is set in motion."[16] Thus, in an era when nationalism was reconceiving historical research as a quest for authenticity and indigeneity, a largely forgotten project emerged in the human and natural sciences to focus attention instead on the complexity of cultural flows.[17]

Although the term "monument of nature" [*Naturdenkmal*] was not yet in common use, the analogy between nature preservation and historic preservation was implicit in these "whole-state" surveys. This is well illustrated by the work of the plant geographer Anton Kerner, who set out to write a botanical history of the Habsburg lands, from alpine flowers to the grasses of the Hungarian steppe. Kerner suggested that it was the duty of the imperial-royal scientist to advance modernization and, at the same time, to salvage as quickly as possible what was left of traditional natural-cultural landscapes: "to preserve these last remains of authenticity [*Ursprünglichkeit*] in image and word."[18] Kerner took it upon himself to teach a German-speaking public to see the beauty of these "peripheral" landscapes, and he did so with lyrical prose and delicate brushwork.

"Whole-state" research in the human sciences resulted in a host of cultural productions, from atlases to exhibits to memorials, all of which conditioned

13

Matthew Rampley, *The Vienna School of Art History: Empire and the Politics of Scholarship, 1847–1918* (University Park: Penn State Press, 2013), 84.

14

Peter Stachel, "Die Harmonisierung national-politischer Gegensätze und die Anfänge der Ethnographie in Österreich," in *Geschichte der österreichischen Humanwissenschaften. Bd. 4: Geschichte und fremde Kulturen*, ed. Karl Acham (Vienna: Passagen Verlag, 2002), 323–367; Brigitte Fuchs, *"Rasse," "Volk," Geschlecht: anthropologische Diskurse in Österreich 1850–1960* (Frankfurt: Campus, 2003), ch. 10.

15

Matthew Rampley, "Peasants in Vienna: Ethnographic Display and the 1873 World's Fair," *Austrian History Yearbook* 42 (2011): 110–132.

16

Cited in Cordileone, *Alois Riegl*, 99.

17

Rampley, "World's Fair," 132.

18

Anton Kerner, *Das Pflanzenleben der Donaulaender* (Innsbruck: Wagner, 1863), 28.

Watercolor of a moor on the Hungarian steppe, Anton Kerner, ~1855–60.

a new way of regarding a landscape. Alois Riegl termed it a "view from a distance." He described the subjective effect as a "mood" or "ambience," akin to religious devotion, yet awakened by the new scientific worldview, which saw a world interlinked by causal relations—the desire for "the comforting conviction of the unshakeable rule of the law of causality."[19] According to Riegl's successor Max Dvorak, the quest for the synthetic overview derived from the artistic effort to capture "the diversity of natural phenomena ... so that the full richness of the world down to the smallest wildflower and the most fleeting change in the atmosphere and quality of light has become a source of artistic sensations."[20] Dvorak's description resonates with the aesthetic aims of Habsburg natural scientists, as we might illustrate by means of the famous alpine panoramas of Friedrich Simony.

Simony was the first professor of physical geography at the University of Vienna, and the first to insist that this discipline needed to take the entirety of the Habsburgs' territory as its field of study. He described it as his duty to teach the public to see individual landscapes as part of this greater whole—in his words, to "awaken love and enthusiasm for the singular, beautiful, great fatherland" by means of "a life-like representation, from first-hand observation, of the most interesting and instructive natural phenomena from the diverse regions of the imperial state in word and image." Technically, this necessitated a range of techniques for representing nature across dimensions of time and space: "panoramas and profiles, characteristic landscapes and images of individual natural historical points of interest or objects, then too graphical representations of various kinds," all produced as posters "of the largest possible size" to be suitable for public lectures.[21] In his panoramic landscape paintings, for instance, Simony perfected visual methods for capturing fine-grained detail within compositions that are, nonetheless, clearly legible as foreground, middle ground, and background.

19

Quoted in Bernd Euler-Rolle, "Der 'Stimmungswert' im spätmodernen Denkmalkultus-Alois Riegl und die Folgen," 3, http://www. denkmaldebatten. de/fileadmin/ dateien/Download-Materialien/B._ Euler-Rolle_-_Der_ Stimmungswert. pdf.

20

Max Dvorak, "Einleitung," in *Die Denkmale des Politischen Bezirkes Krems*, ed. Hans Tietze (Vienna: Anton Schroll, 1907), xvii.

21

Quoted in Albrecht Penck, *Friedrich Simony: Leben und Wirken eines Alpenforschers. Ein Beitragzur Geschichte der Geologie in Österreich* (Vienna: Hölzel, 1898), 12.

Painting of a glacier in the Dolomites, Friedrich Simony, 1856. Courtesy of the Bildarchiv der Österreichischen Nationalbibliothek.

In the 1890s, Alois Riegl introduced the concept of age-value [*Alterswert*] to describe the quality of historic monuments that defined their worth. Like *Stimmung*, age-value was linked to a view from afar, now in both the temporal and spatial sense, yet a view concerned with the specificity of individual objects and cultural landscapes. Just as the effect of *Stimmung* arose from an awareness of causal connections in space, age-value corresponded to an equally naturalistic vision of the cycle of universal birth and decay. Riegl imagined it as the view of the Habsburg subject traveling as a tourist within his own state: "the Bohemian searches to satisfy his deep longing for mood in something like a cathedral in Dalmatia, while a person from Styria finds it in Tyrolean wall painting, and the person from Silesia finds it in Salzburg's Italianate architecture."[22] By the 1890s, Habsburg subjects had been trained to take such a spatially and temporally "distant" view of cultural landscapes. Their eyes had grown accustomed to the techniques developed by sciences like climatology, geology, ethnography, and art history for representing fine-grained detail within a large-scale overview. Riegl's concepts of *ambience* and *age-value* should thus be understood as theorizations of a gaze that was cultivated jointly by the natural and human sciences under the patronage of the Habsburg state in the wake of 1848, a gaze associated with an important historical transition in the understanding of climate change.

Today, the climate crisis compels us once again to hone such a multiscalar vision. To do so, as this history suggests, we will need to transcend the political and intellectual constraints imposed by the borders between nation-states. But there is another lesson implicit in the history of the Austrian Problem, and that is the potential for the arts and sciences to reinforce each other's efforts toward a new way of seeing. Rather than assuming a division of labor according to which the sciences provide "explanation" and the arts "understanding," Habsburg scholars across the disciplines adopted a common ideal of *Anschualichkeit* or visualizability, and they held themselves to this high standard when communicating with the public. Thus the goal of climatology was to produce "a maximally life-like [*lebendig*] picture of the interaction of all atmospheric phenomena over a patch of the earth's surface."[23] Jointly producing such a picture would be an equally worthy goal for artists and scientists today. Architects, in particular, are as skilled as climatologists in the creative work of up- and down-scaling.[24] Their collective experience might hold the key to a new intuition for the multiple scales of climate change.

[22]
Quoted and translated in Cordileone, *Alois Riegl*, 276.

[23]
Julius Hann, *Handbuch der Klimatologie* (Stuttgart: J. Engelhorn, 1883), 3.

[24]
Albena Yaneva, "Scaling Up and Down: Extraction Trials in Architectural Design," *Social Studies of Science*, vol. 35, no. 6 (December 2005): 867–894.

Deborah R. Coen is a professor of history at Barnard College, where she teaches modern European history and the history of science. Coen is the director of research clusters and curriculum of Columbia University's Center for Science & Society and a member of Columbia's Committee on Global Thought. She is the author of *Vienna in the Age of Uncertainty: Science, Liberalism, and Private Life* (2007) and *The Earthquake Observers: Disaster Science from Lisbon to Richter* (2013), both published by the University of Chicago Press, and co-editor of *Intimate Universality: Local and Global Themes in the History of Weather and Climate* (Science History Publications, 2006). Her current project is *Climate in Word and Image: Science and the Austrian Idea*, to be published by the University of Chicago Press.

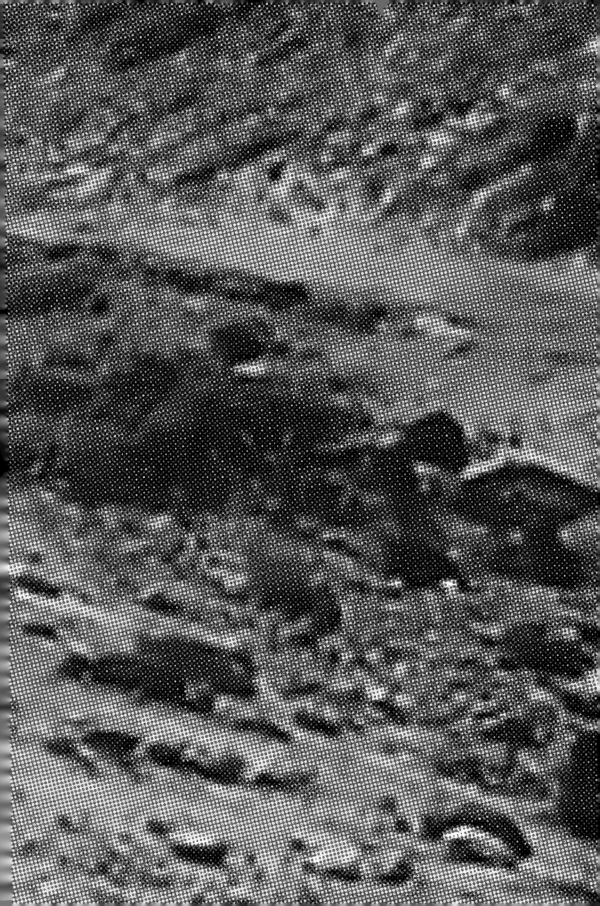

Notes from the Wasteland: Competing Climatic Imaginaries in the Post-Apocalyptic Landscape

JACOB BOSWELL

In his 1977 book *God Is Alive and Well and Appearing in Popular Culture*, John Wiley Nelson argues that "popular culture is to what Americans believe as worship services are to what members of institutional religions believe."[1] According to Nelson, American pop culture carries all the trappings of a traditional religion: a description of worldly problems, a method of deliverance, a picture of the post-salvation world, and a suggested road map toward that future perfection.[2] Russel Nye, the Pulitzer Prize–winning godfather of popular culture studies, confirms this, writing, "The popular artist corroborates values and attitudes already familiar to his (or her) audience ... aim[ing] less to provide a new experience than to validate an older one."[3]

Popular culture is rife with images of environmental apocalypse. From George Miller's *Mad Max* cycle to Margaret Atwood's *MaddAddam* trilogy to Cormac McCarthy's *The Road*, images of climatic dystopia saturate contemporary media. At least superficially, these stories share a common trope: humanity's technological hubris produces a dramatic environmental change. As a culture, we've been telling ourselves this story for a long time, and the reasons for its resurgence of late are obvious. Yet, the future imaginary produced within popular media is not monolithic. While these futures share common themes, they often envisage two conflicting results. Superficially these manifest in scenarios of either too little or too much water—a dry apocalypse or a wet one—yielding either a desert or tropical landscape. While they conveniently coincide with topical anxieties, these visions have their roots in highly differentiated and often competing historical narratives. As such, they act out two very different visions of the future that encompass divergent constellations of society's hopes and fears. This divergence has implications for the way design and planning work to mediate between the now and the imaginary future.

THE DRY APOCALYPSE

The dry apocalypse tells the story of humanity's struggle for survival following some cataclysm, most often a nuclear war. The dry apocalypse is uniformly hot and uniformly arid.[4] Here in the sunburnt desert wasteland, the remnants of humanity search for water, food, weapons, sex, personal salvation, or civilization. Our antihero, a lone drifter, reluctantly aids a helpless survivor of the old world—a caravan, a settlement, a colony—escape the perils of the wasteland.

Examples abound, from *Mad Max* to Electronic Arts' classic 1987 videogame *Wasteland* (and Bethesda Software's series of *Fallout* video games that followed) to Harlan Ellison's series of Vic and Blood tales captured in *A Boy and His Dog* (1969). Collectively this apocalyptic form has been termed *dieselpunk* or sometimes *atompunk* for its fixation with the technology and

1

John Wiley Nelson, *Your God Is Alive and Well and Appearing in Popular Culture* (Philadelphia: Westminster Press, 1976).

2

Bruce David Forbes and Jeffrey H. Mahan, *Religion and Popular Culture in America* (Berkeley: University of California Press, 2000).

3

Russel B. Nye, *The Unembarrassed Muse: The Popular Arts in America* (New York: Dial Press, 1970).

4

The assumption that a nuclear holocaust would result in the desertification of the globe is curious given what science supposes about nuclear winter, let alone what has been revealed by real post-nuclear sites like Chernobyl.

Dry apocalyptic narratives typically center on a single antihero wandering the desert wasteland, as seen in *A Boy and His Dog*, directed by L.Q. Jones (1975), and *Mad Max: Fury Road*, directed by George Miller (2015). Courtesy of Shout Factory and Warner Home Video.

aesthetics of the 1940s, '50s, and '60s.[5] In the dry apocalypse, the devices and symbols of modernity—the car, the airplane, the atomic bomb—figure into a mythologized past. The lost civilization that produced such devices is simultaneously worshiped for its technological prowess and condemned for destroying their perfect "tomorrow-morrow land."[6] The cataclysmic event and subsequent breakdown create a radical discontinuity between pre- and post-apocalyptic worlds, resulting in a society that has reverted to a premodern "savage" state.

In most dry apocalypses, the protagonist eventually delivers the good community from the bad, and in so doing creates a new civilization from within the wasteland. This is an old story, given new life in the twentieth century via the prospect of totalizing nuclear reset. In many ways, dry apocalypses recall the Lucretian narrative of birth and death captured in *De Rerum Natura*—a model later perfected by Virgil in *The Aeneid*. According to Denis Cosgrove, the Lucretian narrative understands culture as "a long process of trial and error, always threatened by a return to brutality."[7] Virgil transforms this process into a perpetual cycle; civilization is born from brutal nature, it proceeds through an Arcadian stage on its way to an imperial golden age, and then on to the inevitable collapse brought about by imperial hubris. Virgil's Aeneas, fleeing the fiery annihilation of Troy, creates a new civilization from his violent conflict with the Latins on the banks of the Tiber. Here, according to Cosgrove, we see the formation of Imperial Rome, "born in the blood of native and colonizer ... Only Aeneas's own heroism and eventual death will produce the proper circumstances to guarantee the return of the gilded age."[8] When viewed in this way, the dry apocalypse sits squarely within the narrative of Western cultural progression, a twentieth-century

5

Mr. "Piecraft" and Nick Ottens, "Discovering Dieselpunk," *The Gate House Gazette* 1 (July 2008), http://www.ottens.co.uk/gatehouse/Gazette%20-%201.pdf.

6

Terry Hayes, George Miller, George Ogilvie, Mel Gibson, Tina Turner, and Bruce Spence, *Mad Max: Beyond Thunderdome* (Burbank, CA: Warner Home Video, 1997).

7

Denis Cosgrove, "Mapping Arcadia," in *Geography and Vision: Seeing, Imagining and Representing the World* (London: I. B. Tauris, 2008), 80.

8

Cosgrove, "Mapping Arcadia," 80.

A clan of feral children collect along the spine of a crashed 747 in *Mad Max: Beyond Thunderdome*, directed by Terry Hayes and George Miller (1985). A symbol of the civilization that was lost, the downed plane is a ritual device within the children's mythologized past and prophesied future. In that myth, the plane carrying their parents crashed in a wasteland. They believe that one day the pilot will return to take them away to "tomorrow-morrow land." Courtesy of Warner Home Video.

redux of the pastoral dialectic so convincingly traced by Raymond Williams in his classic study of the trope *The Country and the City*. Williams's analysis reveals a kind of Western cultural double-think—at once basking in the narrative of progress while simultaneously hoping for a return to simpler times.[9]

George Miller's *Mad Max* films directly parallel the Virgilian narrative. Like Aeneas, Max wanders a savage world, perpetually reinitiating civilization for those, presumably, more innocent than himself. In the 1979 original, Max (Mel Gibson) is set adrift in the nuclear wasteland after his family's seaside idyll is destroyed by a rampaging motorcycle gang. In *The Road Warrior* (1981) Max saves the innocent settlers of an oil refinery from the deprivations of a wasteland warlord. We are told in a narrated epilogue that the settlers

9

Raymond Williams. *The Country and the City* (New York: Oxford University Press, 1973).

Thomas Cole's *The Course of Empire* is a set of five paintings that capture the rise and fall of a single civilization. The paintings—*The Savage State* (1834), *The Arcadian or Pastoral State* (1834), *The Consummation of Empire* (1836), *Destruction* (1836), and *Desolation* (1836)—work as a linear narrative and warning, moving from barbarism to empire. However, in *Wastelands*, Vittoria Di Palma has shown that Cole intended for the paintings to be displayed in a circular fashion with *The Savage State* at the upper left and *Desolation* at the upper right, suggesting, rather than a linear trajectory ending in death, a cycle from birth to death to birth. *The Course of Empire: Destruction*, Thomas Cole, oil on canvas, 1836. Courtesy of the New York Historical Society Collection.

At the end of *Mad Max: Beyond Thunderdome*, the band of feral children escape to the wreckage of Sydney, Australia. In this scene, set several years later, we witness the ritual of a new, more innocent society birthed from the ashes of the old. The children emerge from the technology-fueled death and destruction of the old world to found a new pastoral state that still worships Max as its deliverer. Courtesy of Warner Home Video.

eventually form the great northern tribe. Civilization is reborn. The third and fourth installments in the series reinforce this cycle. In *Beyond Thunderdome* (1985) Max reluctantly saves a band of children from "Auntie" (Tina Turner) and we later see that the children have begun a new civilization within an old seat of mythical power—the collapsing remains of Sydney, Australia. In *Fury Road* (2015), the much more recent latest installment, Max (Tom Hardy) helps a band of female sex slaves escape across the desert in search of "The Green Place," a farming community ruled by warrior women.

Beyond Virgil, George Miller draws on another branch of the Western cannon in the Max films–the denial of Moses. Both characters are denied entry into the new civilization they birth because of their actions in the wilderness. Perhaps here we find the *raison d'être* behind the specific climatic imaginary associated with the dry apocalypses—the desert wasteland. The desert wasteland is both the site of redemption and is itself redeemable. According to art historian Vittoria Di Palma, the term *wasteland* stems from the Old English precursor *weste londe* or *westen*, indicating a place of desolation marked by a harsh climate and lack of sustenance. The biblical wasteland is a place of suffering and atonement, but one through which redemption can be won.[10] Di Palma makes an important distinction between the notion of wasteland and wilderness. While the term *wilderness* connotes an original or primitive state of nature, the biblical wasteland refers instead to a post-apocalyptic landscape—a landscape devastated by God's wrath. The capacity to transform a wasteland, a symbol of God's displeasure with humanity, into a garden was indication of redemption, and of God's favor. Thus the physical climate of the dry apocalypses ties together two powerful Western narrative cycles: the Virgilian cultural cycle of barbarism, innocence, empire, and desolation, and, in Di Palma's words, the biblical cycle of "condemnation, devastation, atonement, and redemption."[11]

THE WET APOCALYPSE

The wet apocalypse also tells the story of humanity's struggle for survival following a cataclysm. Global sea level rise is, predictably, a significant aspect of the wet apocalypse, though the underlying causes are typically contested within the story: ranging from corporate genetic tinkering to bio-terrorism. This event has either wiped out or fundamentally altered vast segments of the human

10

Vittoria Di Palma, *Wasteland: A History* (New Haven: Yale University Press, 2014), 16.

11

Di Palma, *Wasteland*, 19.

population while also producing an uncontrollable animal or vegetal overgrowth that literally swamps civilization in a tropical morass. The incursion of a tropical climate and tropical flora and fauna on the remnants of Western civilization triggers the true hallmark of the wet apocalypse: degeneration.[12] Far from the narratives of the antiheroic drifter, the wet apocalypse is primarily concerned with the slow degeneration of humanity into (or displacement by) a genetically altered race, often of its own creation. In these stories, the remnants of humanity rot from within, dwindling both numerically and morally while battling one another in an effort to seize whatever remains of modernity's material culture. In this way the wet apocalypse does not usually hinge on the actual act of submersion. Rather, it results from the manifold notion of total human alteration: physically, mentally, and socially. More recent examples of the wet apocalypse include Margaret Atwood's *MaddAddam* trilogy, Paolo Bacigalupi's *The Windup Girl* (2009), Jeff Noon's *Pollen* (1995), and J. G. Ballard's *The Drowned World* (1962). This genre is often labeled "biopunk" in reference to the nascent bio-hacker movement's desire to liberate biological knowledge from the confines of institutional and corporate labs.[13] Here again, as was the case with the dry apocalypse, technological hubris has led to a radical disruption in the preexisting order, though often for very different reasons. In each of these novels, overt tropical-ization is the result of some human agency. Within the confines of the tropicalized environment, characters experience a crisis of genetics, often resulting in mental anguish over their reversion to a more primitive nature or the alteration of their bodily form. These novels, as well as comics and films like *Swamp Thing*, *The Toxic Avenger*, and *C.H.U.D.*, deal with latent environmental toxicity, bodily contamination, and mutation—crises posed by the human body's either rapid evolution or slow devolution.

Though inherently related to the dry apocalypse via eschatological implication, the wet apocalypse relies on a paradoxical climatic imaginary—the imposed jungle. A more recent historical lineage situates this imaginary within cultural assumptions that are quite different from those found within the dry apocalypse. The wet apocalypse has its origins in *fin de siècle* rhetoric surrounding degeneration. Degeneration theory stemmed from the dawning realization at the turn of the nineteenth century that the progress narrative seemed to have failed on at least three fronts: industrialization and science had neither improved the welfare of industrialized society's working poor, nor had it improved colonial prospects, nor had it advanced the welfare of colonized peoples.[14] Degeneration theory grew from the convergence of Western medicine's long tradition of Humoralism with Darwinian evolutionary and social theory in order to provide productive responses to this failure of the progress epistemology. Medical theory at the time posited that hot, humid, tropical environments were inimical to white bodies and that exposure to such environments resulted in a kind of devolution.[15]

12

I use the term *degeneration* here as an extension of nineteenth-century degeneration theory, which would have seen the acquisition of animalistic characteristics as subhuman, and thus a mark of degeneracy. Certainly, contemporary popular culture provides us with many examples of humans who, in acquiring animalistic characteristics, actually become superhuman. Spider-Man is a good example.

13

See Meredith Patterson, "A Bio-punk Manifesto," http://maradydd. livejournal.com/ 496085.html.

14

For an excellent discussion of the social theory of degeneration and its effects on late-nineteenth and early-twentieth-century literature, see William Greenslade's *De-generation, Culture and the Novel 1880–1940* (Cambridge: Cambridge University Press, 1994). For a broad discussion of Imperialism's relationship to Western perceptions of race, gender, and sexuality, see Anne McClintock's *Imperial Leather: Race, Gender and Sexuality in the Colonial Contest* (New York: Routledge, 1995).

15

For some of the earliest European writing on this subject, see James Ranald Martin's *The Influence of Tropical Climates on European Constitutions, Including Practical Observations on the Nature and Treatment of the Diseases of Europeans on Their Return from Tropical Climates* (London: Churchill, 1856), or James Lind's *Essay on Diseases Incidental to Europeans in Hot Climates*, 5th ed. (London: J. Murray, 1792).

The first step towards lightening The
White Man's Burden

The White Man's Burden
is through teaching the virtues of cleanliness.

Pears' Soap

is a potent factor in brightening the dark corners of the earth as civilization advances, while amongst the cultured of all nations it holds the
highest place—it is the ideal toilet soap.

*The first step towards lightening The
White Man's Burden is through teaching
the virtues of cleanliness, Pears' Soap
Co., ~1900. Colonial imagery at the
end of the nineteenth century and the
beginning of the twentieth featured heavily
paternalistic images of tropical peoples.
The copy here, a reference the Rudyard
Kipling poem of 1899 of the same name,
suggests that it is the duty of civilized
(white) nations to undertake the moral and
physical salvation of tropical peoples and
belies colonial fears about the tropics,
suggesting that soap might allay some of
the threat of contamination and degeneration by literally sanitizing tropical peoples.
Courtesy of the Advertising Archives.*

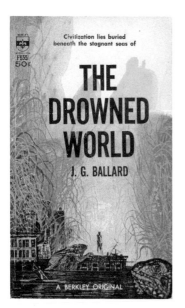

Civilization lies buried
beneath the stagnant seas of

F855
50¢

THE
DROWNED
WORLD

J. G. BALLARD

A BERKLEY ORIGINAL

*The 1962 Berkley Books cover of J. G.
Ballard's The Drowned World perfectly
captures the overwhelming fecundity
described within the wet apocalypse. The
tiny, fading figure of a man and the city
give way before a riotous, hostile, and
ever-encroaching plant and animal world.*

Thus, white settlers and explorers in tropical environments were not inherently predisposed to acts of violence, greed, indolence, or decadence, but, rather, their more pristine nature had been corrupted by the heat, torpid vapors, or "actinic radiation."[16] And such fears were not limited to the tropics. Degeneration could be extended to places that had become, in effect, tropicalized through the imposition of artificial heat, smoke, or congestion. It is through the mechanism of tropicalization that the stagnant tenements of the industrial metropolis became sites of bodily degeneration as well.[17] Tropical people and tropicalized people were made subject to the same series of racist characterizations including innate indolence, lasciviousness, and savagery.

Western cultural products of this era reflect the concern over degeneration and tropicalization, often in distinctly apocalyptic tones. Charles Pearson's *National Life and Character* (1893) cautions against the sheer fecundity of tropical people, suggesting that the "higher races" would become hemmed in and eventually overrun by the swarming hordes.[18] Joseph Conrad's *Heart of Darkness*, published in 1899, is rife with overt allusions to the physical and moral degeneracy of the European ivory hunters who have ventured into the Congo.[19] Within the same decade, H. G. Wells wrestled with the notion of degeneracy and the potential benefits of human plasticity in *The Island of Dr. Moreau*. That Wells set this novel within the confines of a tropical island is not coincidental.[20] William Booth's best-selling *Darkest England and the Way Out* extended fears over tropicalization to urban London, making the direct equivalency: "Darkest England, like Darkest Africa reeks with malaria. The foul and fetid breath of our slums is almost as poisonous as that of the African swamp."[21]

What begins in the 1890s as a social and scientific polemic had become thoroughly ensconced as a cultural trope by the 1950s. By this time the emphasis had shifted, however; science fiction (particularly of British origin) was less concerned with white explorers degenerating in the tropics than it was with England degenerating *into* the

16

Dane Kennedy,
"The Perils of the
Midday Sun: Cli-
matic Anxieties in
the Colonial Trop-
ics," in *Imperialism
and the Natural
World*, ed. John
M. MacKenzie
(Manchester: Man-
chester University
Press, 1990).

17

Rod Edmond,
"Returning Fears:
Tropical Disease
and the Metrop-
olis," in *Tropical
Visions in an Age
of Empire*, ed. Felix
Driver and Luciana
de Lima Martins
(Chicago: Uni-
versity of Chicago
Press, 2005).

18

Charles H. Pear-
son, *National Life
and Character: A
Forecast* (London:
Macmillan and
Co., 1894).

19

Joseph Conrad
and Robert
Kimbrough, *Heart
of Darkness: An
Authoritative Text,
Backgrounds and
Sources, Criticism*
(New York: Nor-
ton, 1988).

20

H. G. Wells and
Leon Stover, *The
Island of Doctor
Moreau: A Critical
Text of the 1896
London First
Edition, with an
Introduction and
Appendices* (Lon-
don: McFarland
and Co., 1996).

21

William Booth
as quoted in
Rod Edmund's
"Returning Fears."

tropics. Novels like John Wyndham's *Day of the Triffids* and J. G. Ballard's *The Drowned World* see a humbled, post-imperial England overrun by hostile plant and animal life.

Within the postmodern violence of Ballard we get our first glimpse of the contemporary wet apocalypse. *The Drowned World* is set within a submerged London, complete with Arthur Conan Doyle–sized Iguanodons. The protagonist, Dr. Robert Kerans, a twentieth-century Marlowe, wrestles with his desire to return to the vestiges of civilization (now located at the poles) but is undeniably drawn to the primal nature he finds within London's tepid lagoons. Kerans is juxtaposed by Strangman—Ballard's homage to both Conrad's Kurtz and Wells's Moreau. Strangman invites Kerans to revel in the decadence of an escape from civilization, to merge with the jungle, fading back into a primeval, pre-moral state. Unlike Moreau, Strangman goes unpunished for his crimes, and Kerans aimlessly drifts south—"a second Adam searching for the forgotten paradise of the reborn sun."[22] Here there is no suggestion of a cycle; the wet apocalypse results in a total state change, flipping the earth and humanity into a new trajectory that is seemingly irreconcilable with the old. *The Drowned World* and earlier novels like John Wyndham's *The Chrysalids* (1955) show a critical shift in wet apocalyptic thought. Rather than signifying the degeneration of humanity into a slovenly or bestial other, this new version saw tropicalization and bodily mutation as a potential escape from the paternalistic constraints of modernity. For instance, *The Chrysalids* explores a post-nuclear world set in a now temperate Labrador. Here, again, the tropics have drifted north, leaving only the northernmost extremes of the planet inhabited by "humans." It is here we find a dystopian pastoral society—dominated by the protagonist's father—bent on protecting its own genetic purity at all costs. The end of the novel juxtaposes the genetically pure, oppressive, society with a hyper-technological society of mutants located in "Zeeland" (New Zealand). The plot strongly implies that rapid bodily mutation, conferred by nuclear fallout, has resulted in a new and better version of humanity. In this version of the wet apocalypse, society achieves salvation not by transforming the wasteland into the pastoral seat of a new empire but by allowing its inherent destructive impulses to be subsumed, giving in to some interior (or sometimes overt) change that enables both survival and a chance to rewrite the course of human history.

More recent wet apocalyptic fiction like Margaret Atwood's *MaddAddam* trilogy (2003–13) and Paolo Bacigalupi's *The Windup Girl* (2009) continue these tropes. Both of these novels struggle with the environmental and bodily consequences of contemporary biotech, represented in both by the prosaic corporate bureaucracies of the food and drug industry (Atwood's CorpSeCorp and Bacigalupi's Calorie Men). Correspondingly, these worlds are both predicated on a biological rather than nuclear apocalypse, and both suggest that within these fundamentally altered environments it is some variant of humanity (Atwood's genetically spliced "Crakers" and "Pigoons" and Bacigalupi's gene-hacked "New People") that are best adapted to survival. Maybe more importantly, these novels suggest that it is through their alteration that these new, perhaps unfallen, forms of humanity may finally be capable of authoring a different outcome—finally breaking the cycle of violent expansion and collapse.

22

J. G. Ballard, *The Drowned World* (London: Fourth Estate Publishing, 2011), 175.

Cover paintings for the classic role-playing game *Wasteland* (1988) and its contemporary sequel *Wasteland 2* (2014) evidence the shift away from the dry apocalypse and toward the wet.

TWO IMAGINARIES
ENTER…

It is apparent that in recent decades we see a shift away from the dry apocalypse and toward the wet as the dominant form in Western culture. Critical and popular acceptance of novels like *MaddAddam* and *Windup Girl* (*MaddAddam* is currently in development as an HBO series directed by Darren Aronofsky) along with the changing morphology of traditionally dry apocalyptic narratives point toward a collective shift in the popular imaginary.[23]

Even the *Mad Max* cycle has deviated, at least partially, from its normally desiccated setting. In the latest installment from 2015, the chrome-plated, uniformly white wasteland survivors bear the unmistakable marks and deformities of a genetically altered humanity—marks noticeably absent from the earlier films in the series.

If, as Nye asserts, popular culture "corroborates values and attitudes already familiar to its audience … aim[ing] less to provide a new experience than to validate an older one."[24] What does it mean that our post-apocalyptic imaginary is shifting? These two imaginaries are undeniably products of the fears of their time. With its concern for the post-nuclear and in its fetishizing of modernist symbols, the dry apocalypse seems to be the climatic imaginary of a past age. Alternatively, the wet apocalypse corresponds more readily with contemporary social concerns. In the wet apocalypse, big pharma, big

[23]

Jeff Dinsmore, "Margaret Atwood Talks HBO's *MaddAddam*," HBOWatch, February 9, 2015, http://hbowatch. com/margaret-atwood-talks-hbos-maddaddam.

[24]

Nye, *The Unembarrassed Muse.*

Many of the lead characters in *Mad Max: Fury Road* bear physical deformities, mutations, or other symptoms of the toxic environment they live in. The rampant evidence of toxicity in *Fury Road* is in stark contrast to earlier films, which make no obvious connection between the nuclear annihilation that has resulted in the wasteland and the physical health of its inhabitants. Courtesy of Warner Home Video.

25

Lawrence Buell,
"Toxic Discourse,"
Critical Inquiry, vol.
24, no. 3 (1988).

26

John McMorrough,
"Design for the
Apocalypse," *Volume*
20 (2009): 40–42.

27

Sheila Jasanoff and
Sang-Hyun Kim,
*Dreamscapes of
Modernity* (Chi-
cago: University
of Chicago Press,
2015), 4.

28

Henry Nash Smith,
"Rain Follows the
Plow: The Notion of
Increased Rainfall
for the Great Plains,
1844–1880," *Hun-
tington Library
Quarterly*, vol. 10,
no. 2 (1947): 169–
193; Marc Reisner,
*Cadillac Desert: The
American West and
Its Disappearing
Water* (New York:
Viking, 1986);
Fazal Sheikh and
Eyal Weizman, *The
Conflict Shoreline*
(Göttingen: Steidl
Verlag, 2015).

29

Alessa Johns,
"Feminism and
Utopianism," in
Gregory Claeys,
ed., *The Cambridge
Companion to
Utopian Literature*
(Cambridge: Cam-
bridge University
Press, 2010), 178.

30

Johns, "Feminism
and Utopianism,"
174.

31

British decoloni-
zation began in
1945—India was
granted indepen-
dence in 1947—and
was in full swing by
the mid-1950s.

agriculture, and big data have replaced isolated, point source disasters like a nuclear attack as the boogeymen of contemporary risk society.[25]

Considering this shift, what are the implications for the ways in which designers conceive of and construct the future? After all, apocalypses are not utopias or dystopias; they are the moment when the world that was ends and the new world begins. Yet utopias and dystopias are often reliant on the apocalyptic as a way to create or explain their critical break with the existing world. John McMorrough has suggested that the "specter of the apocalypse is another version of the modernist tabula rasa, a leveling of the past to make way for the future."[26] Apocalypses are all potential.

In their recent book, *Dreamscapes of Modernity: Sociotechnical Imaginaries and the Fabrication of Power*, Sheila Jasanoff and Sang-Hyun Kim define imaginaries as "collectively held and performed visions of the future, animated by shared understandings of forms of social life and order."[27] Building on this definition, it would seem that design and planning—as cultural and material performances—should mediate between the now and that collectively imagined future. Design has done this in the past. In the nineteenth and early twentieth centuries the transformation of the desert from wasteland to garden emerged as a nationalistic symbol of technological prowess and a rationale for territorial expansion. We see the "successes" of this mythopoeia in the US Bureau of Reclamation's massively scaled infrastructural attempts to alter first the rainfall and then the hydrology of the Western United States, and we see it again in the early-twentieth-century "redemption" of the arid Palestinian landscape by Jewish settlers via the mechanism of modern scientific forestry.[28] Projects like these—ethically and ecologically problematic as they are—nonetheless provide cultural touchstones for the dry apocalypse. They prove, at least in some positivist sense, that the desert is redeemable.

If climatic reclamations were modernity's answer to the dry apocalypse, it would seem that the design community has yet to offer any meaningful contribution toward the future imagined in the wet apocalypse. In a number of ways, the wet apocalypse shares many common tropes with the feminist utopias of the 1970s as advanced by science fiction authors like Ursula K. Le Guin in *The Dispossessed* or Marge Piercy in *Woman on the Edge of Time*. According to Alessa Johns, these feminist utopias share five common themes: (1) a focus on education and intellectual development, (2) a view of human nature as malleable and social rather than determined and individualist, (3) a cumulative approach to history and shared approach to power, (4) an understanding of the natural world as dynamic rather than static, and (5) a pragmatic approach to social change.[29] At least three of these appear within the wet apocalypses of the mid-twentieth century onward: their view of human nature as malleable, their cumulative approach to history, and their understanding of the natural world as dynamic rather than static. Johns argues that these shared structures set feminist utopias apart from their traditional, paternalistic counterparts, which feature fully mapped, centralized, and geometrically manifested seats of power, and state-sanctioned behaviors, disciplines, rituals, and symbols.[30] Perhaps then the sharing of themes between feminist utopias and the wet apocalypse stems from a search for an alternative to this paternalistic course. Wet apocalypses of the 1960s shared a desire for (or anxiety over) the end of the colonial era—particularly the British colonial era.[31] The tropical setting of these tales thus evokes a principle crisis of colonialism: its inability to reconcile itself with a tropical nature. One could therefore see these early-twentieth-century concerns over the maintenance of the white body as a static form within a tropical environment giving way to a postwar inversion—a romanticized mutability of human nature in response to a vastly altered and dynamic environment. Similarly, the idea of the dry apocalypse as a wiping clean of past society in order to make way for a new, ahistorical condition gives way to a cumulative process of social and genetic change, gradually but fundamentally altering the way humans interact with one another and their environment.

If so, the wet apocalyptic imaginary poses distinct problems for traditional design and planning. In many ways the modes of production with these disciplines (organization, structure, scenario) are inherently paternalistic—they map out a given future for the built environment, however contingent or provisional, and they document how people will, or should, live there. But what the wet apocalypse suggests is a readiness to accept not just an altered environment or mode of living but a changed way of being entirely. And this way of being, like the environments and "lifestyles" that came before and will inevitably come after, must also be designed.

Jacob Boswell is an assistant professor of landscape architecture at the Ohio State University's Knowlton School. Jacob's writing attempts to trace the effect of social and scientific imaginaries on the production of designed and vernacular landscapes in the United States and abroad. His design work builds on this fascination with the imaginary, forwarding speculative environmental futures that stem from both historic and contemporary discourse around medicine, climate, and the body. Jacob's written and design research have been published both nationally and internationally. He is the recipient of the 2013 Ohio State University Alumni Award for Distinguished Teaching.

Gaïa Global Circus:
A Climate Tragicomedy

RANIA GHOSN AND EL HADI JAZAIRY

All images in this essay are of *Gaïa Global Circus*, project by Bruno Latour, play by Pierre Daubigny, directed by Frédérique Aït-Touati and Chloé Latour, Compagnie AccenT and Soif Compagnie, at The Kitchen, 2014. Photographs © Paula Court, courtesy of The Kitchen.

WHAT SORT OF STORY IS CLIMATE CHANGE?

Gaïa Global Circus takes aim at the deficiency of our emotional repertoire for dealing with the climate crisis—a condition that this theatrical event's conceiver Bruno Latour describes as the "abysmal distance between our little selfish human worries and the great questions of ecology."[1] This experimental play can be seen as a confluence of two areas of interest: On the one hand, the director and artists sought to reanimate the theater's historic connection with the cosmos, and on the other, the public scholar questioned how he might best address environmental disasters beyond the usual apocalyptic cultural imaginary. These two groups share a sense that the great challenge facing the debate around climate today is one of new forms and forums of eco-political engagement. And both also address a shared concern: If the threats are so serious, if we worry once again that the sky might be falling on our heads, how is it that we are all so little mobilized?

In her analysis and critique of the abstract images produced by experts in the discourse of climate change, Birgit Schneider elaborates on problems of perception as well as of scale. People observe daily weather changes, she notes,

[1]

Laura Collins-Hughes, "A Potential Disaster in Any Language: 'Gaïa Global Circus' at the Kitchen," *New York Times*, September 25, 2014, http://www.nytimes.com/2014/09/26/theater/Gaïa-global-circus-at-the-kitchen.html.

but they do not perceive climate—which is, according to its modern definition, a statistically created, abstract object of investigation with a long-term assessment period. Furthermore, people can experience local weather but not the global effects of climate change, which would require no less of them than to perceive the world as a whole.[2] How do we think about something as intangible and invisible as climate? What are the aesthetics and tone of narrating climate change, and to what ends? If environmental issues are un-representable in their scale, their ubiquity, and their duration, then perhaps it falls to works of art (which are still works of thought) to present them to the senses.[3]

Gaïa Global Circus belongs to the genre of the arts of climate change. This rapidly emerging body of work explores the interplay between climatic knowledge and aesthetic experience to engage with the temporal and scalar dissonances of the issue at stake, and to acknowledge and deal with the effects of environmental processes upon life. Such practices deploy a range of aesthetic formats to explore our chaotic relationship with Gaia, be they Olafur Eliasson's ice installations (the most recent of which was at the 2015 Paris Climate Change Conference), Ursula Biemann's video essays, or the *Climate Changed* graphic novel book by Philippe Squarzoni, to name only a few. Latour and his collaborators envisaged a play that commands a new approach to science, politics, and nature by combining varying tones of tragedy, comedy, and ritual.[4] Theater, by their estimation, is uniquely capable of exploring the dramas and emotions not elucidated in public discourse. Their intention was to make sensible our thing-world by creating a collective aesthetic experience, which in turn implies the possibility of new configurations of climatic publics. Their concerns resonate with Ulrich Beck's "emancipatory catastrophism," the term by which he proposes that we can and should turn the question on climate change upside down—not to ask "what can we do for climate change?" but rather, "what is climate change good for?"[5]

POLITICAL ARTS: FROM ABSTRACT KNOWLEDGE TO COLLECTIVE AESTHETIC EXPERIENCE

Latour proposes that climate change calls for a new worldview, one that includes the figure of Gaïa as a new personage on the theater of the world. In his view, the assumed divide between nature and society—and the accompanying focus on deanimate, disembodied, undisputed reason—has led directly into the current ecological crisis. We do not live on a "Blue Marble," insofar as that famous image of our planet symbolizes an objective, holistic, impersonal earth made visible by our own technological achievements. Such metaphysics of technological progress, Latour argues, should now be countered by a redefined assemblage of *values*, so as to extend beyond the critique of the modern objectification of the Earth to a new ecological belief-system In the embodiment of Gaïa. This carries a scientific as well as a mythological dimension—Gaïa derives from technological processes of modeling and measurement but also incorporates an abundance of mythological connotations, as its name evokes the Greek goddess of Earth. Gaïa is an "odd, doubly composite figure … the Möbius strip of which we form both the inside and the outside, the truly global Globe that threatens us even as we threaten it."[6] Latour cites *The Revenge of Gaïa* (2006), in which James Lovelock discusses positive feedback "tipping points" leading to significant and irreversible climate system changes.[7]

2

Quoted in Antonia Mehnert, "Climate Change Futures and the Imagination of the Global in Maeva! by Dirk C. Fleck," *Ecozone*, vol. 3, no. 2 (2012): 28.

3

Gaïa Global Circus, http://www.bruno-latour.fr/fr/node/359.

4

The collective work was undertaken with Chloé Latour and Frédérique Aït-Touati (directors), Claire Astruc, Jade Collinet, Matthieu Protin, and Luigi Cerri (actors), and Pierre Daubigny (playwright).

5

Ulrich Beck, "How Climate Change Might Save the World," *Development and Society*, vol. 43, no. 2 (2014): 169–183.

6

Bruno Latour, *An Inquiry into Modes of Existence: An Anthropology of the Moderns* (Cambridge, MA: Harvard University Press, 2013), 9f.

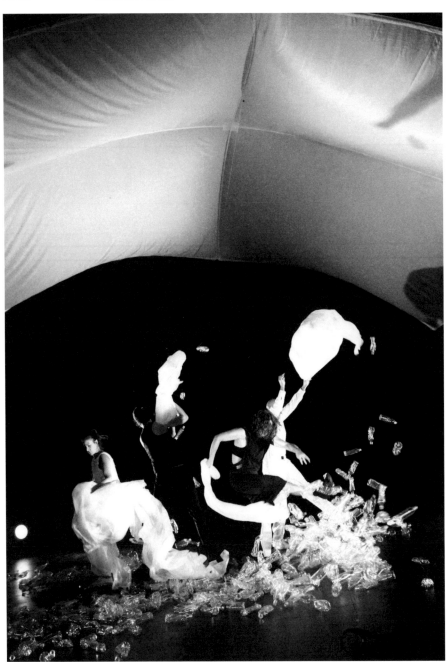

7

Lovelock defines Gaïa as "a biotic-planetary regulatory system. Over 30 million types of extant organisms, descendant from common ancestors and embedded in the biosphere, directly and indirectly interact with one another and with the environment's chemical constituents. They produce and remove gases, ions, metals, and organic compounds through their metabolism, growth, and reproduction. These interactions in aqueous solution lead to modulation of the Earth's surface temperature, acidity-alkalinity, and the chemically reactive gases of the atmosphere and hydrosphere." See James Lovelock, *The Revenge of Gaia: Why the Earth Is Fighting Back and How We Can Still Save Humanity* (London: Penguin Books, 2007).

Beyond the accumulation of scientific knowledge, Gaïa embodies questions of representation, of what the issues are and where we stand vis-à-vis those issues. For Latour, "the Big Picture is just that: a picture. And then the question can be raised: in which movie theatre, in which exhibit gallery is it shown? Through which optics is it projected? To which audience is it addressed?"[8] Beyond the big picture, the absorption of this concept of Gaïa in the public consciousness requires a new and different rhetoric that connects political ecology with the energy of collective aesthetic experience. Latour calls for a new worldview that might "counter a metaphysical machine with a bigger metaphysical machine." He adds: "Why not transform this whole business of recalling

8

Bruno Latour, *Reassembling the Social: An Introduction to Actor-Network-Theory* (Oxford: Oxford University Press, 2005), 187.

9

Latour, *An Inquiry
into Modes of
Existence*, 23.

modernity into a grand question of design?"[9] His response calls for crafting
the "political arts"—an experimental method for conceiving and responding
to the problem of climate change. If politics is the art of the possible, then the
multiplication of the possible requires a reconnection with the many available formats of the aesthetic. The project of the political arts fits into Latour's
broader quest for a new eloquence with which to engage political ecology. In
his books *Making Things Public* (2005) and *Politics of Nature* (1999), both
of which include the word "democracy" in their subtitles, Latour explores the
gap between the importance of the politics of representation in politics and
ecology and the narrow repertoire of emotions and sensations with which we
understand these issues. He asks what would happen if politics revolved instead
around disputed things, atmospheres, natures, and what techniques of representation might help make them public. In his recent book *An Inquiry into
Modes of Existence* (2013), Latour demands nothing less than to overcome
the modern preoccupation with objective scientific truth and to rediscover
the plurality of vastly different modes of existence (like religion, morality, or
law). Latour repeatedly states the reason for which this is needed at this very
moment: "Gaïa approaches."[10]

10

Latour, *An Inquiry
into Modes of
Existence*, 13.

THE THEATER: MAKING CLIMATE PUBLIC

Latour argues that the assembly, the model of political accord organized
according to a very particular architecture (for example, Étienne-Louis
Boullée's Cenotaph for Isaac Newton) has disappeared. Which assembly,
then, are we in now? What spaces could stage a totality, especially when that
whole is opaque, fragmented, contradictory? In *Reassembling the Social*,
Latour outlined the *panorama* as a historical visual practice and space that
stages such a sense of wholeness. From the Greek *pan-* (all) and *-rama*
(spectacle), the panorama is a view of totality. Installed in rotundas, panoramas were immense 360-degree paintings that hermetically surrounded the
observer. From a darkened central platform, the observers found themselves
completely enveloped in visual illusions illuminated by concealed lighting.
These "sight travel machines" transposed the visitors into the image, be
it simulations of distant lands, familiar cities, or catastrophes of nature or
wars.[11] Struck with enchantment in the middle of a magic circle, the spectator is sheltered from unwelcome distractions all while being immersed in
a foreign landscape. Latour found these contraptions quite powerful, particularly as they solved the question of staging totality and nesting a range
of scales, from the micro to the macro, into one another. However, he also
points to their limitations, in that "they don't do it by multiplying two-way
connections with other sites." A panorama designs a picture with no gap in
it, "giving the spectator the powerful impression of being fully immersed in
the real world without any artificial mediations or costly flows of information
leading from or to the outside."[12]

11

Stephan
Oettermann, *The
Panorama: History
of a Mass Medium*
(New York: Zone
Books, 1997).

12

Latour, *Reassem-
bling the Social*,
187.

The limits of the panorama as a form of composing totality led Latour
to explore other modes of representation, particularly those that stage their
own technology and capitalize on their distance from the real. In describing
controversies and scientific evidence, Latour has worked on what he calls "the
theater of proof": how evidence is made *convincing* in the eyes of the wit-
nesses. This is not to jeopardize the actual qualities of the evidence but rather
to show what motivates scientists to develop *effective* evidence. This research

in turn interested Latour in the reverse process: how the stage might help scientists, especially climatologists, follow the threads of what makes convincing proof—a crucial issue at a time when climate skeptics have such influence on public opinion.[13] Hence the idea that he could explore, onstage, all the dissonances of climate change with an "older and more flexible medium than philosophy." For Latour, "only the theater can afford to explore the range of passions corresponding to contemporary political issues."[14]

HOW DO WE TALK WHEN WE TALK IN CLIMATE THEATER?

The Theater is thus the collective aesthetic equivalent of the Parliament or the Congress. It appropriates the technologies of the "image machine" to place the story of climate change, a story that is difficult both to tell and to hear, at the center of the "Theater of the World." The theater is neither theory nor teaching; it is a practice that makes possible through the medium of the stage a thought experiment that is done in public, not just in the head.[15] This form of communication addresses environmental matters by sharing them in full scale and in real time with an audience that is assembled in small collectives. It responds to the accelerationist temporality of climate change, a phenomenon well represented in recent short videos on human-induced climate change. One such example is *Welcome to the Anthropocene*, a three-minute roller-coaster ride through the latest chapter in the story of how one species has transformed a planet. Commissioned by the London Planet Under Pressure conference, *Welcome to the Anthropocene* provides a data visualization of the state of the planet. It opens at the beginning of the Industrial Revolution. As the camera swoops over Earth, viewers watch the planetary impact of humanity: cities, roads, railways, pipelines, cables, and shipping lanes, until finally the world's planes spin a fine web around the planet.[16] Contrary to such representations of acceleration, *Gaïa Global Circus* slows down thought to ground it in the immediacy of the present. Latour's piece also adopts a different narration tone. Rather than a foretold tragedy as it unfolds in disaster movies and short films, *Gaïa Global Circus* is a tragicomedy that blends those opposing but complementary genres with decorum, in order to prevent the listeners from falling into the excessive melancholy of what is at stake.

With monsters, storms, a modern-day Noah, scientists, and divinities onstage, the theater is the setting in which the performance and speech of nonspeaking and nonhuman entities operate as devices of estrangement. *Gaïa Global Circus* counters the familiarity of disaster satellite images that numb the senses into a "feeling of vaguely blasé nonchalance."[17] The piece animates an era when humans recognize their transformation into a climatological entity, all while foregrounding the frictions and dissonance of cross-scalar, multispecies, and intertextual thinking. It is a show that reflects on the tensions between the cacophony of human positions on ecology, our own contradictions in relating to them, and what encompasses and surpasses them. These various threads trace, watch, project, worry, make astonishing discoveries, and knit together the voice of Gaïa—a voice that has many interpretations, because it emanates from a complex and non-unified figure. *Gaïa Global Circus* animates the earth in an era when humans recognize their transformation into a climatological entity, all while hindering the possibility of a simple identification with the characters in the play. It

13

Gaïa Global Circus, http://www. bruno-latour.fr/fr/ node/359.

14

Bruno Latour, A propos de Gaia Global Circus (GGC) Réponses à quelques questions fréquentes (FAQ), http://www.bruno-latour.fr/sites/ default/files/ downloads/FAQ %20GAIAGLO BALCIRCUS_0. pdf.

15

Gaïa Global Circus, http://www. bruno-latour.fr/fr/ node/359.

16

See *Welcome to the Anthropocene*, http://www. anthropocene.info/ short-films.php.

17

Frédérique Aït-Touati and Bruno Latour, "The Theatre of the Globe," *Exeunt*, February 13, 2015, http:// exeuntmagazine. com/features/ the-theatre-of-the-globe.

invites the audience to engage the performed actions and utterances on an aesthetic and cognitive plane, rendering them astonishing in intellectually challenging and sometimes frightening ways.

Faced with this inaudible speech, the theater intervenes with its proper tools: thought experiments in the form of scenic and mental images are active fictions of a world yet to come. This model of the theater resonates with Donna Haraway's concept of "worlding" as a process of actively reimagining a non-anthropocentric world. "These knowledge-making and world-making fields," Haraway observes, "inform a craft that for me is relentlessly replete with organic and inorganic critters and stories, in their thick material and narrative tissues."[18] The model of the world that *Gaïa Global Circus* projects moves away from the dominant discussion of technical fixes for the climate, which focus on the improvement of technology, information, and policy incentives as means to "manage" or even "reverse" climate change. Rather, it proposes to advance new hypotheses and cultivate thinking about what current technologies, theories, or habits can't yet solve. It is not "the job of theatre to find a solution," Latour notes, but to play with "the dialectic between philosophical reasoning and theatrical experiment ... It is a dance, rather than an argument."[19]

A NEW PERSONAGE HAS ENTERED THE THEATER OF THE WORLD

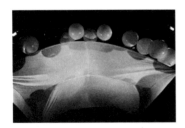

In his article titled "La Non-invitée au Sommet de Copenhague"—roughly translated as "Who Wasn't Invited to Copenhagen?"—Michel Serres points to the one empty seat at Copenhagen's Parliament of Things: that of Gaïa. He wondered how to make it possible for her to sit, speak, and be represented. What is the Gaïa equivalent of Thomas Hobbes's Leviathan's frontispiece? The challenge of governing the climate is that we are addressing the global without a world state, requiring a form of representation to think through the new geopolitics of climate change.

18

Donna Haraway, "SF: Science Fiction, Speculative Fabulation, String Figures, so Far," Pilgrim Award Acceptance (2011), http://people.ucsc.edu/~haraway/Files/PilgrimAcceptanceHaraway.pdf.

19

Jonas Tinius, "'All the World's a Stage?' A Review of Bruno Latour's *Gaïa Global Circus*," March 3, 2015, http://allegralaboratory.net/all-the-worlds-a-stage-a-review-of-bruno-latours-Gaïa-global-circus.

Gaïa Global Circus responds to this provocation by borrowing from techniques of the Baroque theater. It deploys the ancient theater of shadows and more contemporary optical machines to imagine a *theatrum mundi* for our time. The scenography makes sensible the scalar dissonance between the human and nonhuman, and explores a possible relationship with the environment in which the human is no longer the center. The play takes place in a circus tent, with the audience occupying one part of the arena on stepped rows of seats. Both actors and spectators are under a canopy on which different atmospheres are projected—similar to other world representations like a geodesic dome or planetarium. The stage becomes an actor in its own right. It seeks to capture the issue of an environment that no longer surrounds us because it has become a player on the world stage. The centerpiece of the décor is a translucent canopy floating in the air and suspended by helium balloons. This mainsail device (measuring some 20 by 25 feet) enables the actors to transform the stage area at every moment, as it can be moved like a canopy over any portion of the theater. When a storm from what seems like the end of the world rumbles through, the floating canopy envelops the audience, as a comfort object or a security blanket. Both a model of the world and a wonder object in itself, the "flying tent" is both an effort to put the world onstage and an attempt to question our perception of Nature. Mobile, changing, and unpredictable, this décor-actor is a living object moved by the actors, which transforms the stage and constantly produces atmospheres and climates. At every performance, this flying machine seeks a collective experience of another relation to our common world, at the scale of the theater. "In a way," Latour notes, "this canopy screen is the lead actor in the play."[20]

20
Bruno Latour,
Frédérique
Aït-Touati, and
Chloé Latour,
"Material for Stage
Writing Within
the Framework of
the Project: *Gaïa
Global Circus,*"
trans. Julie Rose
(May 2011),
http://www.
bruno-latour.
fr/sites/default/
files/downloads/
KOSMOKOLOS-
TRANSLATION-
GB.pdf.

GAÏA, THE URGENCY TO THINK AND FEEL

Just as a geologist can hear the clicks of radioactivity, but only if he is equipped with a Gcigcr counter, we can register the presence of morality in the world provided that we concentrate on that particular emission. And just as no one, once the instrument has been calibrated, would think of asking the geologist if radioactivity is "all in his head," "in his heart," or "in the rocks," no one will doubt any longer that the world emits morality toward anyone who possesses an instrument sensitive enough to register it.

—Bruno Latour, *An Inquiry into Modes of Existence* [21]

Why is Gaïa the lead actor in the play? Because global warming, the most important event concerning us (according to climatologists and environmentalists) is also the symptom of the emergence of this new controversial figure called Gaïa. *Gaïa Global Circus* appeals to affective, aesthetic, and media practices in an effort to address the cognitive dissonance between the scale of the issues to be addressed and that of the set of emotional and experiential states that are associated with the task. It is one appeal for an aesthetic practice to engage the contemporary pressing matters of the world. "If theatre is to become, once again, the theatre of the globe," Latour observes, "then it must re-learn, like Atlas, how to carry the world on its shoulders, both the world and all there is above it."[22] It must relearn the pleasure of a collective aesthetic experience of connecting our individual dynamics of hope, fear, and desire to a larger scale of environmental, planetary, and ultimately cosmic dynamics of the same order. At the core of *Gaïa Global Circus*, you find a fundamental question about the fabric of reality, the forms of knowledge that frame that reality, and the impossibility of ever fully knowing or comprehending it. Yet, to quote Isabelle Stengers, a philosopher and longtime interlocutor of Latour's, Gaïa has the urgency to induce thinking and feeling in a particular way.[23]

[21] Latour, *An Inquiry into Modes of Existence*, 456.

[22] Aït-Touati and Latour, "The Theatre of the Globe."

[23] Isabelle Stengers, "Gaia, the Urgency to Think (and Feel)," *Os Mil Nomes de Gaia* (September 2014), https://osmilnomesdegaia.files.wordpress.com/2014/11/isabelle-stengers.pdf.

Rania Ghosn and El Hadi Jazairy are partners of DESIGN EARTH. The practice engages the geographies of technological systems to open up a range of aesthetic and political concerns for architecture and urbanism. DESIGN EARTH's work has been recognized with several awards, including the 2015 Jacques Rougerie Foundation's First Prize.

Ghosn and Jazairy hold doctor of design degrees from Harvard University Graduate School of Design, where they were founding editors of the journal *New Geographies*. They are authors of the recently published *Geographies of Trash* (Actar, 2015), for which they received the 2014 ACSA Faculty Design Award. Some of their recent work has been published in *Journal of Architectural Education*, *MONU*, *San Rocco*, *Bracket*, *Perspecta*, and *Topos*.

The Magical Ecology of *Laudato Si'*

CATHERINE SEAVITT NORDENSON

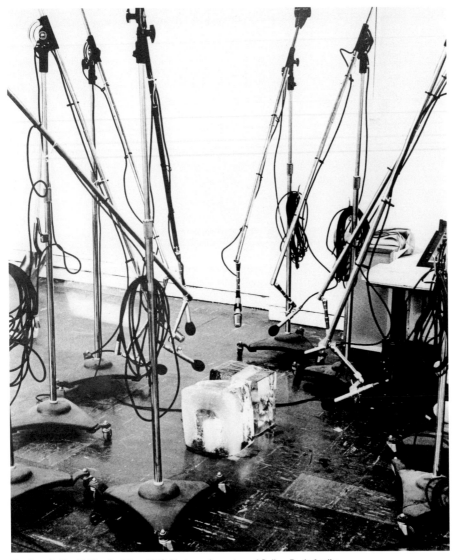

The Sound of Ice Melting, Paul Kos, 1970. Photograph courtesy of Gallery Paule Anglim.

> Many years later, as he faced the firing squad, Colonel Aureliano Buendía was to remember that distant afternoon when his father took him to discover ice...
> —Gabriel García Márquez, *One Hundred Years of Solitude*

The planet and the environment are the topics of the most recent encyclical letter of Pope Francis, the former Jorge Mario Bergoglio, released by the Vatican on June 18, 2015. The letter, *Laudato Si'* (Praise Be to You, from a canticle by Saint Francis of Assisi), was released by the Vatican just a

Pope Francis's motorcade in Central Park, September 25, 2015.

few months before his wildly popular September visit to the United States. Cruising the streets and avenues of New York City in his little black Fiat Cinquecento, he was a fascinating sight. That white-robed arm waving at us from the open window of the backseat, in contrast to the flashing red lights of the New York City Police Department's escort, spoke volumes. The choice of Central Park as the site for Francis's Friday papal motorcade was a brilliant one, evoking the elastic power of Olmsted and Vaux's Greensward Plan to provide a graft of the natural world as the setting for this twenty-first-century version of the 1960s "Be-In." Too bad for Francis that his view of the tree canopy's biomass was seen through a plastic-bubbled Popemobile, though he didn't appear to mind. Even the massive traffic disruptions didn't seem so bothersome to typically grumpy New Yorkers. Need to get around with the Pope in town? We just walked—it took longer, but it's better for the environment anyway, and the weather was great.

Many reactions to Francis's letter are concerned with the perceived correctness (or incorrectness) of the Pope's climate science. Indeed, the references and sources are almost entirely drawn from theological precedent, but there are secular references as well—most notably two United Nations documents, the 1992 Rio Declaration from the Earth Summit and the 2000 Earth Charter. Clearly the Pope was hoping to influence the forthcoming United Nations Climate Change Conference, COP21, held in Paris in December 2015, and one could now argue he was successful. But it is the literary quality of the poetic writing and Francis's choice of language in *Laudato Si'* that is most interesting to me. A massively important aspect of climate change is communication, and Francis has a fantastic ability to tell a story. Scientists, unfortunately, don't always do such a good job. Of course, the letter was written in Spanish; I read the English translation as published by the Ignatius Press of San Francisco, and perhaps it would be better to read the original. This seems to be a faithful translation, however, as one would expect from the Vatican translators. Many of the poetic chapter subheadings capture the imagination: "Nothing in this world is indifferent to us." "Weak Responses." "The Common Destination of Goods." "Justice Between the Generations." "Beyond the Sun." My personal favorite: "Ecological Conversion." Despite the ecclesiastical nature of the text, Francis's prose has the ability to engage a reader, no matter his or her religious convictions. I suspect the Pope imagined that the letter's readership would extend well beyond the list of those in its subtitle—Bishops, Priests, and Deacons, Men and Women Religious, and the Lay Faithful—and he acknowledges this in Chapter Two, the one that addresses the "convictions of believers" most explicitly. Francis wants to start a conversation, not shut it down, and he sees religion as a productive part of a dialogue:

> I am well aware that in the areas of politics and philosophy there are those who firmly reject the idea of a Creator, or consider it irrelevant, and consequently dismiss as irrational the rich contribution which religions can make towards an integral ecology and the full development of humanity. Others view religions simply as a subculture to be tolerated. Nonetheless, science and religion, with their distinctive approaches to understanding reality, can enter into an intense dialogue fruitful for both.
> —Chapter Two, "The Gospel of Creation," Paragraph 62

Certainly, some of Francis's thoughts read like a sensible list of best practices; in Paragraph 211 he evokes "ecological citizenship," suggesting we reduce our use of paper and plastic, conserve water, plant trees, separate our trash for recycling, use public transportation, and not waste food. But he also importantly addresses a point that is sometimes overlooked: climate change is not just an environmental problem; it is also a social one. Francis notes the unequal effects that changes will bring to the poor, a concern that is one of the hallmarks of his papacy:

> Today, however, we have to realize that a true ecological approach *always* becomes a social approach; it must integrate questions of justice in debates on the environment, so as to hear *both the cry of the earth and the cry of the poor.*
> —Chapter One, V, "Global Inequality," Paragraph 49

However, where the Pope really gets it right is on the topic of biodiversity. And this, I find, is where he reveals his South American roots, where environmentalists have been most profoundly articulate in their critique of *desenvolvimento* (development) and the devastating effects of species loss. Francis makes several references to the Amazon basin and deforestation, explaining that production and profit should not justify the clearing of a forest. Losses of biodiversity must be factored into the equation of profit. Francis's discussion of biodiversity is particularly insightful—he describes a kind of "intrinsic value" that each species possesses, beyond any discussion of profit or loss. This intrinsic value, and the message to us as fellow earthlings, is what is at risk.

> Each year sees the disappearance of thousands of plant and animal species which we will never know, which our children will never see, because they have been lost forever. The great majority become extinct

for reasons related to human activity. Because of us, thousands of species will no longer give glory to God by their very existence or convey
their message to us. We have no such right.
—Chapter One, III, "Loss of Biodiversity," Paragraph 33

Pope Francis's strongest message in the letter, I would argue, is not the
acknowledgement of the Church's acceptance and adaptation to science, or
even the endorsement of climate science, but rather the insistence that we
need to readjust the relationship between human and nonhuman biota, as
well as between humans and the planet. He evokes a sense of stewardship and
responsibility by quoting Genesis 2:15, the notion of "tilling and keeping the
garden of the world." In fact, the Pope almost seems to approach a Parliament of Things—might Francis have Bruno Latour's books on his bedside
table at the Vatican Guesthouse?

> Although it is true that we Christians have at times incorrectly inter-
> preted the Scriptures, nowadays we must forcefully reject the notion that
> our being created in God's image and given dominion over the earth
> justifies absolute dominion over other creatures.... This implies a rela-
> tionship of mutual responsibility between human beings and nature.
> —Chapter Two, II, "The Wisdom of the Biblical Accounts,"
> Paragraph 67

The literary influences that emerge in Francis's writing are most compelling.
As a Jesuit priest in Santa Fe, Argentina, Francis taught secondary school
boys literature and creative writing, and is a professed admirer of Dostoevsky,
Hölderlin, and his fellow Argentine Jorge Luis Borges, whom he met as a young
man. There even seems to be some magical realism in the Pope's narrative style.
Not in a dreamy Brazilian *telenovela* sense of magic, but in the way that Gabriel
García Márquez describes magical realism, as the incredible and fantastical stories told to him as a child by his grandmother. She delivered these unbelievable
stories without changing her expression, with a serious stone face. Márquez
reflects on this skill in a 1981 interview with Peter Stone for the *Paris Review*:
"That's a journalistic trick which you can also apply to literature. For example, if
you say that there are elephants flying in the sky, people are not going to believe
you. But if you say that there are four hundred and twenty-five elephants flying
in the sky, people will probably believe you. *One Hundred Years of Solitude* is
full of that sort of thing. That's exactly the technique my grandmother used."
Francis, likewise, captures our imagination, employing everyday metaphors and
evocative imagery—sheep in hair curlers, for example—to convey deep meaning. And this makes us listen, imagine, and pay attention. It's an effective way to
deliver a message, particularly one about the importance of the unseen aspects
of our environment. The Pope even takes us on a dive into the ocean, plunging
us down into this great aquatic forest to show us our invisible impact.

> In tropical and subtropical seas, we find coral reefs comparable to
> the great forests on dry land, for they shelter approximately a million
> species, including fish, crabs, mollusks, sponges, and algae. Many of
> the world's coral reefs are already barren or in a state of decline. "Who
> has turned the wonderworld of the seas into underwater cemeteries
> bereft of color and life?"
> —Chapter One, III, "Loss of Biodiversity," Paragraph 41

Consider Francis's discussion of biodiversity. What exactly does he mean by "approximately a million"—is that a Jesuit koan? I am reminded of the wonderful taxonomy presented by Borges, with his intriguing and compelling categorizations of animals in the 1942 essay "The Analytical Language of John Wilkins" (*El idioma analítico de John Wilkins*). He sorts them into just fourteen categories. There are those that belong to the emperor; mermaids; those that tremble as if they were mad; those that have just broken the flower vase; those that, at a distance, resemble flies. They all seem unique, and vastly important. Francis also gives us a list of the invisible in Paragraph 34: "The good functioning of ecosystems also requires fungi, algae, worms, insects, reptiles, and an innumerable variety of microorganisms." We feel a Buddhist sense of "save all beings" moving through Francis's discussion of loss and inherent value, and given his Jesuit training and his namesake Saint Francis of Assisi, this comes as no surprise. Our scientific minds acknowledge the reality of the planet changing. We must change too, Francis gently reminds us, and like Borges, subverts our rational notions of order. Amen to ecological conversion, Pope Francis.

Catherine Seavitt Nordenson is an associate professor of landscape architecture at City College of New York and principal of Catherine Seavitt Studio. Her research focuses on design adaptation to sea-level rise in urban coastal environments and explores novel landscape restoration practices, given the dynamics of climate change.

Views from the Plastisphere:
A Preface to Post-Rock Architecture

MEREDITH MILLER

Plastiglomerate sample, 2013. This and the following plastiglomerates were collected through a collaboration between Kelly Jazvac, geologist Patricia Corcoran, and oceanographer Charles Moore. Photograph by Jeff Elstone, courtesy of Kelly Jazvac.

WHAT DOES ARCHITECTURE HAVE TO DO WITH GLOBAL CLIMATE CHANGE?

You might respond, "it has everything to do with global climate change" and then further articulate this sentiment with a list of the building industry's culpable features, such as the carbon footprint of steel production, or all the emissions from transporting construction materials, or the large share of energy consumption by buildings, or at an urban scale, wasteful mobility habits caused by the horizontal expansion of cities.

Okay. A fine response. But let me ask the same question again: "What does architecture have to do with global climate change?" As in, how does design as a cultural practice bear on the extensive processes of atmospheric modification and the social, scientific, and political circumstances through which we have come to know about, and have attempted to manage, this slow inevitability? To some, this version of the question might overreach or sound too theological coming from an architect. It's like asking, "Where does my design fit within the cosmic order of the universe?" But perhaps speculating on the tangible links

This text parallels a research and design project I am currently working on with Thom Moran, which speculates on plastiglomerates as a future building material. Called "Post Rock," this project is funded by the Research Through Making Program at Taubman College of Architecture + Urban Planning, University of Michigan, and was exhibited in March 2015.

between the immediate experience of architecture and its planetary milieu is exactly where we might look for design's significance right now. Escaping the nowhere of abstract metrics and summary figures that characterize climate's representation, and pursuing instead a cosmology of physical things, where might we end up?

It is in search of this cosmological perspective that I want to first offer a contemporary parable. On the surface, the parable involves land, water, and garbage moving about the earth's surface in biblical proportions. But beneath these large maelstroms of matter and energy, there is, I hope, a small lesson for architecture and about climate.

A CONTEMPORARY PARABLE FOR ARCHITECTURE AND ABOUT CLIMATE

Three travelers, having walked the greater part of the day, come upon a kind of monument made of what appears to be stone. The exterior surfaces of the open-air structure are very smooth and similarly colored, as if each large block had been cut from a much larger boulder or outcropping and then polished. The interior surfaces, on the other hand, are rough and uneven, with bulges and dimples of varying shapes and sizes. The travelers cannot determine whether this unevenness resulted from someone carving away purposefully at the stone or if these are the eroded edges of the original boulder or outcropping turned inward. From apertures in the ceiling, light falls upon these pockets and lumps to reveal the orange curve of a family-size detergent bottle here, a scattering of seashells and rubber oyster tubing there, and throughout, marbled stalagmites of misshapen ropes, plastic mesh, and minerals. Despite these fragmentary indications to the contrary, the overall effect of the space is entirely monolithic, as if the whole thing has been quarried directly from the earth. The travelers then wonder if the crenulations are too colorful, the crystalline specks too bright, and the cave-like forms too reminiscent of garbage for this material to be strictly geological...

On second thought, let me start again:

Three figures, two women and a man, are walking along a remote seashore. The beach is punctuated with black volcanic rock, but its most distinctive feature is a layer of colorful trash occupying a wide margin of sand. They look past the loose pieces of plastic and glass pushed about in the surf and inspect various nooks in the volcanic rock. Stooping over, one of them dislodges a softball-size object from a cluster of rocks, driftwood, and trash. Mostly smooth and granite-like in texture, the object is clearly geological in origin, except there is a bit of yellow nylon rope protruding from one end and a marbled neon-pink vein down the center.

These people happen to be an earth scientist, an artist, and a marine scientist. Defying categorization as either geological material or manufactured product, this rock-like object, and the many similar ones they collect along Hawaii's Kamilo Beach, prompts them to come up with a new system for classifying this hybrid material. In June 2014, the *Geological Society of America* published their report announcing a stone: "plastiglomerate." According to the article, a plastiglomerate is "an indurated, multi-composite material made hard by agglutination of rock and molten plastic."[1] This rock-like substance results from plastic waste of various sizes and types accumulating in the world's oceans and beaches. Much of this plastic breaks down into

1 Patricia L. Corcoran, Charles J. Moore, and Kelly Jazvac, "An Anthropogenic Marker Horizon in the Future Rock Record," *Geological Society of America Today*, vol. 24, no. 6 (June 2014): 4–8.

smaller bits called microplastics (less than 5 mm in diameter); some remains on the ocean surface; some of it drifts down and away from the ocean surface; some is ingested by marine life; and some, these researchers discovered, fuse with sand, shells, stone, glass, and other marine debris to form strange, heterogeneous rocks. In other words, a plastiglomerate is an emergent product of human and geological processes.

Given the inherent durability of both stone and thermoplastics, plastiglomerates are likely to last for a very long time. If geology is the study of the earth's history, plastiglomerates represent an unusual historiographical conundrum. Without a long record of existence (the mass production of synthetic thermoplastics began in the twentieth century), their geological classification relies instead on their likelihood of remaining in the earth's surface far into the future—probably well beyond the human race.[2] Thus, plastiglomerates are yet another indication that the Anthropocene has arrived.[3]

Within this pair of parables, the large stones (speculatively) encountered in an architectural structure and the rocks (actually) collected on the beach, are objects that mobilize concepts and territories outside their immediate moment of encounter. The rocks and stones, with visible remnants of fishing apparatus and seashells fused with smaller multicolored polymer fragments, tell a reverse history of the near and distant materials brought together by the heat of sunshine, the scattering of winds, the churn of ocean currents, the toss of a hand, the stamp of a thermoset mold, and the chemical daisy-chaining of synthetic polymers. While each is unique in its exact composition of parts, they all fall within a bracketed range of hardness, density, and hybrid materiality that places them

2

By plastic I am referring to thermoplastics or synthetic polymers; there are of course other resins that occur in nature. Billie Faircloth's recent book thoroughly dissects the many types of polymers, identifies where they appear in construction products, and reflects on the conceptual limitations placed on plastic's presence in architecture. Billie Faircloth, *Plastics Now: On Architecture's Relationship to a Continuously Emerging Material* (New York: Routledge, 2015).

3

Corcoran, et al., "An Anthropogenic Marker," 4.

Plastiglomerate sample, 2013. Photograph by Jeff Elstone, courtesy of Kelly Jazvac.

4

The researchers distinguished two types of plastiglomerate found on Kamilo Beach: "an in situ type, in which plastic is adhered to rock outcrops, and a clastic type, in which combinations of basalt, coral, shells, and local woody debris are cemented with grains of sand in a plastic matrix." They cite the manner of plastic's adherence to rock ("molten plastic had infilled vesicles in volcanic rock, thereby forming plastic amygdales") and the density of the samples ("Bulk density of the clastic fragments ranged from 1.7 to 2.8 g/cm^3, with the highest values determined from fragments rich in basalt pebbles. The measured bulk densities show that plastiglomerate has greater potential to become buried and preserved in the rock record than plastic-only particles, which typically have densities in the range of 0.8–1.8 g/cm^3"). Corcoran, et al., "An Anthropogenic Marker," 4.

5

The aim here is not to reify the auratic object or speculate on its impenetrable ontology.

6

This is a reference to James Corner, "Agency of Mapping." Many other examples of the system's position in landscape are possible here, such as Corner and Stan Allen's entry to the Downsview Competition in 2000 and its influence captured in publications like *Large Parks*, ed. Julia Czerniak, George Hargreaves, and John Beardsley (New York: Princeton Architectural Press, 2007). In architecture, a significant example from this time period is the work of Jesse Reiser and Nanako Umemoto, who also describe their geometric surface modeling and its effects through an encompassing logic of landscape.

within this new geological series.[4] In this way, the qualities of each plastiglomerate sample, from its colors to its particular proportions of heterogeneous components, are inseparable from the dispersed geographies, energetic inputs, and consumer or commercial refuse that contributed to its making.

What do these physical qualities tell us that abstractions cannot? This essay will position such objects as points of access to broader knowledge formations, in particular the fraught epistemologies of global climate. By focusing on the literal qualities of a thing,[5] this essay looks at the potential of materials for creating new subjectivities in an era of climate anxiety and information overload. As an alternative to the abstracting tendencies of data-focused practices, this claim implies an approach to architecture. It acknowledges that architecture has a particular capacity to work on and through its physical, material specificity in order to make sensible and immediate those ideas that are more abstract and distant. The mediation of climate knowledge through the aesthetic qualities of things will be considered here as a kind of cosmology of subject, object, and environment: a means of apprehending the world by way of the here and now.

ABSTRACTION VS. EVIDENCE

At the turn of the millennium, architecture experienced a renewed interest in the informational, and diagrammatic techniques expanded into a broader descriptive field linking ecological processes with spatial form. Landscape became the envy of building designers; the allure was not in the dirty stuff of soils, plant matter, and hydrology but in the open-ended way in which designers could referee a complex set of circumstances toward an imagined future. The "agency of mapping" placed authorship at a remove from the matter meant to be authored. Notational systems of representation took precedence over measured drawings or experiential images to demonstrate the design's networks of relationships and their open range of possibilities.[6] This widespread shift to the paradigm of landscape was also significant for the scale and scope of architecture's purported capacities: programs, habitats, ecosystems, economies, were all seen as equal subjects for design's management. Anticipating effects and outcomes rather than specifying them, this architecture became more and more infrastructural and, thus, less and less material.

Today's design discourse has clearly benefited from this brand of systems thinking and a broader awareness of an ecological or even planetary context for design. These theories have been influenced by 1960s systems thinkers and environmental designers—figures like Ian McHarg, Buckminster Fuller, and John McHale, among others—evidenced in the representational techniques that support recent landscape-focused practices (network diagrams, energy budgets, data flowcharts). Yet it is instructive to note what representational practices did not carry over from that era of global consciousness. Collaging existing urban paradigms with visionary and formally distinct proposals (think Fuller and Shoji Sadeo's domes and pyramids) gave way to imaging emptiness (think James Corner's Fresh Kills) and demonstrating programmatic indeterminacy. This latter mode of representation is not without an aesthetic, of course, but the emphasis is on behaving like something and not looking like anything.

With some distance from the landscape paradigm's first appearance, it is time to recognize the presiding habits of mind that have developed from this influential chapter in our discipline's recent history. Of particular concern is the idea that the physical and aesthetic qualities of architecture take a backseat to the mutable circumstances that they frame, or that form simply serves to make visible the diagrammatic relations that are the real substance of the work. Either scenario easily argues for design as a passive instrument of information, which in turn implies an impossible neutrality. But we might instead see that architecture (framed here as a material practice) can be an active participant in the construction of, or challenge to, new knowledge formations. The physical and aesthetic qualities of architecture can create visceral cues, sensible reminders of the elsewheres and elsewhens that encompass and support that architecture's existence (and our own).

Moreover, systems practices that perpetuate a modernist concept of Nature's alterity often do so by privileging certain aesthetic categories over others. The scientism that is inherent to a modernist, managerial approach toward the earth's systems breeds a kind of false consciousness, disclaiming the considerable role that aesthetics play in shoring up certain ideological positions. Perhaps turning off that false consciousness would free us up to really "see" the aesthetic categories that often go ignored but that are intrinsic to the many "naturecultures" that constitute our planetary environment. (I am using Donna Haraway's term "naturecultures" here to refer to conditions that have fully dismantled any remaining possibility of a binary separation of culture from nature.) Design that enables us to "see" these "other" aesthetic categories is a step toward understanding the types of cultural and epistemological work they do. As designers we do not just document existing conditions—we can put things together in new ways, adding value through form, image, coloration, organization, and so on. However, this cannot happen from within autonomous bubbles of a "disciplinary" practice. It involves a more inclusive attention to the naturecultures that comprise architecture's contingencies and that condition architecture's reception.

In other words, the data-landscape project is not the only way for architecture to engage matters outside disciplinary boundaries. And conversely, the form-aesthetics project is not limited to discourses of autonomy. The geodesic dome was one techno-utopian image whose proliferation circulated a transforming set of political and social affiliations, while enduring as the aesthetic of technology's empowerment to individuals.[7] (It not only *behaved like* something, it also *looked like* something.) The ideas affiliated with that form evolved from the designer's original intentions; the dome is eminently recognizable and yet, it remains open to appropriation and discourse. Moving from abstraction toward evidence offers a model for architecture's capacity to mobilize ideas and associations outside its immediate material limits. One benefit of this model is the focus on architecture's primary domain of knowledge, which addresses the question of where form, material, and aesthetics can actually have effects in the world.

7

Felicity Scott, *Architecture or Techno-Utopia: Politics After Modernism* (Cambridge, MA: MIT Press, 2013).

Plastiglomerate sample, 2013. Photograph by Jeff Elstone, courtesy of Kelly Jazvac.

MATTERS OF FACT, MATTERS OF OPINION, MATTERS OF CONCERN

> It is not unanimous among scientists that [climate change] is dispropor-
> tionately man-made. What I get a little tired of on the Left is this idea
> that somehow science has decided all this so you can't have a view.
> —Jeb Bush, interview with Fox News, 2011[8]

Climate change is not simply an atmospheric phenomenon. It is a multitude
of competing narratives that shape what we know about climate change and
what we are willing to put at risk in response to that knowing. Among these
narratives is the perennial debate around the scientific evidence for climate
change's anthropogenic causes. In the seven years between the Fourth and
Fifth Assessments authored by the Intergovernmental Panel on Climate
Change (IPCC), the consensus around the reliability of climate models as a
source of data shifted. The Fifth Assessment Report from 2014 included risk
scenarios and projected outcomes that were developed through data models.
Still, public opinion on climate models remains uneven, as Paul N. Edwards
discusses in his history of climate science, *A Vast Machine: Computer Models,
Climate Data, and the Politics of Global Warming*.[9]

   Edwards begins his history with the 1968 portrait of Earth from Apollo
8. He emphasizes that the timing of this new vantage point corresponded
with numerous scientific activities and cultural movements already in motion
(the 1957–58 International Geophysical Year, the United Nations, Cold War

8

Suzanne Golden-
berg, "Jeb Bush
May Be 'The Smart
Brother'—But
He's as Much of a
Climate Denier as
Any Conservative,"
the *Guardian*,
December 15,
2015, http://www.
theguardian.com/
commentisfree/
2014/dec/16/
jeb-bush-climate-
denier-republican-
presidential-
candidate-2016.

9

Paul Edwards,
*A Vast Machine:
Computer Models,
Climate Data,
and the Politics of
Global Warming*
(Cambridge, MA:
MIT Press, 2013).

"closed world" discourses, the "One World" movement, and orbiting satellites with funny names dazzling Western and Soviet audiences). A planetary imaginary was in place, as were the international instruments for thinking and managing a global environment.[10] What is significant about this image, for Edwards, is how it exposes the tremendous gap between the simple immediacy of a single atmosphere as a concept and the intricate, layered, and multi-scaled composition of people, systems, and infrastructures involved in making that atmosphere "knowable" as climate. While computing, infrastructures, and measurement protocols are highly technical and essential components of this construction of climate knowledge, its effective aggregate is a socio-technical system, one that includes the scientists, their habits and errors of judgment, monitoring stations, organizations, and communication systems. It is a "vast machine."[11]

Asking "how did the world become a system?" Edwards frames a detailed account of how early, distributed forms of weather observation became incrementally consolidated into climate science. Localized practices, varying instruments, and uneven material conditions were absorbed into a coordinated network of data, through arduous institutional oversight and the labor-intensive task of reconstructing historical data sets. This system makes it possible to think of the Earth's climate as a "knowable entity" and its climate as something "conceivably managed by deliberate intervention." The incomprehensibly large and complex entity of global climate is rendered "knowable," but its key representation is unstable—and that merging of multiplicity into a single model has been at the heart of debates over the "truth" of climate change.[12] One objection is that science can only produce truths through empirical data or experimental evidence. The problem with this criticism, Edwards points out, is that it wrongly assumes that data has a greater degree of objectivity or autonomy than the sociotechnical system that generates it. Furthermore, the world's climate processes are clearly too complex and too many to reproduce experimentally. Edwards makes the case that the existence of climate models is what makes climate data possible. In other words, the model precedes the data it represents. Without this form of representation, much of climate history would remain in "shadow."

If matters of fact are inseparable from the sociotechnical apparatus that produces them, climate narratives that hang on science's objectivity are equally inseparable from those that appeal to public opinion. (The former Florida governor's insistence on having his own "view" is one of many examples of this popularized distrust in expertise.) Geographer Mike Hulme blames the slow public acceptance of climate change on a failure of communication. The deficit model of communication supposes that if the public is not convinced of a theory, it is due to a lack of information. Hulme argues that the problem is not an information deficit—advances in climate science and the unified front of the IPCC prove climate data's abundance and internal consistency. Instead, he claims, it is a problem of popularizing the information's *message*.[13] He proposes alternative models, including "deliberation," where communication between citizens and the scientific community would move two ways, exchanging the sentiments, beliefs, and histories of those who participate.[14]

The previous two examples expose the counterintuitive ways in which climate data becomes more meaningful when wrapped up with less objective modes of representation. Following these perspectives, it is clear that climate discourse needs a "powerful descriptive tool," as Bruno Latour articulated a

10
Edwards, *A Vast Machine*, 3.

11
Edwards, *A Vast Machine*, 3. The term "vast machine" is taken from a John Ruskin passage he quotes as the book's epigraph: "The meteorologist is impotent if alone; his observations are useless; for they are made upon a point, while the speculations to be derived from them must be on space... The Meteorological Society, therefore, has been formed not for a city, nor for a kingdom, but for the world. It wishes to be the central point, the moving power, of a vast machine, and it feels that unless it can be this, it must be powerless; if it cannot do all, it can do nothing. It desires to have at its command, at stated periods, perfect systems of methodical and simultaneous observations; it wishes its influence and its power to be omnipresent over the globe so that it may be able to know, at any given instant, the state of the atmosphere on every point on its surface. —John Ruskin (1839)." Edwards defines the vast machine of climate science as: "a sociotechnical system that collects data, models, physical processes, test theories, and ultimately generates a widely shared understanding of climate and climate change."

12

In 1988, James Hansen testified before the assembly of the US House Energy Committee, with a graphic analysis of projected temperature ranges based on a model attesting to a 99 percent rate of reliability.

13

Mike Hulme, *Why We Disagree About Climate Change: Understanding Controversy, Inaction, and Opportunity* (Cambridge: Cambridge University Press, 2009), 217.

14

What this looks like is unclear, but it would require various media to convey evolving messages to different subjects. Hulme, *Why We Disagree About Climate Change*, 218–221.

15

"The mistake we made, the mistake I made, was to believe that there was no efficient way to criticize matters of fact except by moving away from them and directing one's attention toward the conditions that made them possible. But this meant accepting much too uncritically what matters of fact were. This was remaining too faithful to the unfortunate solution inherited from the philosophy of Immanuel Kant. Critique has not been critical enough in spite of all its sore-scratching. Reality is not defined by matters of fact. Matters of fact are not all that is given in experience. Matters

of fact are only very partial and, I would argue, very polemical, very political renderings of matters of concern and only a subset of what could also be called states of affairs. It is this second empiricism, this return to the realist attitude, that I'd like to offer as the next task for the critically minded." Bruno Latour, "Why Has Critique Run Out of Steam? From Matters of Fact to Matters of Concern," *Critical Inquiry*, vol. 30, no. 2 (Winter 2004): 231–232.

decade ago, in "Why Has Critique Run Out of Steam?" While Latour helped create a discipline out of investigating the conditions in which scientific knowledge is produced, this essay laments the use (or misuse) of similar constructivist tactics to debunk the science of climate change. Rather than a full reversal, Latour attempts to find another "powerful descriptive tool" that does away with "matters of fact" and allies instead with what he calls "matters of concern."[15] This approach takes root interrogating the materiality of science not as a series of "objects"—which are factual and undesigned—but rather to attend to them as "things," which he frames as a "gathering" of meaning and intents.

While climate is typically considered an atmospheric phenomenon or a representational problem of data, plastiglomerates are physical markers of climate's ongoing transformation. They are the sum of various inputs. Even if their existence cannot be ascribed human authorship, these things are crafted by a more complex composition of industrial and consumer activities, thalassic and riparian forces. Their physical qualities are signatures of this complex composition, or natureculture, that authored them. It would also be possible to say these objects are "post-natural," existing outside a modernist division of civil society from pure nature. They represent a possible avenue for thinking about a more

Plastiglomerate sample, 2013. Photograph by Jeff Elstone, courtesy of Kelly Jazvac.

literal version of architecture's connection to larger milieus. Architecture's capacity to link an aesthetic to a world of ideas, so clearly evidenced by plastiglomerate samples, starts with materiality.

Imagining the geocentric arrangement of a Ptolemaic cosmology, suppose that the physical encounters that make up an architectural experience correspond to the nested rings layered around the central subject. These nested encounters with physical things act like membranes through which subjects (individuals, collectives, publics) develop a new awareness of surrounding milieus at various scales. These are not conclusive encounters: the strange rocks and stones of my two-part parable leave much to the imagination while evoking some history outside that moment, a human and natural history. In this way, perhaps an architectural cosmology of things can re-enchant us with the nearby world or provide new perspectives on the "wicked problems" of contemporary life.

NAVIGATING THE PLASTISPHERE

The annual global production of plastics is currently estimated to be 245 million metric tons (270 US tons). According to one study, this amount "represents 35 kg of plastic produced annually for each of the 7 billion humans on the planet, approximating the total human biomass."[16] It is vivid and staggering to picture that each year, the earth's surface is populated with new plastic whose combined bulk is roughly equivalent to that of all the human bodies that also populate the earth's surface—and that year after year, another total-human-biomass's worth of plastic is added.

Of that 245 million metric tons, only 0.1 percent is believed to end up in one of the five subtropical gyres, the vast islands of floating debris that have now been well measured and documented.[17] Still, plastic has become the primary source of marine pollution in the sixty years of its manufacture. A new report by a group of marine chemists and biologists documents the microbial communities that are flourishing on fragments of floating plastic. This study found that a variety of "heterotrophs, autotrophs, predators, and symbionts" are concentrated on these plastic fragments at a density and diversity much greater than that of the surrounding ocean water. Plastic waste has become a substrate for "novel ecological habitats in the open ocean," and one that given plastic's long half-life, guarantees a stable alternative to indigenous substrates found at sea.[18]

16
"Plastic accumulates not only on beaches worldwide, but also in 'remote' open ocean ecosystems. Drifter buoys and physical oceanographic models have shown that surface particles such as PMD can passively migrate from Eastern Seaboard locations all the way to the interior of the North Atlantic Subtropical Gyre in less than 60 days, illustrating how quickly human-generated debris can impact the gyre interior that is more than 1,000 km from land. Plastic debris in the North Atlantic Subtropical Gyre and North Pacific Subtropical Gyre is well-documented and models and limited sampling confirm that accumulations of PMD have formed in all five of the world's subtropical gyres." Erik R. Zettler, Tracy J. Mincer, Linda A. Amaral-Zettler, "Life in the 'Plastisphere': Microbial Communities on Plastic Marine Debris," *Environmental Science and Technology* 47 (2013): 7,137–7,146.

17
Marcus Eriksen, Laurent C. M. Lebreton, Henry S. Carson, Martin Thiel, Charles J. Moore, Jose C. Borerro, Francois Galgani, Peter G. Ryan, Julia Reisser, "Plastic Pollution in the World's Oceans: More than 5 Trillion Plastic Pieces Weighing over 250,000 Tons Afloat at Sea," *PLoS ONE*, vol. 9, no. 4 (December 2014).

18
"Plastisphere communities are distinct from surrounding surface water, implying that plastic serves as a novel ecological habitat in the open ocean. Plastic has a longer half-life than most natural floating marine substrates, and a hydrophobic surface that promotes microbial colonization and biofilm formation, differing from autochthonous substrates in the upper layers of the ocean." Zettler et al., "Life in the "Plastisphere," 7,137.

WELCOME TO THE "PLASTISPHERE"

The "plastisphere" joins an assortment of other "-spheres" that encircle the planet with distinct but interrelated material conditions: atmo-, bio-, hydro-, litho- and so on. Each is defined by its particular components and by its vital function within the planetary system; and each outlines a distinct knowledge category comprised of the disciplines and institutions that attend to its study and management. By selecting the term "plastisphere" to describe the totality of garbage-surfing microbes and their nearly imperceptible ecosystem, these scientists underscore the phenomenon as a pervasive materiality and a global infrastructure of life. The proliferation of plastic waste represents the entropic flipside of the world's industrial system while revealing the adaptive capacities of life worlds outside our own. The balancing of ecological gains and losses according to a static idea of "nature" begins to feel like a futile motivation for environmentalist action. From this perspective, change—to the atmosphere, to the biosphere, to the lithosphere, to the financial sphere— becomes less an indicator of nature out of balance. Change appears instead as a consistent property of environment and a reminder of the conceptual limits to technocratic models of sustainability. This is not to argue that architecture has nothing to do with climate; rather, it is to modify Latour's and Hulme's call for descriptive tools or new mediums through which different perspectives, multiple views, and alternate sensibilities can be shared, in order to begin assessing which forms of change, what methods of adaptation, and whose burdens of responsibility are acceptable.

> The planet will be here for a long, long, LONG time after we're gone, and it will heal itself, it will cleanse itself, 'cause that's what it does. It's a self-correcting system. The air and the water will recover, the earth will be renewed. And if it's true that plastic is not degradable, well, the planet will simply incorporate plastic into a new paradigm: the earth plus plastic. The earth doesn't share our prejudice toward plastic. Plastic came out of the earth. The earth probably sees plastic as just another one of its children. Could be the only reason the earth allowed us to be spawned from it in the first place. It wanted plastic for itself. Didn't know how to make it. Needed us. Could be the answer to our age-old egocentric philosophical question, "Why are we here?"
>
> Plastic, asshole.
> —George Carlin

A PREFACE TO POST-ROCK ARCHITECTURE

Returning us to a cosmological perspective, George Carlin's punch line points out the absurdity of humanist logic in the context of geological and climatic transformations. What if "earth plus plastic" is a new paradigm of lithospheric materiality? What if plastic, or a plastic-rock hybrid, is the answer to the most basic existential questions? While this scenario may resonate with "post human" discourses circulating today, for me the humor is key here; as a speculative device, it offers a possible technique for reconfiguring persistent frameworks of environmental thinking and the subject-object relationships those frameworks support. It relieves the proprieties of a modernist idea of

Collection of identifiable plastic objects found by Noni Samford on Kamilo Beach and along the nearby coastline. Photograph courtesy of Kim De Wolff.

environmental design and its aesthetics. Learning from the qualities of plasti-glomerates as material and medium, perhaps a post-rock architecture might be formulated:

Post rocks resist abstraction. They are neither symbolic nor instrumental. They embody the trajectories of materials and forces, rather than diagram them; they give physical presence to the entropic processes and cultural tendencies behind the plastic's production and eventual removal from a system of value. However, to describe this relation to process as indexical is not quite right either. Post rocks' physical appearance does not *index* their formation, a process that involves degrees of complexity and many agents acting at different scales. It's a process that is impossible to repeat precisely. As a "thing," post rock makes sensible those scattered inputs and distant geographies without *explaining* their contingencies. Carrying that external history, the aesthetics of post rocks both arrest with familiarity and resist easy categorization.

Now insert the word *architecture* after "post rock" in that last paragraph. Both literally and as a model for practice, what is envisioned here is a way of engaging architecture's milieus—atmo-, hydro-, bio-, plasti-, or other—not by emulating the abstract logic of the system but by authoring tangible things of the here and now.

Meredith Miller is an architect and an assistant professor of architecture at the University of Michigan's Taubman College of Architecture and Urban Planning. Through design research, writing, and collaborations, she explores the interactions of architecture and environmental thinking.

Deep Mapping

JEANNE GANG
WITH CLAIRE CAHAN AND SARAH KRAMER

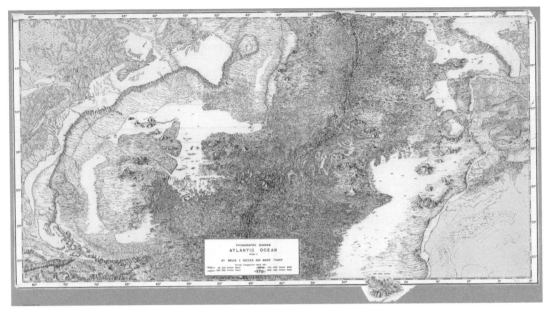

Physiographic Diagram of the North Atlantic, Marie Tharp and Bruce C. Hezeen, 1957. © Marie Tharp 1980, reproduced by permission of Marie Tharp Maps, LLC.

It must have been frustrating for an explorer with such tenacity to be so soundly defeated at measuring the depth of the ocean. Sailing west across the Pacific in 1521, on a voyage that would eventually circumnavigate the globe—and prove indelibly the spherical shape of the earth—Ferdinand Magellan lowered a weighted line to measure the fathoms below.[1] When the line dropped 750 meters without contact, he ordered it pulled up, declaring the ocean immeasurable. It was utterly apparent his tools were no match for the ocean's scale and vastness, its impenetrable watery territory. As technology has advanced into the twenty-first century, scientists and ocean explorers have continued to develop new and better methods to measure these depths—from longer weighted lines to echo and sonar, to satellite imagery and submersibles outfitted with GPS. But although humanity has now compiled vast amounts of data into maps, diagrams, and digital models, the full scope of the ocean—the hydraulic, topographic, geologic, and biological systems that comprise more than 70 percent of the Earth's surface—have yet to be fully comprehended.

Nearly five centuries of marine discovery notwithstanding, the ocean remains largely unknown, unmeasured, and unmapped.[2] This is problematic for many reasons: the ocean is responsible for the production of food, water, and, most importantly, oxygen; as a site of carbon sequestration, it is one of our major defenses against climate change. And yet—because much oceanic territory remains unexplored, humans struggle to grasp the connection between our survival and a healthy marine ecosystem. Compounding this situation are glaring misconceptions about the ocean that remain embedded

1

Although Magellan's voyage is commonly thought of as providing proof that the world is round, a spherical Earth was already an accepted belief in his time. The main impetus behind Magellan's voyage, acknowledged both within his time and ours, was the pursuit of a new trade route west to reconnect Spain with the resources of the Spice Islands after the Treaty of Tordesillas placed the more practical eastern route under Portuguese control. See the History Channel, http://www.history.com/topics/exploration/ferdinand-magellan.

2

What we can and do measure and map of the ocean is its acidity and carbon content—which we believe to be increasing to potentially danger-ous levels—as well as what we believe to be dwindling populations of certain species and the destruction of certain habitats. Yet it remains difficult to assess, and then to convey to the general population, just how much damage human activities have wrought. Institu-tions contributing key research and knowledge in this area include the National Oceanic and Atmospheric Administration (NOAA), the United States Environmental Protection Agency (US EPA), National Geographic, and the Scripps Institution of Oceanography at the University of California, San Diego.

in public consciousness. Our inability to experience the ocean as we do land has contributed to a view of the ocean as an eternally bountiful, self-sustaining entity too vast and mysterious to require our protection.

Maps can be one of our most powerful tools in changing humans' under-standing of the ocean. By depicting the Earth as an interconnected ecosystem, with land and ocean mutually dependent, they can effectively reveal that biodiversity, geology, and the need for resource protection do not stop at the shore. To do so, however, mapmakers need to navigate a tricky balance of data, accuracy, and politics. Two maps in particular demonstrate this point, exhibiting two extremes of a continuum: the 1957 map of the Atlantic Ocean floor by Marie Tharp and Bruce C. Heezen, and a less widely circulated, more recent map of the Clarion-Clipperton Fracture Zone (CCZ) in the Pacific Ocean, published by the International Seabed Authority (ISA) in July 2014.

Tharp and Heezen's map, a physiographic diagram, is revolutionary for depicting the topography of the ocean floor in three dimensions for the first time. It also unearthed a discovery that upended previous scientific beliefs and fundamentally altered the way we understand the planet today. The map was the first to locate the Mid-Atlantic ocean ridge and its rift, providing evi-dence in support of the theory of continental drift, now known as the science of plate tectonics. It is also the first map to identify fracture zones—linear ridges and troughs in the ocean's crust that evidence past plate movements. Tharp and Heezen's deliberate graphical choices—chiefly, their decision to apply the physiographic style of land-based diagrams to the ocean, which allowed them to show explicit topographic connections between the ocean floor and terrestrial landscapes, and therefore depict traces of the forces of plate tectonics—made the inaccessible accessible to a general audience, all the while hinting at the mechanisms by which the world was formed.

Compared to maps made with contemporary technologies, Tharp and Heezen's hand-drawn map from 1957 is a "low-resolution" image of the ocean floor. Nevertheless, it provides a key framework for the many ways we use the ocean today, even in spite of the restrictive context in which it was created. National claims to the ocean have a complex history, bound up in issues of State power and the control of resources. World War II remapped the European continent and its colonial extents, sparking a renewed focus on claiming oceanic territory and resources. It was during this time that Tharp and Heezen began their work, and the form of their map reflects these tense postwar political conditions—in particular, rising Cold War paranoias. Fears of Soviet submarine warfare led the US government to make precise ocean depths classified information. Tharp and Heezen surmounted this challenge not only by employing a physiographic style for their map but also by using a hachuring technique, exaggerated at a scale of 40:1, to emphasize a sense of depth while obscuring factual information.

Decades before Tharp and Heezen ever met, in 1921, a German scientist named Alfred Wegener put forth the theory of continental displacement, or conti nental drift. Veering radically from accepted beliefs of the time, Wegener's theory postulated that the continental landmasses were "drifting" across the Earth. Hav-ing studied the fossils of ancient reptiles and plants as well as rocks and mountain ranges across the world, Wegener was convinced that all of Earth's continents were once part of an enormous, single landmass, an *Urkontinent*, commonly known today as Pangaea.[3] Most scientists strongly rejected Wegener's theory; more than a half-century later, Tharp and Heezen would meet similarly heated resistance against the bold ideas revealed in their map of the ocean floor.

3

"On the Shoulders of Giants," NASA Earth Observatory, http://earthobservatory.nasa.gov/Features/Wegener.

The first two-dimensional map of the Atlantic Ocean floor was created from soundings collected between 1872 and 1876 by a team of British scientists aboard the HMS *Challenger*, a refitted Royal Navy warship. Although a wircline sounding machine had been recently developed by Irish physicist and engineer Sir William Thomson, one of many iterative advances in this method of data collection made throughout the late nineteenth century, the *Challenger* employed various hemp-line sounding machines.[4] The wireline machine and other improvements would soon make the sounding process faster and more accurate, but for the *Challenger* crew, it remained an arduous process and fairly approximate: the ship stopped every 200 miles, at which point its scientists used the sounding machines and other equipment to lower a fathometer (a weighted line marked by flags) to record the measured distance between the water's surface and the ocean floor. In 713 days at sea, the HMS *Challenger* collected 492 soundings.

As people came to use the ocean increasingly to connect continents via transportation and communication networks, the desire to develop new ways to map its myriad features grew at an unprecedented pace. The speed and quantity of oceanic data collection became a public priority after the sinking of the *Titanic* in 1912. Following the end of World War I, advanced sonar technology developed during the war soon replaced the weighted lines of manual soundings to measure the ocean's depth. The establishment of the International Hydrographic Bureau (today the International Hydrographic Organization) in 1921 compelled a significant increase in the speed, quantity, and, crucially, the accessibility of oceanic data collection, as the Bureau began compiling data from all of its member nations into the General Bathymetric Chart of the Oceans (GEBCO). Through GEBCO, the Bureau began standardizing the nomenclature and terminology of such information, incentivizing oceanic data collection and bringing the work of the international scientific community to the public on a much larger and broader scale.[5] One mission of particular significance during this period was that of the *Meteor*, a German survey vessel, which traversed the southern Atlantic thirteen times between 1925 and 1927, collecting 67,338 sonar soundings that provided the data for a much more accurate depiction of marine topography.

It was in the wake of this period of intense data collection that Tharp and Heezen began work on their map of the Atlantic Ocean floor while at Columbia University's Lamont Geological Observatory under Dr. Maurice Ewing. Between 1947 and 1952, Ewing, Heezen, and Ewing's students collected their own sounding records across the Atlantic. Because the paths of the ships were inconsistent, the soundings were incomplete. Tharp developed a process to piece them together, "splicing about three thousand feet of sounding records into six complete paths … representing about a hundred thousand miles of travel."[6] Using these compiled soundings, she then translated the data into topographical profiles, or cross-sectional views, of the ocean floor.

Plotting the sounding data onto two-dimensional graphs that would ultimately form her profiles, Tharp was able to confirm the location of a "wide medial ridge that had been surmised by oceanographers since the late nineteenth century." This was not exactly revelatory; oceanographers on the *Meteor* had produced similar profiles twenty years prior. But probing further, Tharp noticed something startling: a rift, or, in her own words, "a deep notch near the crest of the ridge." As she would later write, "If there was such a thing as continental drift, it seemed logical that something like a mid-ocean rift valley might be involved. The valley would form where new material came

4

T. H. Tizard, R. N., et al., "Narrative of the Cruise of HMS *Challenger*," *Report of the Scientific Results of the Voyage of the HMS* Challenger *During the Years 1873–76* (London: Longmans & Co., 1885).

5

"GEBCO General Bathymetric Chart of the Oceans," International Hydrographic Organization (IHO) and the Intergovernmental Oceanographic Commission (IOC) of UNESCO, www.gebco.net.

6

Hali Felt, *Soundings: The Story of the Remarkable Woman Who Mapped the Ocean Floor* (New York: Picador, 2012).

up from deep inside the Earth, splitting the Mid-ocean ridge in two and pushing the sides apart. That, in turn, would move the continents on their various
tectonic plates."[7]

7

All quotations
in this paragraph
from Felt,
Soundings.

To prove the accuracy of her analysis and thus the existence of the rift to
a skeptical Heezen, who saw it as a dangerous subversion of scientific ortho-
doxy, Tharp began converting the profiles into a single drawing that could
more realistically communicate the data. Working on a related project, they
"had found a definite association of topography with seismicity." Plotting
the known locations of earthquake epicenters along with the data from the
soundings, and then interpolating and extrapolating between known data
points, Tharp identified the extent of the rift valley. She noticed that the
epicenter locations "led south through the Atlantic, around the African cape,
north into the Indian Ocean, west into the Gulf of Aden, and then made land-
fall in the form of the East African Rift." A few decades prior, seismologists
Beno Gutenberg and Charles F. Richter had first noted a relationship between
earthquake epicenters and the oceanic ridge. But Tharp could trace her rift
nearly continuously across the world, not only discovering a "forty-thousand
mile long underwater structure, quite possibly the largest geologic feature on
Earth," but proving its relationship to seismic activity, lending credence to the
theory of continental drift.[8]

8

All quotations
in this paragraph
from Felt,
Soundings.

Contemporary reactions to Tharp's discovery, and the map that she and
Heezen developed to graphically communicate it, were far from favorable—
their ideas and their map were fervently criticized as works of imagination,
both by the scientific community and the public at large. It would take Jacques
Cousteau and his Troika camera to turn the tide, screening physical proof of
the ridge and its rift valley at the 1959 International Oceanographic Conference. Cousteau's video led to expanded support for Heezen's data collection
and Tharp's mapping, resulting in a spate of new oceanic breakthroughs.

But even with filmic evidence, and in spite of many other technological
advances made over subsequent decades, achieving agreement about the
ocean, and the data collected about and within it, remains far from easy.
Consensus proves especially difficult when viewing such data through the lens
of largely arbitrary human-made laws and systems developed for terrestrial
landscapes. For instance, the United Nations Convention on the Law of the
Sea (UNCLOS), which defines the boundaries of national jurisdiction along
the ocean floor as well as the adjudication of international waters, was opened
for signatures in 1982 after a decade and a half of deliberations. It was not
ratified until 1994—all told, involving nearly twenty-six years of deliberations
"by more than 150 countries representing all regions of the world, all legal
and political systems and the spectrum of socio/economic development."[9]
(Significantly, as of this writing, the United States has yet to officially recognize the convention, despite participating in the negotiation of its terms.)

9

*United Nations
Convention on
the Law of the
Sea of December
10, 1982*, United
Nations Division
for Ocean Affairs
and the Law of the
Sea, http://
www.un.org/depts/
los/convention_
agreements/
convention_
overview_
convention.htm.

In addition to its clear establishment of the "rights, duties, and jurisdic-
tions of maritime states ... the limits of a country's 'territorial sea' ... rules
for transit through 'international straits' and [the boundaries of] 'exclusive
economic zones,'" the convention also establishes international waters—the
High Seas, which are beyond the claim of any nation and cover 63 percent
of the ocean's floor.[10] As technological advancements and infrastructure
have come to dominate life in new and unanticipated ways, the High Seas
are being reconsidered, with various countries laying claim to territories
previously declared unclaimable. Current technologies (including remotely
operated deep sea submersibles, LiDAR, and GIS) are also allowing us to

10

*United Nations
Convention on the
Law of the Sea*.

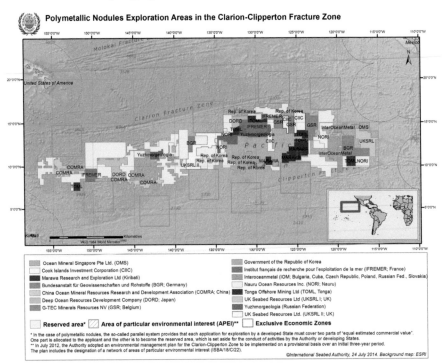

Exploration areas in the Clarion-Clipperton Fracture Zone, International Seabed Authority, 2014.

reach new depths and discover more of the ocean floor. They have enabled scientists, private and public corporations, and other entities to identify areas where they believe valuable mineral deposits are located, spurring new claims to the seabed to mine these deposits for the minerals used in the manufacturing of cell phones, computers, and other devices.

To oversee exploration, mapping, and resource management in the High Seas, the UN Convention formed the International Seabed Authority (ISA) in 1982. The map of the Clarion-Clipperton Fracture Zone that the organization released in 2014, the result of much international collaboration and advanced technical knowledge, is disappointingly little more than a prospector's map. It shows proposed claims made by different countries on a remarkably mineral-rich area, located on the deep ocean floor in the middle of the Pacific, that stretches as wide as the continental United States and lies outside any national jurisdiction.

Like all maps, Tharp and Heezen's physiographic diagram of the Atlantic Ocean floor and the ISA's image of exploration areas in the CCZ are examples of data organized in a deliberate graphic style to communicate a specific point of view. As graphic tools, they have strategic viewports and scales chosen to articulate a clear message to an intended audience. In the case of Tharp and Heezen's map, the process of visualizing the remote ocean floor led to a great shift in our understanding of the dynamic geological events that continue to shape the planet. The ISA's map looks to imminent events, geological in nature but this time human-made, hinting at a future in which the ocean is viewed fully as a privatized, parcelized commodity. Both maps mobilize standards of geological representation to create a specific image of the ocean—ultimately, rendering it as an exploitable territory.

Despite ever-more sophisticated sonar technology, to date, a mere 5 percent of the ocean has been mapped in high resolution.[11] Compare this to the

11
National Ocean Service, "How Much of the Ocean Have We Explored?" National Oceanic and Atmospheric Administration, US Department of Commerce, http://oceanservice.noaa.gov/facts/exploration.html. As noted by Jon Copley, the ocean floor has been 100 percent mapped at approximately 5-meter resolution and partially mapped at approximately 20-meter resolution. See Jon Copley, "Mapping the Deep, and the Real Story Behind the '95 Percent Unexplored' Oceans," October 4, 2014, http://moocs.southampton.ac.ukoceans/2014/10/04/mapping-the-deep-and-the-real-story-behind-the-95-unexplored-oceans.

12

Mars, the Moon, and Venus have been mapped 100 percent, 100 percent, and 98 percent, respectively, at approximately 100-meter resolution. Copley, "Mapping the Deep."

13

David E. Sanger and Eric Schmitt, "Russian Ships Near Data Cables Are Too Close for US Comfort," *New York Times*, October 25, 2015.

14

Svati Kirsten Narula, "A Blueprint for Protecting the World's Oceans," the *Atlantic*, April 25, 2014.

15

Richard W. Spinrad and Ian Boyd, "Our Deadened, Carbon-Soaked Seas," *New York Times*, October 16, 2015, http://www.nytimes.com/2015/10/16/opinion/our-deadened-carbon-soaked-seas.html; Don Hinrichsen, "Appropriating the Water: A Human Thirst," World Watch Institute, http://www.worldwatch.org/node/527.

16

L. M. Wedding, et al., "Managing Mining Of The Deep Seabed: Contracts Are Being Granted, But Protections Are Lagging," *Science*, July 10, 2015.

17

Wedding, et al., "Managing Mining of the Deep Seabed."

100 percent high-resolution mapping of the surfaces of Mars and the Moon, and the implications of this disparity vis-à-vis humanity's ability to respond to climate change begin to come into view.[12] If colonizing outer space seems more viable than looking to the ocean, where climate change can be mitigated here on Earth, then that disparity places an arbitrary limit on the scope of our imaginations and ability to act in response to the climate crisis.

Oceans, and their territorial ambiguity, have long been used as a tool for political, economic, and social manipulation, with some of the most recent fearmongering headlines warning of an impending Russian plot to dismantle Internet cables on the ocean floor.[13] With little land now left to parcel out and plumb for resources, governments have turned to the space of the ocean, contesting the precarious boundaries between Exclusive Economic Zones and asserting new claims in the High Seas. Because the majority of the ocean lacks sovereignty, its territory has no representation in, and no safeguard against, any of the policies, politics, and resulting consequences to which it is subjected; and those areas of the ocean that are claimed without dispute are not necessarily protected from harm. Forty-two percent of the ocean is "claimed" by national interests (through Exclusive Economic Zones). Only 2 percent of the ocean is actively safeguarded as "marine protection areas."[14] As it has throughout history, the ocean provides a link between sovereign spaces, but it does not share the same protections that have been established for terrestrial environments. Despite land-based conservation efforts, upland, tidal, and freshwater resources are nearing the brink of collapse, calcifying this shift in focus to the oceans' depths as a critical future site for resource extraction.[15]

Areas outside the internationally recognized Exclusive Economic Zones are considered part of the "common heritage of man" and placed under a trusteeship managed by the International Seabed Authority.[16] As of July 2015, however, with support from the ISA, a twenty-first-century gold rush is set to begin on the ocean floor—specifically, within the CCZ. Singularly biodiverse (it is home to a range of habitats and species found nowhere else in the world), the CCZ is also rich with many of the minerals and metals used to manufacture the infrastructure that makes modern life possible. Smartphones and computers are included in this list as well as construction materials, wind turbines, and a host of common household appliances. With its excavation set to be carried out by an internationally governed, multinational team of experts mobilizing state-of-the-art technologies, the CCZ's imminent strip-mining makes clear that the ocean's deepest layer—long considered beyond human experience and comprehension—is "open for business."

The magnitude of the damage to the 6-million-square-kilometer zone of the CCZ reserved for resource extraction cannot be fully anticipated, as the scale of mining is unprecedented and the ecosystems at this depth of the ocean are not fully understood. A cross-disciplinary team of oceanographers, biologists, ecologists, and environmental advocates recently addressed this threat in an article published in *Science* magazine for which our office, Studio Gang, provided imagery. Weaving their research findings with a call for a new type of conservation protocol, they implored the ISA to consider developing "a process to establish regional environmental management plans as part of the framework for governing both explorations and exploitations of deep seabed minerals that includes a network of no-mining areas."[17]

These urgent requests to study the life in the fracture zone's marine environment are graphically represented in both the ISA's and our image as

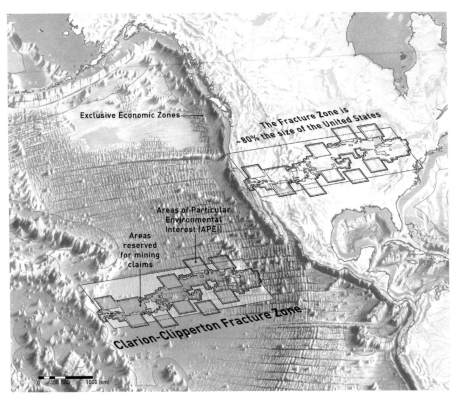

Labels on map: Exclusive Economic Zones • The Fracture Zone is ~80% the size of the United States • Areas of Particular Environmental Interest (APEI) • Areas reserved for mining claims • Clarion-Clipperton Fracture Zone • 0   300 500   1000 (km)

From Wedding et al., "Managing Mining of the Deep Seabed," *Science* 349, July 10, 2015, 144–145. The vast extent of mining exploration claims and areas reserved for mining in the CCZ in the abyssal Pacific Ocean. Image courtesy Studio Gang Architects; base map courtesy Heinrich Berann, World Ocean Floor Panorama, based on Tharp and Heezen's map and data, © 2015 National Geographic Creative.

loosely defined, disconnected green squares. Lacking official representation, they hover awkwardly outside the zone of national claims, like uninvited guests without seats at the table. Thus far, those who have invested the most in visualizing the CCZ are those seeking to profit from extracting and selling its natural resources—corporations, governments, and other organizations, like the ISA itself, that are intent on gauging the value of the materials found in the seabed. Using advanced remote sensing methods such as LiDAR (a pulsed laser system capable of accurately and precisely mapping the ocean floor), they have gathered a tremendous amount of three-dimensional data from the CCZ. The images they have developed from their data, however, clearly show the narrow scope of their interests. Intent on delineating existing mineral and metal deposits in order to most efficiently mine them, and to territorialize these deposits in order to negotiate the eventual flows of capital that buying and selling them will produce, their two-dimensional "floor plans" map and delimit zones with orthogonal boundaries that treat the ocean as though it were surface alone. Below the colorful parcels denoting reserved areas in the ISA's map, the ocean is a largely featureless blue plane. Nothing about the map helps us visualize the scale of the mining set to take place, the particularities of this ecosystem, or the relationship of this area to the rest of the planet. It certainly does not pique the human imagination as Tharp and Heezen's map did.

Although the plethora of data available today has created a widespread sense that the planet is a completely known, "open source" entity, the deployment of this information and those enriched by its analysis remain highly restricted. This condition is reflected in the ISA map's lack of scale—the continents of North America and Asia that lie just beyond the extraction zone are not shown, removing any reference to familiar land-based features and inhabitants. Furthermore, by eliminating any large-scale landforms from the map, it is impossible for the viewer to grasp the scale of the CCZ and to intuit how mining in this area may affect human life on land.

The reality of this scale is startling—the parcels approved for resource extraction (simply labeled "Reserved Areas" on the map) are the equivalent in total surface area of almost 80 percent of the contiguous United States. But with the US cropped out, viewers are not able to make this association without independent inquiry. In contrast, Tharp and Heezen's map draws the viewer's attention to the heavily shaded rift valley in the center, in part by framing it with familiar land masses and rendering it as an extension of the topographic features visible on land. While the valley may have once been highly controversial, technology has enabled us to intuit and verify its presence—today it is indisputable. The ISA's map, in contrast, crops out the land, placing all of the viewer's focus on a single feature of the ocean—the fracture zone. What is there to dispute? All context has been removed, presenting a view of the ocean that is difficult to interpret and a mining project whose ramifications are difficult to judge. While many people throughout the world rely on technologies built with the materials believed to be abundant in the CCZ, given our collective dependence on the ocean, societies at large should be demanding to know more details about this enormous project.

In the 1950s, political and social forces restricted what could be shown on a map as well as who could contribute to scientific inquiry. Because of her gender and superstitions involving women at sea, Marie Tharp was kept off the Observatory's ships during research expeditions. Perhaps this distancing fired up her imagination, driving her to use the tools that were available to her to interpolate and layer the information she was not allowed to gather firsthand. Rather than investing her energies in collection, she employed her own spatial sensibilities and an open mind to synthesize the data gathered by Heezen and formulate conclusions.

To improve humanity's understanding of the ocean, we now face a different challenge. As new tools help us accumulate more and more data, we face an overload, rather than a lack, of information—a vastness of data, like the vastness of the ocean itself, that has the potential to impede deeper knowledge and insight. Geopolitical realities, further complicated by corporate and commercial interests, industry, manufacturing, and transportation, will continue to play a role in the way information is shared or withheld. Design practices should begin to use our tools and techniques in resourceful ways to develop and circulate cross-disciplinary work in the rising field of marine spatial planning. Seeing information layered in insightful ways will make what exists in the space of the ocean more visible, thereby contributing to more developed scientific and lay understandings of its complex conditions.

This design challenge entails an engagement process with the ocean's many users, including representatives of governments, industries, conservation initiatives, and even recreational bodies. Architects and urban designers, with their abilities to draw information out of diverse stakeholders and delineate spaces, are distinctly suited to lead this kind of spatial inquiry. While a public process to sustainably manage the resources of the ocean has been developed, at least in a nascent form, design thinking currently plays no role in its operation. Designers could help to address what are essentially spatial

and organizational conflicts by translating these highly complex, interconnected issues into visual resources. New kinds of communicative drawings
could emerge, analogous to those that organize terrestrial environments, but
with an inherent temporal dimension. Though some may argue that mapping
is not central to design, especially when it is focused on the ocean, this kind
of deep, cross-disciplinary mapping is essential for understanding the climate
changes ahead and our ocean's potential role in addressing them.

Jeanne Gang is an architect, MacArthur Fellow, and founding principal of Studio Gang, an
architecture and urban design practice in Chicago and New York. Author of *Reveal* (2011),
Reverse Effect (2011), and *Building: Inside Studio Gang Architects* (2012), Jeanne leads
the Studio's design work as well as its cross-disciplinary collaborations. She is currently
engaged in design on major projects throughout North America, including the expansion
of the American Museum of Natural History in New York City and a strategic plan for the
National Aquarium in Baltimore.

Architect Claire Cahan is design director at Studio Gang. Claire has led and collaborated on
many of the Studio's architectural projects, including the National Aquarium strategic plan and
the design phase of the American Museum of Natural History. Claire worked closely with Jeanne
and the Center for Ocean Solutions to produce the Studio Gang image featured in *Science*.

Writer and editor Sarah Kramer is publications director at Studio Gang. She works closely
with Studio leadership and project teams to communicate architectural and design ideas
through various web- and print-based publishing projects.

Securing Adjustable Climate

FELICITY D. SCOTT

Endcap View of Cylindrical Colony with Suspension Bridge, Don Davis, 1975. Courtesy of NASA Ames
Research Center.

On May 13, 1974, front page headlines in the *New York Times* read, "Proposal for Human Colonies in Space Is Hailed by Scientists as Feasible Now."[1]
The article was illustrated with a rather prosaic diagram modeled after
eighteenth-century mathematician Joseph Louis Lagrange's hypothesis on
celestial mechanics. It identifies Lagrange Libration point 5 (or L5), a point
of stable equilibrium between the sun, the earth and its moon, as an ideal site
for the first space colony, since the colony could retain its position within this
celestial orbit without having to accelerate. Walter Sullivan, a prominent science journalist, reported on an event convened three days earlier at Princeton
University by physics professor Gerard K. O'Neill, a renowned high-energy
particle physicist. Liberated not only from gravity and friction but also from
inhospitable climates, material scarcity, "large scale governments," and other
Earthly threats, O'Neill's space colonies were imagined to take the form of
giant, rotating, man-made habitats (initially in cylindrical form) that would
replicate, or so he insisted, the most beautiful parts of Earth, exemplified for
him by Carmel Bay, California, along with the Grand Teton mountains in
Wyoming, the island of Bermuda, and "attractive villages in Italy and Southern France."[2] His space colonies were represented at this time by technical
diagrams, supplemented by a powerful and distinctly neoliberal narrative.
With an abundance of material goods, endless sunshine, "virtually unlimited"

1

Walter Sullivan,
"Proposal for
Human Colonies in
Space Is Hailed by
Scientists as Feasible Now," *New
York Times*, May
13, 1974, 1, 23.

2

Gerard K. O'Neill,
"A-III. The
Colonization of
Space," Appendix
A, Proceedings
of the Princeton
Conference on
the Colonization
of Space, May
10, 1974, in
*Space Manufacturing Facilities
(Space Colonies),
Proceedings of the
Princeton/AIAA/
NASA Conference, May 7–9,
1975*, ed. Jerry
Grey (New York:
American Institute
of Aeronautics
and Astronautics,
Inc., 1977), A7,
A10. O'Neill
speaks of their
"independence
from large-scale
governments" on
page A7 and likens
them to "attractive
villages in Italy and
Southern France"
on page A10.

Construction Crew at Work on Bernal Sphere Colony, Don Davis, 1976. Courtesy of NASA Ames Research Center.

Cutaway View of Torus Space Colony, Rick Guidice, c. 1975. Courtesy of NASA Ames Research Center.

resources harvested from outer space, freedom to travel, and, as he repeatedly underscored, "independence from large-scale governments," pioneering colonists were promised attractive, self-sufficient, profitable, Earth-like environments.[3] Yet, unlike Earth, there would be no unproductive workers, no pollution, no limits to energy consumption, no garden-destroying pests. Fresh strawberries would be available throughout the year.[4]

Confidently pitching science-fiction-like narratives as the most rational scientific solution to the world's problems, O'Neill offered truly fantastic figures of emigration rates, population growth, and (through an avowedly "bootstrap" plot) the rapid self-replication of space communities. Starting with a small, higher-density Model 1 colony (soon to be called Island One and from which the others would be fabricated), he estimated that by 2074 "more than 90% of the human population could be living in space colonies" such as his Model 4.[5] Although not necessarily desirable, there would be room, he claimed, to expand the human population by a factor of twenty thousand. Here was an exponential growth curve speaking not to imminent doomsday, as with neo-Malthusian systems dynamics studies like Jay W. Forrester's *World Dynamics* of 1971 and *The Limits to Growth*, published by the Club of Rome as an intervention to the UN's 1972 Conference on the Human Environment in Stockholm.[6] O'Neill's diagrams indicated Earth's population decreasing as that in outer space spiraled upward on account of unlimited resources. With industry and populations relocated to

3

O'Neill, "A-III. The Colonization of Space," A7.

4

O'Neill, "A-III. The Colonization of Space," A10.

5

O'Neill, "A-III. The Colonization of Space," A6–A7.

6

On this aspect of the Stockholm Conference, see Felicity D. Scott, *Outlaw Territories: Environments of Insecurity/ Architectures of Counterinsurgency* (Brooklyn: Zone Books, 2016).

7

O'Neill, "A-III. The Colonization of Space," A7.

8

Gerald Feinberg cited in Sullivan, "Proposal for Human Colonies in Space Is Hailed by Scientists as Feasible Now," 23.

9

Sullivan, "Proposal for Human Colonies in Space Is Hailed by Scientists as Feasible Now," 1, 23. O'Neill expands on the bootstrap logic in "The Space Manufacturing Facility Concept," in Space Manufacturing Facilities (Space Colonies), 7–11.

10

O'Neill, "A-III. The Colonization of Space," A11.

11

See, for instance, "'Is the Surface of a Planet Really the Right Place for an Expanding Civilization?': Interviewing Gerard O'Neill," in Stewart Brand, ed., Space Colonies (New York: Penguin Books, 1977), 22.

12

See W. Patrick McCray, The Visioneers: How a Group of Elite Scientists Pursued Space Colonies, Nanotechnologies, and a Limitless Future (Princeton, NJ: Princeton University Press, 2013); and Gerard O'Neill, "The Colonization of Space," Physics Today, September 1974, 32–40.

outer space, as Sullivan reported, Earth would be left with "few permanent residents. It would be 'a worldwide park, a beautiful place to visit for a vacation.'"[7]

As indicated in the New York Times, this rosy vision was haunted by a constellation of contemporary anxieties: Columbia University physicist professor Gerald Feinberg, Sullivan reported, "said that in a world threatened by nuclear devastation or catastrophic pollution effects, colonies in space would provide insurance for the continuity of the human race and other life forms." Life itself, that is, was at stake. Feinberg, too, mobilized the Jeffersonian appeal to self-sufficiency and self-government, drawing analogies to the colonization of the Americas to suggest that space colonies would "tend to be independent" and "could provide a haven for dissidents and would offer the advantages of small, independent political units."[8] Sullivan concluded by alluding to a lingering doubt: "Within the solar system, Dr. O'Neill pointed out, there is plenty of room for colonization 'without shooting any Indians.'"[9] "In contrast to our experience with expanding civilizations on Earth," O'Neill had explained at Princeton, distancing himself from the specter of colonial violence, "in space colonization there would be no destruction of indigenous primitive populations; nothing corresponding to the Indian wars of 19th century America."[10] Space colonization was repeatedly and ambiguously likened to the European discovery of the New World and the ideology of manifest destiny associated with the nineteenth-century American frontier: at a moment when US expansion and economic growth seemed threatened by resource scarcity, environmental degradation, nuclear fallout, or political pressures both at home and from developing countries, including the oil-rich nations of OPEC, space colonization suggested continuity in US supremacy and pioneering know-how.

The New York Times coverage proved pivotal. O'Neill had struggled to gain support for his ideas in the preceding years, eventually gaining seed money for the 1974 event from Michael Phillips, president of the Point Foundation, the California agency through which Stewart Brand channeled the immense profits of his alternative lifestyle initiative, the Whole Earth Catalog. O'Neill's space colony obsession began somewhat by chance in fall 1969 when, in the wake of the euphoria of the Apollo moon landing, and seeking to counter growing disenchantment with science and engineering among the country's youth arising from the violence of the US-led war in Vietnam, he posed the question to his freshman physics students, "Is the surface of a planet really the right place for an expanding technological civilization?"[11] As detailed by W. Patrick McCray in The Visioneers, O'Neill became increasingly convinced by his findings, and increasingly frustrated by their rejection among the scientific community, finally gaining an audience when the popular magazine Physics Today published "The Colonization of Space" in September 1974.[12]

In addition to outlining technical and scientific details behind his evidently inflated claim that self-sufficient space colonies were achievable in the next few decades, O'Neill's Physics Today article underscored that colonization held the promise of solving not only the US's but the world's major problems by offering an abundant clean energy supply, protection of the biosphere, the expansion of living space (lebensraum) and even equalizing living standards. Indeed, adding the question of security to that of scarcity, territory, and population, he claimed nothing less than world peace to be at stake.

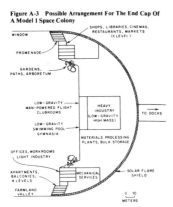

Figure A-3   Possible Arrangement For The End Cap Of A Model 1 Space Colony

Possible Arrangement for the End Cap of a Model 1 Space Colony, c. 1974.

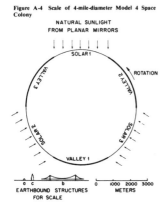

Figure A-4   Scale of 4-mile-diameter Model 4 Space Colony

Diagram showing scale of a 4-mile-diameter Model 4 Space Colony, c. 1974.

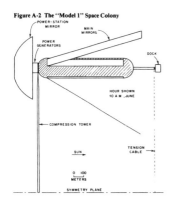

Figure A-2   The "Model 1" Space Colony

Diagram of the Model 1 Space Colony, c. 1974. Images from *Space Manufacturing Facilities (Space Colonies): Proceedings from the Princeton/AIAA/ NASA Conference May 7–9, 1975.*

I hesitate somewhat to claim for space-colonization the ability to solve one other problem, one of the most agonizing of all: the pain and destruction caused by territorial wars. Cynics are sure that humanity will always choose savagery even when territorial pressures are much reduced … Yet I am more hopeful; I believe that we have begun to learn a little bit in the past few decades. The history of the past 30 years suggests that warfare in the nuclear age is strongly, although not wholly, motivated by territorial conflicts; battles over limited, nonextendable pieces of land.[13]

It was powerful rhetoric. Picking up momentum from the *Times* article, and the popular reception of O'Neill's ideas in *Physics Today*, the US National Aeronautics and Space Administration (NASA), its funding then in decline in the wake of the Apollo missions and the winding down of Skylab, recognized an opportunity. NASA gave $12,000 toward, and co-convened, a second, larger conference at Princeton in May 1975, the Conference on Space Manufacturing Facilities (Space Colonies). In addition to convening physicists from Princeton, Columbia, and MIT (as in the first event), it brought together experts from large corporations; the US government and military agencies; as well as from legal, diplomatic, social-scientific, cultural, and management realms.[14] Additionally, NASA contributed $100,000 for the NASA/Ames-Stanford University Summer Study on Space Colonization that year and funded O'Neill's book-length manifesto, *The High Frontier: Human Colonies in Space* of 1977.[15] It also contributed to visual marketing. Hence, when on July 23, 1975, O'Neill testified about the benefits of space colonization to the US House of Representatives' Committee on Science and Technology, he arrived armed with a large model, seductive renderings by California artist Donald Davis and Frank Guidice of the NASA-Ames laboratory, and even a short film produced by NASA in association with the National Public Affairs Center for Television and Dolphin Productions, New York.

13

O'Neill, "The Colonization of Space," 36.

14

I have written elsewhere on the contribution of MoMA's Ludwig Glaeser to this event and the relation of O'Neill's work to the period's architecture more generally. See Felicity D. Scott, "Earthlike," forthcoming in *Grey Room*.

15

Gerard O'Neill, *The High Frontier: Human Colonies in Space* (New York: Morrow, 1977).

Wavering between the language of plenty and economic development and the threat of scarcity, insurrection, and war, and with characteristic rhetorical flair, O'Neill launched his congressional testimony at the nexus of "American know-how" and appeals to freedom, cast in distinctly neoliberal terms. The moon landing, he proposed, was better understood not as a scientific venture but as a "prospecting survey" for space colonization, much as a mining company might undertake. He ended by recalling that, on a recent trip to Alabama, he was greeted by a large group of young people waving placards and shouting not in protest but in an enthusiastic embrace of his new techno-scientific developments. As demonstrated in the beautiful renderings, space colonies would establish productive, profitable, isolated, normative, passive workforce communities living "in comfort, even in some luxury, within a large enclosed volume having a climate where flowers, trees, birds and animals could flourish."[16] As I argue elsewhere, drawing on De Witt Douglass Kilgore's *Astrofuturism*, O'Neill's promise of a lush, protected suburban lifestyle was code for racial segregation in America.[17] Here I want to try to refract this "promise" through a different lens, for the proclaimed "benefits" and the violent reshufflings implied by his space colonization enterprise were directed not only toward forces within the US, wherein they might serve to contain insurrection and dissent, but also globally. Indeed, at stake was their potential to help foster the country's dominant role in processes of economic globalization and in scripting the global management of environmental resources, the contours of which were then being articulated simultaneously through business and institutions like the United Nations and World Bank. Whether domestically or globally, within this shifting geopolitical landscape space colonies were to operate in the interest of corporate profit.

O'Neill told Congress that space colonies were key to the US's economic and resource security now that "both the oil-consuming nations and the underdeveloped third world are vulnerable to the threat of supply cutoff from the Middle East."[18] Promising to transmit solar satellite energy back to earth via giant microwave beams, US energy independence would be assured without the political backlash caused, domestically, by strip-mining and nuclear proliferation, and at a scale far beyond the Alaska pipeline. Given the scale of the marketplace for the primary product—energy—the payoff for investors would also be enormous. "We can put the Middle East out of business!" he recounted one of collaborators as having exclaimed.[19] Moreover, taking lessons from the development sector, he rehearsed the argument that what was good for the US was good for the world, claiming the US would be able to supply cheap energy to developing countries or even provide it as humanitarian aid, thereby overcoming growing hostility to the US as "exploiters of scarce resources."[20] Additionally, through promoting development, this energy supply would even reduce population growth in the global south and with it, as was widely feared at this moment, threats to political stability.

The fantasy that no one was exploited during processes of territorial expansion was an old trope, manifest in the European habit of imagining new worlds as empty territory, or in claims that modernization or religious conversion of indigenous populations was in their interest, as in the so-called civilizing mission. By the 1970s, development initiatives were increasingly forced to account for the rights of indigenous and formerly colonized peoples, and as with the period's developmental ideology, space colony enthusiasts thus turned to casting their work as humanitarian aid. Under the auspice of aid, poorer countries might be remade as a degraded image of the

16

Gerard K. O'Neill, testimony at *Future Space Programs 1975*, Hearings before the Subcommittee on Space Science and Applications of the Committee on Science and Technology, US House of Representatives," July 23, 1975, 170. The main body of O'Neill's testimony is reproduced in Brand, ed., *Space Colonies*, but that version elides the preamble, images, and question-and-answer period.

17

See Scott, "Earthlike."

18

O'Neill, testimony at *Future Space Programs 1975*, 129.

19

O'Neill, testimony at *Future Space Programs 1975*, 134.

20

O'Neill, testimony at *Future Space Programs 1975*, 135.

Model of Torus Colony, Don Davis, c. 1975. Courtesy of NASA Ames Research Center.

free world, replete with massive debt and burdened with technologies that, whether appropriate or not, ensured their ongoing dependence upon multinational corporations who were invariably the primary beneficiaries.

In a 1976 *Penthouse* interview, O'Neill again claimed space colonization for liberalism, calling it "a natural continuation of greater freedom, a greater amount of diversity and control over the environment."[21] Pointing to warlike tension resulting from the 1973 oil crisis and "pressure on land area as populations increase," as well as rising nuclear threats, he posited, "The unattractive alternatives seem to be a more tense situation in which nations are increasingly threatening each other in order to get the raw materials they need or some massive type of conflict that will result in a global dictatorship."[22] Space colonies, by contrast, offered a panacea, a way to "de-escalate that situation on Earth." There would be "much less reason for warlike activities than [between] countries on the Earth" not only on account of self-sufficiency but also since "their boundaries would be their own choice: if they don't like the neighboring colony, they could move somewhere else. If they don't like the land area they have, they could build more very easily without encroaching on anyone else's space."[23] The need for political negotiation, that is, would simply be eliminated. Yet within such a limitless, supposedly borderless place, space colonies would proliferate a new generation of border, even if freed from geographical constraints. In a world increasingly interconnected by communication and travel, the surfaces of space colonies sought to operate less like a border between sovereign states—which came with international protocols—than a police or even private security checkpoint that could regulate flows and movements of people more tactically, managing the distribution of populations following a "rational" metric of productivity and profit. As a counterpart to the ambiguous territorial logics of space colonies, many were left wondering if Earth would become a privileged site for those who could afford to live in nature, supported by energy from outer space, or if, rather, it would become the prison for those without the means, education, or work ethic to leave.

OUTLAW AREA

It is not surprising that O'Neill first obtained support for his space colony initiative from the Point Foundation. An offshoot of the Portola Foundation, which published the *Whole Earth Catalog*, the Point Foundation was established in late 1971 by Stewart Brand and Richard Raymond, buoyed by enormous profits from the *Last Whole Earth Catalog*. Noting that it served as "an activist arm" for Brand, Andrew Kirk argues in *Counterculture Green*, "Point was an active experiment fostering the design science revolution."[24] That is, we might say, concomitant with Brand's avowed indebtedness to R. Buckminster Fuller, Point served as another mechanism through which to promote revolution not through politics but by design. In 1974 Brand launched *CoEvolution Quarterly* under the auspice of the Point Foundation, using it to sermonize on his rising fascination with space colonization. While O'Neill was in town for the summer study session at Stanford two years later, Brand and Phillips interviewed him in what they termed, trying to maintain the semblance of a countercultural edge, a "ghetto apartment in San Francisco." After discussing O'Neill's early struggles for support, the conversation turned to Point and the "famous $600" grant. "So the *Whole Earth Catalog* is

21

"Penthouse
Interview with
Dr. Gerard K.
O'Neill," *Penthouse*
(August 1976),
175.

22

"Penthouse
Interview with
Dr. Gerard K.
O'Neill,"176.

23

"Penthouse
Interview with
Dr. Gerard K.
O'Neill,"176.

24

Andrew G. Kirk,
*Counterculture
Green: The Whole
Earth Catalog
and American
Environmental-
ism* (Lawrence:
University Press
of Kansas, 2007),
119.

responsible for the colonization of space," Brand blithely exclaimed, alluding to his savvy ability to script emergent cultural imaginaries. Phillips recalled his idea of putting the grant in Princeton's name, which O'Neill concurred served very well in harnessing the university's publicity apparatus. This strategy facilitated Sullivan's article and the ensuing "media flash."[25] Formerly a director of marketing and planning for Bank of America and vice president of the Bank of California, and with an economics degree from the University of Chicago, Phillips at this time effectively engineered a dramatic turn toward commercial entrepreneurship and free market ideals among hippies and the counterculture, the bible for which was his 1974 publication, *The Seven Laws of Money*. The seventh law, Kirk recalls, submits "you can never really receive money as a gift."[26]

When Brand anthologized *CoEvolution Quarterly* entries on the subject as *Space Colonies*, he attributed his conversion "from mild interest in the Space Colonies to obsession" to O'Neill's 1975 lecture at the World Future Society convocation in Washington, D.C., a few weeks before the professor's congressional testimony.[27] Brand's interest in the libertarian potentials of outer space in fact predated this encounter by a number of years. In January 1970, "The Outlaw Area" supplement to the *Whole Earth Catalog* included "The space out," an inconspicuous note that, in retrospect, seems to have haunted Brand's thinking throughout the decade. Citing British physicist Freeman Dyson, it reproduced part of a text from the December 1969 issue of *The Futurist*. The answer to the contemporary threat of permanent extinction of the human race on Earth following a nuclear holocaust, Dyson explained, was not found in colonizing planets like Mars—terraforming would not increase living space very much—but in "isolated city states floating in the void" and possibly attached to comets. "Above all they provide an open frontier, a place to hide and to disappear without trace, beyond the reach of snooping policemen and bureaucrats," Dyson argued. "Space is huge enough so that somewhere in its vastness there will always be a place for rebels and outlaws … Perhaps most important of all for man's future, there will be groups of people setting out to find a place where they can be safe from prying eyes, free to experiment undisturbed with the creation of radically new types of human beings, surpassing us in mental capacities as we surpass the apes."[28] Resonating with countercultural and libertarian ideals, and with the alternative lifestyle promoted in the *Whole Earth Catalog*, Dyson's thesis was mirrored in Brand's editorial for promoting "outlaw areas," described as testing grounds beyond the domain of the law, or "state-of-the-art frontiers whose languages are still foreign to lawmakers."[29] Despite international laws put in place following the launch of Sputnik, Brand listed space among "present outlaw areas."

O'Neill's presentation at the World Future Society, which so captured Brand's attention, had a particularly Fulleresque tone. After refuting the premise that human activity, as with material and energy resources, be confined to Earth, O'Neill rejected the assumption that "any realistic solutions to our problems of food, population, energy, and materials must be based on a zero-sum game, in which no resources can be obtained by one nation or group without being taken from the other."[30] It was such beliefs, he objected, that had "driven most observers to the conclusion that long-term peace and stability can only be reached by some kind of systematic global arrangement, with tight constraints to insure the sharing, equable or otherwise, of the limited resources available." Repeatedly insisting that he was avoiding

25

O'Neill in "'Is the Surface of a Planet Really the Right Place for an Expanding Civilization?'" 25.

26

See Michael Phillips, *The Seven Laws of Money* (Menlo Park, CA: Word Wheel, 1974).

27

Stewart Brand, editorial note to Gerard O'Neill, "The High Frontier," in *Space Colonies*, 8.

28

"The Space Out," *Whole Earth Catalog* supplement, "The Outlaw Area" (January 1970): 48. The note is citing "Garbage Disposal Seen as Benefit of Space," the *Futurist* (December 1969): 148, which in turn published excerpts from Freeman Dyson, "Human Consequences of the Exploration of Space," *Bulletin of the Atomic Scientists* (September 1969): 8–13.

29

Stewart Brand, "Apocalypse Juggernaut, Hello," *Whole Earth Catalog* supplement, "The Outlaw Area" (January 1970): 21. See Scott, *Outlaw Territories*.

30

O'Neill, "The High Frontier," 8.

prophecy and speaking only of realistic possibilities, O'Neill posited that the frontier "can be exploited for all of humanity, and its ultimate extent is a land area many thousands of times that of the entire Earth."[31] Like Fuller's World Game, O'Neill rejected the political mandate of any such a "systematic global arrangement"—presumably a reference to the United Nations—in favor of technical solutions in line with the evolution of capitalism. Moreover, following what Brand called Fuller's "wealth sanction," not only would these solutions help overcome famine, war, and disease; like an earlier phase of colonization they promised enormous economic profit for nations who get there first.[32] "The human race," O'Neill proclaimed of the urgency to try, "stands now on the threshold of a new frontier, whose richness surpasses a thousand fold that of the new western world of five hundred years ago."[33] "It would be naïve to assume that its benefits will be initially shared equably among all of humankind," he acknowledged, reassuring potential investors of his intent, "The world has never worked that way."[34]

That Brand's conception of the new social formations possible within such outlaw areas remained premised on political exceptionality and militarism was evident in the scenarios he offered to readers of the *Whole Earth Catalog* in 1970. As Kirk recalls, Brand's initial concept for the use of the vast profits from the catalog was in fact to purchase vast tracts of land to realize "Mountain Fantasy" as a "permanent encampment" or "proving ground" to foster "social invention."[35] In 1975 Brand asked O'Neill what had come out of the second Princeton conference. O'Neill acknowledged "an interesting paper on Space Law," presumably that of Edward R. Finch, a permanent NGO representative to the UN Committee on the Peaceful Uses of Outer Space. The presentation, O'Neill recalled, brought three constraints to his attention:

> First thing, it's got to be non-military. The second, that if anything interesting, new research, comes out of it, like information about the surface composition of the moon … that it does have to be made available through the United Nations … And last is that, at least in some nominal form, the community has got to be under the jurisdiction of the nation or group of nations which establishes it. You cannot, at least deliberately, send people out to be absolutely on their own.[36]

Finch, who preferred the term "space station" to "colony," was in fact also an advocate for commercialization of space research, later playing a major role in a business advocacy group, the National Space Society.[37] While Finch was largely concerned with reading international law to allow for the excavation of

31
O'Neill, "The High Frontier," 8.

32
See Thomas Albright, "The Environmentalists: The Whole Earth Catalog," *Rolling Stone*, December 13, 1969, 30–33.

33
O'Neill, "The High Frontier," 8.

34
O'Neill, "The High Frontier," 11.

35
Kirk, *Counterculture Green*, 120.

36
O'Neill in "'Is the Surface of a Planet Really the Right Place for an Expanding Civilization?'" 26.

37
See also Edward R. Finch and Amanda Lee Moore, *Astrobusiness: A Guide to Commerce and Law of Outer Space* (New York: Praeger, 1984); and Finch and Moore, "Outer Space Law and the Global Community," *The International Lawyer*, vol. 8, no. 4 (October 1974): 752–771.

Gerard K. O'Neill testifying before the Subcommittee on Space Science and Applications of the Committee on Science and Technology, United States House of Representatives, July 23, 1975. Photograph by Punky Crow from Stewart Brand, ed., *Space Colonies.*

materials on the moon, and other vital elements of O'Neill's vision, his out-lining of the relevant UN resolutions and treatise made it evident that outer space was no longer beyond the law or the responsibilities attending national jurisdiction. Indeed, from *Resolution 1721* of 1961, which stated that inter-national law and the UN charter applies to outer space, and that territory in outer space could not be subject to national appropriation, to the 1967 *Treaty on Principles Governing the Activities of States in the Exploration and Use of Outer Space, Including the Moon and Other Celestial Bodies*, and its updat-ing, it was evident that prospective colonies fell firmly within the domain of international law. O'Neill's retort, resisting UN constraints, was that "tech-nology and scientific advancement can be retarded if international law does not keep step with the progress of science," a notion of progress serving the interests of financial investors.[38]

When O'Neill testified to the US House of Representatives, he found him-self again confronted with UN protocols, appearing after Peter Jankowitsch, Austrian ambassador to the United Nations and chairman of the thirty-seven-nation Committee on Peaceful Uses of Outer Space. He accounted for limits born of international law, not by embracing them but modifying his language to ensure that his scheme remained "realistic." By 1977, following repeated critiques of his proposal for strip-mining the moon, O'Neill decided to have a voice in such matters, establishing the Institute for Space Studies at Prince-ton, a nonprofit corporation that sought to become a recognized NGO by the UN. "We want to be able to make an input to the UN deliberations on such things as treatise about the Moon. We don't want things to be bargained away which we may very much want to be able to use later on," he explained to Brand. Citing a precedent in the troublesome Law of the Sea, he remarked, "Nobody's mining the sea because of the arguments."[39] Brand, by contrast, continued to seek a domain beyond the law. Introducing *Space Colonies*, he wrote, "for those who long for the harshest freedoms, who believe with Buckminster Fuller that a culture's creativity requires an Outlaw Area, Free Space becomes what the oceans have ceased to be—Outlaw Area too big and dilute for national control."[40] Fuller had long celebrated the maritime power born of mastering the high seas, regarding it not only as the key to European expansion but to technological invention. Aligning himself with Fuller, Brand, too, celebrated practices seeking to operate beyond national borders and out-side the law as giving rise to spaces wherein one could try anything. Hence, although, as he noted—"the term 'Space Colony' has been expressly forbid-den by the US State Department because of anti-colonial feelings around the world"—he would be sticking to it. Returning us to O'Neill's claim, Brand added, "It's more accurate. This time there's a difference in that no space natives are being colonized." Noting that some things "went well in previous colonizations," Brand concluded, "If we're lucky we may enact a parallel with what happened in Europe when America was being colonized. Intel-lectual ferment—new lands meant new possibilities; new possibilities meant new ideas."[41] When, in 1988, as founder of the Global Business Network, he interviewed Dyson for *Wired*, the elderly physicist refused Brand's techno-logical determinism by invoking the presence of international laws, to which Brand responded, hopefully, "Won't overseas labs that don't care about such matters show up soon and do all the forbidden things?"[42]

Buoyed by the rhetoric of a better and more open future and no doubt visually seduced by the spectacular pastoral images produced to illustrate O'Neill's ideas, many within the counterculture and environmental movement

38
Edward R. Finch, "International Law and Outer Space Stations," in *Space Manufacturing Facilities (Space Colonies)*, 192.

39
Gerard O'Neill, in Gerard O'Neill, Stewart Brand, and Jane McClure, "Instead of Fric-tionless Elephants; Talking with Gerard O'Neill," in *Space Colonies*, 153.

40
Stewart Brand, "The Sky Starts at Your Feet," in *Space Colonies*, 6.

41
Brand, "The Sky Starts at Your Feet," 5. See Scott, *Outlaw Territories*.

42
"Freeman Dyson's Brain," accessible at http://archive. wired.com/wired/ archive/6.02/ dyson.html.

embraced *CoEvolution Quarterly*'s celebration of space colonization as the next frontier. But, as recorded in the magazine following Brand's solicitation of commentary, many of his long-standing interlocutors expressed doubts and even an outright rejection of his new obsession. Some recognized the impossibility of simulating ideal landscapes: as John Holt suggested, the environment would be closer at best to the lobbies of Las Vegas hotels and luxury ocean liners, but more likely to military barracks and troopships; alternative technology celebrity Steve Baer offered an even more compelling image:

> Once on board, in my mind's eye I don't see the landscape of Carmel by the Sea as Gerard O'Neill suggests … Instead, I see acres of airconditioned Greyhound bus interior, glinting, slightly greasy railings, old rivet heads needing paint—I don't hear the surf at Carmel and smell the ocean—I hear piped music and smell chewing gum. I anticipate a continuous, vague low-key "airplane fear."[43]

43

John Holt and Steve Baer, commentary in *Space Colonies*, 64, 40.

Even more tellingly than rejecting the visual sales pitch, "biological designer" John Todd questioned the scientific claims upon which the agricultural and landscape vision was premised. Co-founder of the New Alchemy Institute, Todd was then working to complete the Ark on Prince Edward Island, an experimental "bioshelter" designed to simulate an almost closed ecosystem, and then the closest experimental test site to O'Neill's vision. Todd pointed out that ecological systems were far from simple to replicate in artificial environments, current understanding of whole systems being entirely "primitive" compared to nature's complexity. "When I read of schemes to create living spaces from scratch upon which human lives will be dependent for the air they breathe, for extrinsic protection from pathogens and for bio-purification of wastes and food culture," he scoffed, "I begin to visualize a titanic-like folly born of an engineering world view." Citing statistics derived from Howard Odum's research, he suggested that Island One would be more appropriate to sustain forty rather than ten thousand people.

Beyond those refuting the aesthetic, scientific, and technical basis of O'Neill's arguments, others rightfully questioned its political underpinnings. Even neo-Malthusian Garrett Hardin, best known for his problematic diatribe "The Tragedy of the Commons," had doubts, recognizing that the Brave New World envisioned would likely be subject to totalitarian rule or that it would manifest as an expanded domain of hermetic religious cults. "The principle attraction of the Space Colony proposal is that it apparently permits us to escape the necessity of political control," he proffered, adding, "But, as we have just seen, this is only an apparent escape. In fact, because of the super-vulnerability of the spaceship to sabotage by tribal action, the most rigid political control would have to be instituted from the outset in the selection of the inhabitants and in their governance thereafter."[44] Indeed, by 1977 O'Neill was willing to acknowledge that, as with a sailing ship in open waters, the most effective governance structure for an isolated group might be far from democratic: "a dictatorship is what works," he noted in a later interview with Brand, since "there's nothing that produces conflict more than an ill-defined situation of authority."[45] With conflict comes the need for political negotiation within a democratic framework, hence dissensus had to be banished from the homogenous communities isolated in space. If space colonies were cast as a utopian multiplicity of potential domains, in which groups could maintain autonomy and diversity and assert their distinctions,

44

Garrett Hardin, commentary in *Space Colonies*, 54.

45

O'Neill in "Instead of Frictionless Elephants," 151.

the identitarian structure (which they cast as diversity—not within a colony but between them) and selection process implicitly evacuated the possibility of opening democratic forms of political space.

That O'Neill's politics remained aligned with economic paradigms driving colonialism and, more proximately, with the neocolonial ambitions of multi-national corporations and neoliberal policies informing the US-led process of globalization was evident to others. Ridiculing Brand's suggestion that a democratic processes would prevail, as evident in Brand's claim that "voters will be interested enough to approve the requisite $100 billion," Jan Bronstein insightfully responded by pointing to forces driving capitalism's long-standing expansionist ethos.

> Since when do voters, or congress for that matter, appropriate money for those kinds of projects? They are pushed through by the folks that profit from huge government expenditures (enterprising capitalists and corporations) and passed by the people (government officials) who profit from the profit. Who stimulated European settlement of the Americas? The British East India Company, the Dutch East India Company, the gold seeking Spanish royalty. So realistically, the space colonies will get started when the Exxons of the future decide to monopolize this energy resource too.[46]

Novelist, environmental activist, and farmer Wendell Berry offered the most insightful dissent, pointing to many interconnected facets of O'Neill's elaborate and cynical apparatus. Mobilizing "every shibboleth of the cult of progress," he argued, O'Neill's vision was entirely conventional in its "lust for unrestrained expansion, its totalitarian concentrations of energy and wealth … its exclusive reliance on technical and economic criteria … its compulsive salesmanship." Here was a plan to strip-mine the moon presented as care for Earth's environment. "Anyone who has listened to the Army Corps of Engineers, the strip miners, the Defense Department, or any club of boosters will find all this dishearteningly familiar," he lamented of the "thug mentality of the technological specialist." With unchecked chauvinism and mindless of the neocolonial violence it implicitly condoned, O'Neill's public relations exercise was, as Berry put it, referring to the physicist as a "professional mind-boggler," "superbly attuned to the wishes of the corporation executives, bureaucrats, militarists, political operators, and scientific experts who are the chief beneficiaries of the forces that have produced our crisis."[47] What bothered him most, however, was that Brand had finally revealed himself to be an enemy masquerading as a friend.

A closer look at Brand's activities seeking to mediate between the counterculture and the US establishment suggests that such an ambiguous identity had long facilitated a less than progressive agenda under the language of social and technological innovation; his later founding of the Global Business Network was far from out of character.[48] Indeed, Brand's real savvy was in understanding that radical and reactionary agendas sometimes intersected or looked the same, often drawing from the same cultural and technical imaginary. Hence countercultural ideals were easily redirected toward cynical ends, as Brand demonstrated so well in taking countercultural celebrities to perform an "environment yes, politics no" worldview at the 1972 United Nations Conference on the Human Environment in Stockholm, or, as in Phillips's sponsoring of entrepreneurial attitudes toward money, the

46
Jan Bronstein, commentary in *Space Colonies*, 35.

47
Wendell Berry, commentary in *Space Colonies*, 36.

48
See Scott, *Outlaw Territories*.

realization that hippies' interpersonal skills could be monetized. But such an ambivalence does not mean that distinct agendas should be neatly collapsed. It is equally important to remember that semantic and political revalencing can, in turn, operate in the other direction: critical voices might infiltrate and even produce an interruption or redirection of mainstream ideologies as well. Hence the importance of ironic practices or counter-discourses that seek not to follow technocratic ideals to their logical conclusions but to unpack their political underpinnings and inherent paradoxes, to render the apparatus through which they operate more legible.

I want to come back, in concluding, to the visual logics at play here. Unlike tabulated data and technical diagrams, and the perhaps more spectacular use of data visualization and satellite images that require expertise to interpret, NASA's space colony images appear to partake of a more archaic representational and media logic, to be an archaism with a contemporary function. Underscoring the role of visual media within the space colony initiative, Brand posited with typical entrepreneurial flair,

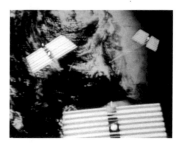

Stills from *Space Colonization*, NASA, 1975.

> Now is the time for NASA to encourage people besides engineers to get into the act. The program needs administrators who are not afraid of excellent artists, novelists, poets, film-makers, historians, anthropologists, and such who can speak to the full vision of what's going on. And their voice needs to be a design voice, not just advisory. America (and Russia) were in Space for ten years before they bothered to get a photograph of Earth. That's pretty arid thinking.[49]

On account of his 1966 campaign, "Why haven't we seen a photograph of the whole Earth yet?" Brand, or so the story goes, was himself catalytic in NASA's release of Apollo mission images of Earth from outer space, images that catalyzed popular anxieties about limited resources and environmental devastation and were widely mobilized by a range of parties, from President Nixon

49
Brand, commentary in *Space Colonies*, 73.

to environmental activists and Brand himself. Space colonies, too, lent themselves to such an opportunity; seductive images were instrumental in sponsoring public support and economic investment. Cast as a vanguard, sometimes even thought to be avant-garde, futuristic visions could slip seamlessly from alternative to libertarian to neoliberal ideals and function all too effectively at the forefront of free-market capital. Resonating between an uncannily familiar environment and a spectacular otherness—a world quite literally turned inside out—renderings of O'Neill's visions, such as those produced in the mid-1970s, were powerful tools in garnering support across the social and cultural spectrum.[50]

While NASA's spectacular images of space colonization were put to work to mobilize support for O'Neill's purportedly "utopian" visions of a neoliberal future, with its claims to ensuring American-style freedom and diversity, power was shifting from technical know-how embodied in the agriculture and industrial machinery depicted in the NASA renderings, toward a less visible apparatus of information and management. This might help us understand why O'Neill was so disturbed by the Systems Dynamics model of Jay Forrester and the Club of Rome, which not only spoke to Earth's limited resources (in their case in the service of a racially inflected panic about population growth) but prioritized the benefits of computer-driven, social-science-informed management schemes as the techno-scientific infrastructure of new forms of governance, forms that in part displaced the priority of physicists and engineers. By 1976, O'Neill was willing to concede that the first colonies would "be much more like a Texas-tower oil rig, or a construction camp on the Alaska pipeline, or like Virginia City, Nevada, in about the year 1875" than the Carmel coastline or the South of France.[51] Compared to a future of exacerbated, post-planetary segregation and a return to an entirely privatized form of governance over colonial territories (think the East India Company), the violence inherent to techniques of power informing nineteenth- and early twentieth-century colonial and industrial paradigms might indeed have come to seem more limited in their reach.

In 1976 the US Congress, having lost interest in outer space exploration, amended the Space Act to allow NASA to engage in Earth science research and climate monitoring, in effect switching the agency's focus back to the planet with the mandate of expanding knowledge of Earth. At a moment marked by racially charged anxieties about population growth in the developing world, resource scarcity, and environmental catastrophe, such knowledge was critical to the maintenance of political power, whether in outer space or on Earth. Hence the importance of focusing not only on technical questions related to abstract formulations of human comfort (manifest in tables and diagrams) as well as environmental management and control, let alone the claims to

50

Fred Scharmen has offered a compelling archaeology of the visual culture informing O'Neill's space colony designs, from science fiction novels dating back to the 1950s and the research of Wernher von Braun, as well as films such as Stanley Kubrick and Arthur C. Clarke's *2001: A Space Odyssey* of 1968, to the phenomenon of architectural megastructures, which enjoyed an enhanced visibility on account of the 1967 Expo in Montreal. Unlike many other visions of occupying outer space, he notes, O'Neill saw space colonies not as rooms in space but as new ground upon which to erect a new architecture. That architecture was not entirely new by 1974. The environments in the NASA film *Space Colonization* certainly resonate with Moshe Safdie's Habitat in Montreal in its aesthetics of a modular diversity, with any last vestige of irony or counter-conduct long discarded in favor of an efficient, flexible, architectural machine. Here was a vision in which certain trajectories of experimental practice, those engaging scientific and technical

research under the rubric of complexity, and always and already operating at the nexus of the pragmatic and the utopian, were put to work as R&D for rethinking capitalist expansion. Fred Scharmen, "The High Frontier, the Megastructure, and the Big Dumb Object," paper presented at the 101st ACSA Annual Meeting, San Francisco, 2013. Available at http://apps.acsa-arch.org/resources/proceedings/indexsearch.aspx-?txtKeyword1=%22Scharmen%2C+Fred%22&ddField1=1.

51

Gerard O'Neill, commentary in *Space Colonies*, 70. An editorial note explains that the text is derived from "remarks before the Senate Subcommittee on Aerospace Technology and National Needs on January 19, 1976, and his keynote address at the annual national convention of the American Institute of Aeronautics and Astronautics in Washington, D.C., on January 30, 1976."

bucolic earthlike environments that were the subject of visual representation. At stake was intervening within emergent techniques of power and hierarchical systems of global governance that were being instituted by US-driven neoliberal economic paradigms, under the rubric of technocratic forms of management and exemplified in the UN's development decade. When confronted by the seductive NASA renderings, the question is thus not what a space colony environment looked like, or even how pleasant its idealized climates might be, but how we might read the political and economic agendas they served to advance, agendas embodied within and beneath their spectacular surfaces. Such logics are reappearing in different form today in the haunting images depicting the waning of earthly beauty and stranded polar bears for NASA's current Global Climate Change initiative. All such images, and their mobilization, speak to the political nature of the period's rising interest in climate and environment as it has developed over the last five decades; they remind us that it was not natural processes (or even nature's relation to humans as a species) that was at stake so much as how questions of climate, nature, weather, and resources were framed as socioeconomic and political concerns, and hence participated in what Foucault termed "the calculated management of life."[52]

52

Michel Foucault, "Right of Death and Power over Life," in *The History of Sexuality: An Introduction*, trans. Robert Hurley (New York: Random House, 1990), 139–140.

Felicity D. Scott is associate professor, director of the PhD program in architecture (history and theory), and co-director of the program in Critical, Curatorial, and Conceptual Practices in Architecture (CCCP) at GSAPP. Her most recent books include *Disorientation: Bernard Rudofsky in the Empire of Signs* (Sternberg Press, 2016), and *Outlaw Territories: Environments of Insecurity/Architectures of Counter-Insurgency* (Zone Books, 2016).

Poli
Ecol

Exploring the complex relationships between humans and climate allows us to ask new questions of our political systems and subjectivities, and to revisit age-old questions in new, and newly affective, ways. Interrogating the universal "we" and other normalizing representational tropes so often employed in climate change discourse troubles many inherited Enlightenment notions that continue to govern political processes and institutions, from the municipal to the transnational scale. How might architecture inform these decidedly not-smooth (and, in some cases, non-anthropocentric) recastings of the planetary political body? A number of preliminary yet provocative lines of inquiry emerge as familiar territories are made unfamiliar in the process of their remapping.

Examining the sympathetic flows of people and aerosols in Africa's Sahel region, Adrian Lahoud suggests how an uncertain architecture, lying in wait, might subvert outdated modes and expectations of practice. Neyran Turan digs into disciplinary history to reveal how climate change prompts a reappraisal of architecture's negotiation of the material and representational, while Emily Eliza Scott unpacks the work of artist Amy Balkin to offer a compelling strategy in the struggle to visualize its effects. Analyzing the hydrologic dam system of Southeastern Anatolia, Zeynep Akıncı and Pelin Tan probe architecture's deployment as an ecopolitical technology. Marion Weiss and Michael Manfredi survey infrastructure, calling for systems with programmatic thickness. Jesse Keenan lays out what's at stake in the discourse of "resilient" cities, whose futures Kate Orff and Adam Sobel explore in a conversation about how designers, scientists, and policymakers might better address climatic vulnerabilities. Assessing the possible shifts of power made when humans "delegate to the biosphere," Saskia Sassen sketches a "third space," bridging urban and natural with a close reading of the work of the architect and urbanist Vanessa Keith. Ross Exo Adams examines the project of bodily control embedded in Rebuild by Design's trans-scalar urban vision. Finally, drawing on contemporary protest movements, diaspora studies, and Internet history, Carson Chan asks how architecture can support sociopolitical change by accommodating the reimagination of collective identities.

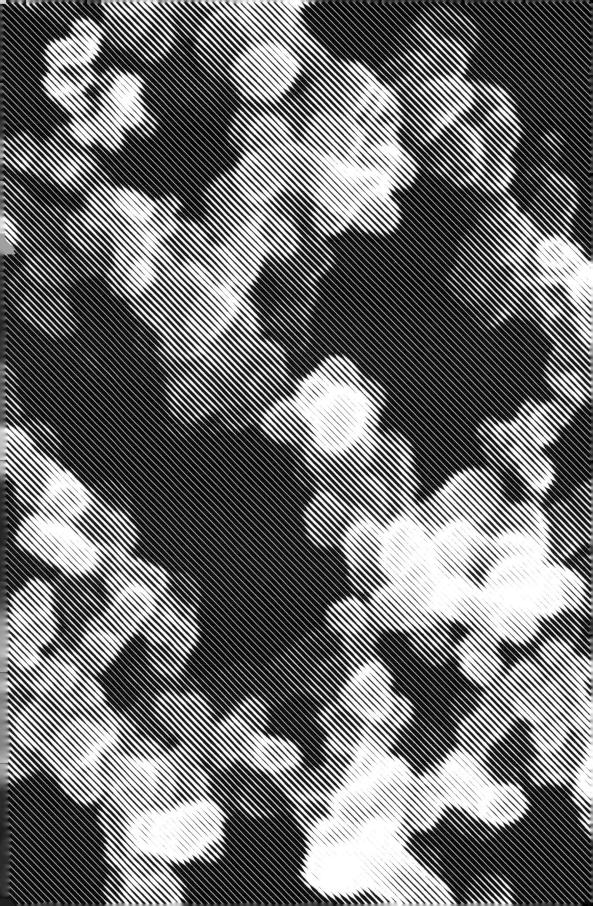

Scale as Problem, Architecture as Trap

ADRIAN LAHOUD

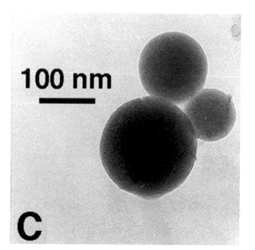 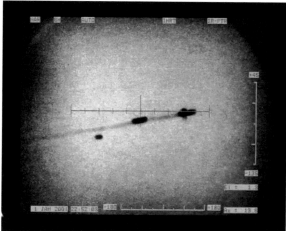

Left, a carbon dioxide molecule under the microscope; right, satellite imaging of human migration.

There is a strange sympathy between the atmospheric particles that float through the sky and the human beings who migrate across the ground and then across the sea. Each body sets the other into motion—a pattern of movement and countermovement. The particle bodies flow from north to south; the human bodies move from south to north. The difference in the kinds of bodies is apparent in the models used to grasp their character: on the one hand, there are the physical and chemical explanations used to model the climate system, and on the other hand there are the anthropological and psychological explanations used to model human character. Thus, the contact between human bodies and atmospheric bodies is the contact between these different kinds of models, the consequences of which are delineated in this story.

The global climate model is actually a series of submodels that are refined until they capture the causal structure of the climatic problem to be reproduced. In climate modeling, as in other forms of simulation, the trick is to somehow reproduce enough information to catch what is relevant in a problem—no more, no less. A working model of causality is really just a reliable explanation of a problem that has been formalized. In terms of scale, this form of modeling operates like a mesh, the apertures of which must be tuned to catch elements of just the right size. Do you calibrate it to catch the weather pattern, the cloud formation, or the water droplet? It depends on what you want to explain. Scale is what organizes this objective relationship between the problem and the model.

Histories of explanation and representation solidify around problems, meaning that we inherit conventions in knowledge production, perhaps none more important than the division of knowledge production itself into scalar categories. These categories emerge both as an objective reflection of the phenomena in question and the social, political, and economic decisions to orient scientific attention in one direction over another.

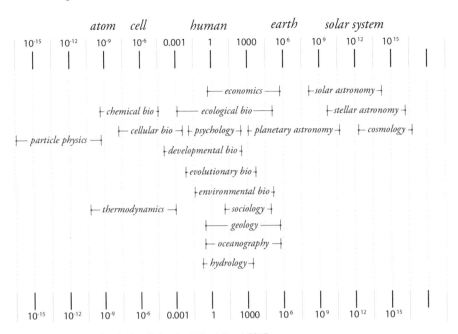

Diagram illustrating scales of scientific inquiry, Adrian Lahoud, 2015.

New kinds of problems—like climate change, for instance—pose special challenges insofar as they bring together the large and the small, the near and the far, the fast and the slow, the weak and the strong, making a mess of existing scalar conventions. In the study of climate, signs must be extracted from a vast sea of scalar variability—this sea of cycles and oscillations span from a nanosecond quick flicker of infrared to the half-million-year passage of our astronomical seasons with all of the endless flux in between. Climate science has to extract individual rhythms from this cacophony. It has to work out what makes a rhythm, what makes it switch tempo, play more insistently, with more syncopation, or just plain out of time.

Scientists obviously set out to explain a wide range of problems through climate models. Even if the ground of verification differs, we find something similar when we look at models of human subjectivity. Here, the aim has always been to use the signs of external conduct to construct a model of internal motivation, to understand the way that conduct emerges out of natural dispositions—or, as Michael Feher has beautifully put it, according to a schema of conflict between good and bad propensities, such as charity and greed, passion and reason, shame and self-worth. Models of subjectivity are supposed to explain what causes human beings to behave the way they do. What happens, then, when environmental models intercept models of human character? How do social modeling and scientific modeling inflect each other? Examining changes across a band of arid land in central Africa may help to demonstrate the inextricability of social and scientific modeling, and the consequences of this encounter.

"Desertification" refers to the process by which arable land becomes unproductive, usually as a result of poor land management. For scientists like Jule Gregory Charney, the term became a kind of conceptual paradigm through which to understand the case of the Sahel and the severe drought that beset the region in the 1970s and 1980s. Essentially, what later became known as the Charney Hypothesis claimed that indigenous mismanagement

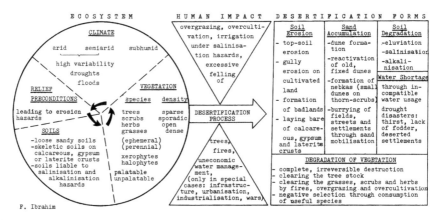

F. Ibrahim

Desertification schema from "Anthropogenic Causes of Desertification in Western Sudan," Fouad Ibrahim, 1978.

of land was leading to a loss of soil and vegetation, and, more importantly, that this loss was changing the reflectivity of the earth's surface, resulting in less rain. The account took hold of the Western scientific imagination; in their view, the people of the Sahel had become weather-makers of the most self-destructive kind.

The two kinds of models described above can be used to understand this account. First there is a generalizing and reductive explanation that attributes land degradation to the people's character. The literature from the period refers to the farmers' inability to reason, to plan, to calculate properly—that is, to their irrationality—but also to their inflexibility, to a lack of capacity to adapt their actions to changing circumstances. The other model of explanation involves the behavior of the environment in response to their actions: the irreversible damage to the earth's surface triggered positive feedback in the atmosphere, which then caused drought. The intersection of these two models gives us a specific geopolitical paradigm: man-made desertification. The consequence of this paradigm was a disastrous legacy of ill-conceived aid packages, reforms, and interventions.

In the last decade, however, climate science has finally confirmed a saying of the Zaghawa people in Chad: The world dies from the north. The severity of the Sahelian drought made it a perfect object to train generations of climate models upon. What climate scientists finally deduced—counter to the prevailing narratives—was that the *oceans* were driving the drying of the Sahel. The first clue in this detective story was a set of models that correlated ocean temperature to precipitation, indicating that the heat in the ocean was acting like a pacemaker for the monsoon.

However, the relationship between heat and rain was still only part of the picture—something was missing. The models could reproduce the signal, but not its strength. Something else was driving the ocean temperature. Was it global warming? Answering this question was complex. What scientists found was that drying depended on a balance between two forces. The first was the temperature difference between cold and warm water in the tropics; the second was the temperature of the troposphere. Instability in the first would tend to more rain, while stability in the second would tend to more drying. The victor would drive precipitation patterns.

The balance of power was poised until an unlikely protagonist tipped the scales. As if in homage to Lucretius, fate would be decided by the infinitesimal

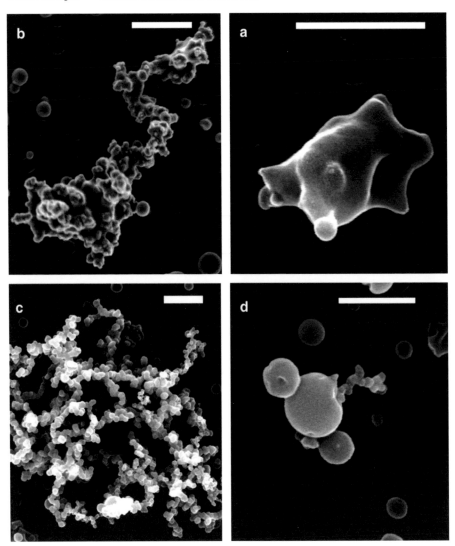

Microscopic images of aerosol particles, each with its own identifiable form.

swerve of a particle. Most people associate global warming with carbon dioxide. But fossil fuels produce another byproduct: aerosols. For some time science has been aware that European and American aerosol emissions were changing the temperature of the oceans. Scientists hypothesized that this was weakening the temperature gradient that was so crucial to precipitation patterns. Aerosol particles are unlike carbon dioxide in that carbon dioxide is long-lived and disperses evenly, which is why we can talk about parts per million as a global concentration. Aerosol particles are short-lived. They get lifted up in air currents, carried through the atmosphere, and then deposited. Their effects, therefore, are far more localized. Unlike carbon dioxide molecules, which are identical, every single aerosol particle is individual.

This character is what makes them so intractable scientifically but also so revealing politically. Individuality is what allows science to identify the source of the particle. Like a fingerprint on a crime scene or a tracing in an atmospheric bloodstream, aerosols tell us a great deal about the structure of the great aerial oceans and the way that their currents redistribute the consequences of human and nonhuman actions. For example, aerosol deposition in the Amazon tells us that, astonishingly, a small, dried-out lake bed named Bodélé in Chad has been secretly supplying the rainforest with nutrients for years. Saharan dust carried dozens of kilometers over the Atlantic has been fertilizing an Amazonian garden. In the case of the Sahel, it does something less metaphysically but more politically suggestive.

Is aerosol dispersion caused by European and North American industrialization significantly contributing to a drying in the Sahel? We don't really know. But the question itself is an unusual one, and its unusualness reveals blind spots in the way climate change is thought of and discussed. It's not, "What is the acceptable average temperature?" It's not, "What is the global concentration of carbon dioxide?" It's not a we-are-in-it-together kind of question, because it's not posed at the scale of the planet or the globe. It suggests that actions in one limited part of the world affect actions in another limited part of the world, even over great distances. It's a different kind of question because it's posed at the scale of specific people and their fate, which is the scale of lived histories, not their negation into a universal humanity. And therefore it's a paradigm of a differently scaled politics.

Interestingly, as the model of the environment changed, so did the models used to understand human behavior. The first shift was to recognize the incredible cultural and linguistic diversity of the region. The second was to recognize the diversity of forms of life: For example, conceiving of a gradient of sedentary and nomadic life where East to West differences yielded similar territorial practices, but different clans and North to South differences yielded both different territorial practices and ethnic groups. As Alex de Waal has shown in his account of famine in Darfur Sudan, looking at the same socioecological diagram from the perspective of the herders, one sees a checkerboard of new nomadic opportunism, in which pastoral routes are used to avoid farms.[1] These models rely on an understanding that what we sometimes take as an essentialist ethnic marker is usually a marker of activity: i.e., you can be an Arab by being a nomadic pastoralist. Indeed there is a certain pragmatic fluidity but also a precarious tension between these livelihoods that is exacerbated under environmental duress. Finally, these models are informed by an awareness of indigenous environmental concepts: for example, the absence of statistical forms of reasoning about precipitation, and in their place something more like a qualitative rather than a quantitative model. Together these factors radically invert the previous concept of the Sahelian farmer plagued by irrationalities and a kind of stubborn ancient inflexibility, instead offering an opposite notion in which the autonomy and flexibility of indigenous farmers and the intelligence of their land use strategies has come to be acknowledged, albeit in a context of extreme global environmental stress. While using these different models in concert has produced a more nuanced and accurate understanding of the social and climatic changes in the region, it does not alter the fact that the Sahel remains caught in a situation not of its own making. Social stress is still being exacerbated by nonlocal environmental factors, and a great diasporic movement of people is still heading to the major cities on the West African coast and north toward Europe and the Mediterranean.

The southward trajectory of aerosols and their effect on the climate of the Sahel and the northward trajectory of migrants attempting to flee sub-Saharan Africa and enter Europe aligns in a pernicious geometry. What gets emitted as a particle returns as a refugee; what is received as a refugee gets returned as a particle. Movement and countermovement. In this drama, Edmond Locard's principle that every contact leaves its trace, the very

1

Alex de Waal, "Sudan, the Sahel, the Sahara: The 99% Principle," *Les Dossiers du CERI* 3 (Summer 2001), 9.

cornerstone of modern forensics, still applies, but with a complicated catch. The contact and the trace drift apart. Environments loosen the bonds between cause and effect, obscuring the link between attribution, responsibility, and—potentially—justice.

What, then, does "justice" mean here? During the 2009 Copenhagen Climate Conference, the public debate was framed by two simple questions: Would an accord be signed or not? And what would be an acceptable average temperature increase? Would it be 1.5 degrees, 2 degrees, or 2.5 degrees? Unbeknownst to the G77 group of developing nations, the Danish had been circulating a secret text exclusively among the G20 that proposed a 2-degree global average temperature increase. When it was discovered, the Sudanese diplomat representing the G77, Lumumba Di-Aping, called an impromptu press conference, saying, "We've been asked to sign a suicide pact." He called it the colonization of the sky, and finally declared it "climate genocide." The proposed average of 2 degrees meant 3.5 degrees in many of the nations that Di-Aping represented, which would be a catastrophic result for sub-Saharan Africa. Di-Aping's claim was an attempt to reestablish the proper political scale within a debate about temperature increase in order to prevent specific populations from being effaced by the coarseness of the model.

The conference was a failure. What it revealed, however, was a certain calculus of life and death concealed in every model. If images belong to a kind of evidentiary paradigm and a symptomatology of signs, then perhaps models belong to a speculative paradigm, a kind of etiology of causes. The model is an observation and a hypothesis. It's both the data that constitutes it and a claim made from that data. But insofar as the model is the means by which scenarios can become rehearsed, it's more than a representation of possible futures. They become a medium through which the present produces the future, though somewhat speculatively.

The Conference of Parties (COP) framework sits at the summit of a vast collective scientific endeavor, but the evidentiary paradigm that forms its foundation presupposes too much. First, it assumes a good faith in the forum and in the way political change flows from understanding. It also presupposes a common regime of intelligibility—shared stakes, institutions, and protocols. In other words, it presupposes so much of what the world lacks. So where does this leave architecture and design? The way that models of the environment intercept models of human subjectivity is a crucial area of inquiry, and one in which architecture and design can intervene. We need to find a way to think of these two things—environment and subjectivity—together, but in light of the failures of COP 15 and the more recent, but nonetheless extremely limited, agreements borne of COP 21, can we imagine an intersection that doesn't presuppose as much faith in the functioning of the forum? That doesn't presuppose a common regime of intelligibility but instead starts with a proposition that embodies the asymmetry of the situation?

The anthropologist Alfred Gell proposes a thought experiment that might offer a useful way of answering this question and, importantly for us, for understanding the role of design within this paradigm. It involves two figures: the hunter and the trapper. Evidence always requires an interlocutor, and the hunter, who is sensitized to science, stands in for this evidentiary regime, for, in a sense, discursive power. Science also requires a figure able to reconstruct a spatial and historical explanation from a series, so the hunter is the first to narrate an event. (Of course Nimrod, the biblical figure of the hunter, as we know, is responsible for the Tower of Babel.) Finally this narration requires

A giraffe trap from "Vogel's Net: Traps as Artworks, Artworks as Trap," Alfred Gell, 1996.

a forum—like a book or a conference—in which the story can be retold. The power of the hunter is an explanatory power, grounded in the interpretation of signs within a space of shared intelligibility. This is the paradigm through which climate science has primarily been pursued.

The second figure is the trapper. Gell's text on "Vogel's Net" is really an extended thought experiment in which he mounts an imaginary exhibition of animal traps. Traps embody what I want to call an asymmetry of legibility. At the very least, the trap is an object that works best when it is unintelligible to its subject. At most, the trap embodies something absolutely essential, which is a kind of fundamental inequality between beings. To work well, the predator must

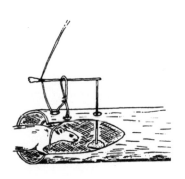

A rat trap from "Vogel's Net: Traps as Artworks, Artworks as Trap," Alfred Gell, 1996.

understand the behavior of the prey. The most successful trapper is the one who is closest to the animal itself—the trapper who grasps the behavior or the prey in all the dimensions of its character and makes a device that embodies the concept of its character.

The trap rarely embodies the form of the prey, though sometimes it does, as in the humorous giraffe trap. More often, it embodies something essential in the prey's character, such as the propensity of the rat to burrow. The trap expresses the ambition of the predator to catch the prey. Therefore, the trap is a designed object in which the character of the prey intersects with the intention of the predator. The chimpanzee trap, for example, solicits the natural curiosity and intelligence of primates. "The thread is very thin and the chimpanzee thinks it can get away," recalls a member of Cameroon's Fang community, Ze. "Instead of breaking the thread, instead of pulling the thread, it pulls on it very gently to see what will happen then. At that moment the bundle with the poisoned arrow falls down on it, because it has not run away like a stupid animal, like an antelope would."[2] The trap preys on the primate's ability to balance its instinct with its intelligence. A small thread captures something essential in the character of its subject, better than an image.

The animal trap turns the personality of the animal against it. This is why Gell is right to think that traps always have a tragic dimension. If we train ourselves to look at architectural and environmental traps, we might be able to extract a portrait of the character of the prey or, more precisely, a portrait of the point of intersection at which the character of the prey and the intention of the predator meet. In architecture, power never touches the human directly. It addresses the human through the life world in which the human subject exists as a set of alternative, conflicting potentials. Environmental determinism would suggest a direct correspondence between climate or geography and the character of the human being, as if an unbroken line of causality chained the human character to the strength of the sun.[3] But in environments, we never find lines of causality without fields of uncertainty. This uncertainty is not only what tempers the strength of our predictions or what qualifies the veracity of our claims. It is the terrain of political struggle itself. Architecture's role in this struggle might be to contribute a particularly uncertain kind of trap, lacking in virtue and good faith, patient, malevolent, living and residing in our blind spots, the amoral, anti-Enlightenment object par excellence.

2

Pascal Boyer, *Barricades mysterieuses & pieges a pensee: Introduction a l'analyse des epopees fang* (Nanterre, France: Société d'ethnologie, 1988), referenced in Alfred Gell, "Vogel's Net: Traps as Artworks and Artworks as Traps," *Journal of Material Culture*, vol. 1, no. 1 (1996): 15–38.

3

Vitruvius, for instance, writes, "Though, however, the southern nations are quick in understanding, and sagacious in council, yet in point of valor they are inferior, for the sun absorbs their animal spirits." *The Architecture of Marcus Vitruvius Pollio in Ten Books*, trans. Joseph Gwilt (London: Priestley and Weale, 1836), 168.

Adrian Lahoud is dean at the School of Architecture, Royal College of Art London.

Measure for the Anthropocene

NEYRAN TURAN

Climate implies environment.
—F. Kenneth Hare, *The Concept of Climate*[1]

The city gives the illusion that the earth does not exist.
—Robert Smithson, *Sedimentation of the Mind: Earth Projects*[2]

Introducing his recently published 2009 book *Why We Disagree About Climate Change*, geographer Mike Hulme writes, "Climate change is not 'a problem' waiting for 'a solution.' It is an environmental, cultural, and political phenomenon that is reshaping the way we think about ourselves, about our societies, and about humanity's place on Earth … Rather than catalyzing disagreements about how, when, and where to tackle climate change, we must approach the idea of climate change as an imaginative resource around which our collective and personal identities and projects can and should take shape."[3]

Moving the attention away from the widespread frameworks of the managerial techno-fix, apocalyptic scenarios, or nostalgia, Hulme's work provides important clues for the role of imagination and representation in relation to climate change. Hulme argues that we have to reconsider our discourses about climate change by understanding the mythical stories that come with it:

> First is the Edenic myth, which talks about climate change using the language of lament and nostalgia, revealing our desire to return to some simpler, more innocent era. In this myth, climate is cast as part of a fragile natural world that needs to be protected … Next, the Apocalyptic myth talks about climate in the language of fear and disaster … and reveals our endemic worry about the future … Then there is the Promethean myth, named after the Greek deity who stole fire from Zeus and gave it to mortals … which talks about climate as something we must control, revealing our desire for dominance and mastery over nature but also that we lack the wisdom and humility to exercise it. Finally, the Themisian myth, named after the Greek goddess of natural law and order, talks about climate change using the language of justice and equity. Climate change becomes an idea around which calls for environmental justice are announced, revealing the human urge to right wrongs.[4]

For Hulme, the more we can understand these myths as stories that reflect our own contradictions, the more we will come to terms with our limitations to deal with the climate change and will be in a better position to assess

1

F. Kenneth Hare, "The Concept of Climate," *Geography*, vol. 51, no. 2 (April 1966): 100.

2

Robert Smithson, "Sedimentation of the Mind: Earth Projects" [1968], in *The Writings of Robert Smithson: Essays with Illustrations*, ed. Nancy Holt (New York: New York University Press, 1979), 82.

3

Mike Hulme, "Why We Disagree About Climate Change," *The Carbon Yearbook 2010: The Annual Review of Business and Climate Change* (September 2010), 41–42. Hulme is professor of climate and culture in the Department of Geography at King's College London.

4

Mike Hulme, "Climate Change: No Eden No Apocalypse" [2009], in *Exploring Climate Change Through Science and in Society: Understanding Controversy, Inaction, and Opportunity* (London: Routledge, 2013), 205.

its potentials. He writes, "There are different ways of portraying this re-imagination of climate change: as a cultural turn, as a crisis in representation or by suggesting that climate change has become a synecdoche ... a term which 'stands-in' for something else."[5]

Hulme's work is all the more noteworthy if one considers that almost all of these myths have infiltrated architectural and urban discourses recently. So how exactly are we to reimagine climate change beyond these myths? And more specifically, what can design possibly contribute toward these urgent concerns? Rather than limiting the role of climate change for design to an external reality to master or solve, might we see it as an opportunity to prompt a renewed understanding of *realism* for architecture? What follows is a brief elaboration on this question and the disciplinary and cultural potentials of such a provocation.

CLIMATE AS THE NEW *CONDITION*

> Humankind cannot bear much reality.
> —T. S. Eliot, *Four Quartets*[6]

In his 1974 article titled "Architecture and Modesty," published in *Casabella*, Andrea Branzi, one of the members of the Italian group Archizoom, wrote:

> With the development of the electronic media and mass-culture, architecture has become something of a minor art ... Once considered the most complete and noble of the arts, architecture has lost its pre-eminence not only because of the external difficulties of a political and economic nature that it has always encountered, but because of a deep internal crisis now afflicting it as a result of modifications in the mechanisms of cultural production and of the urban function itself ... Today the city is no longer a cultural "place" but a "*condition*" ... The quality that we ask of the city today has nothing to do with form or composition, but only with the quality of social services and the market.[7]

For those who are familiar with Branzi's work, this is not a surprising choice of words. One thinks of the No-Stop City project of Archizoom in 1969, which declares the demise of architecture and replaces it with total urbanization, i.e., the limitless horizontal carpet of the capitalist urban *condition*. Conceptualized both as a critique of capitalist development and mass consumerism and as an attack toward the progressive and utopian objectives of late modern and neo-avant-garde architectures of the 1960s, No-Stop City's endless repetition of the big-box structures (the interior climates of the supermarket and the factory) was a demystification of the contemporary metropolis's new logics of production and consumption through the representational technique of overload.[8] Perhaps more importantly, by canceling the distinctions between the interior and the exterior—or work (factory) and consumption (supermarket), city and capitalism, architecture and urbanism—No-Stop City also suggested another kind of collapse, that between *reality* and *realism*. That is, by the relentless repetition of the capitalist city with extreme exacerbation, the realism of the project was nothing more than the excessive and nullifying depiction of external realities. For the No-Stop City, "the city no longer exists any longer outside the system itself: the whole

5

Hulme, "Climate Change: No Eden No Apocalypse." For more on this point, see Mike Hulme, "Climate and Its Changes: A Cultural Appraisal," *Geo: Geography and Environment*, vol. 2, no. 1 (January–June 2015): 1–11, http://onlinelibrary.wiley.com/doi/10.1002/geo2.5/full.

6

T. S. Eliot, *Four Quartets* (Mariner Books, 1968), 14.

7

Andrea Branzi, "Radical Notes 20: Architecture and Modesty," *Casabella* 396 (December 1974): 8. Emphasis added.

8

For a thorough reading of the No-Stop City project in relation to the political climate of Italy, see Pier Vittorio Aureli, *The Project of Autonomy: Politics of Architecture Within and Against Capitalism* (New York: Princeton Architectural Press, 2012), and Pier Vittorio Aureli, "Manfredo Tafuri, Archizoom, Superstudio, and the Critique of Architectural Ideology," in *Architecture and Capitalism: 1845 to the Present*, ed. Peggy Deamer (London and New York: Routledge, 2013), 132–147. See also Andrea Branzi, "City, Assembly Line of Social Issues: Ideology and Theory of the Metropolis," *Casabella* 350–351 (1970): 22–34.

visual relationship with reality loses importance as the distance between the subject and the phenomenon collapses. The city no longer 'represents' the system, but becomes the system itself."[9]

Twenty years after Branzi's announcement of architecture's demise within the relentless market-driven "conditioning" of the city, another kind of "condition"—this time *field conditions*—would prompt a new call for realism. Reacting to the representational concerns of postmodern architecture and shifting focus to the material systems of the city, the attention to conditions in the nineties was promising for its identification of a new realism in architecture. An important difference between these two formulations is that while Branzi's emphasis on conditions had declared the *demise* of architecture, the newer notion aimed to reclaim architecture's agency. Here, the distinction between realism and reality was once again abolished, but this time with an almost opposite premise of instrumentality (instead of critique as it was with Branzi), a realism that aspires to intervene into the very material reality of the city:

> Postmodernism in architecture is usually associated with a rediscovery of architecture's past ... Postmodernism responded not only to a call to re-inscribe architecture into history, it also responded to a contemporary demand for *meaning* in architecture ... Nevertheless, an architecture that works exclusively in the semiotic register and defines its role as critique, commentary, or even "interrogation" ... has in some fundamental way given up on the possibility of ever intervening in that reality. Under the dominance of the representational model, architecture has surrendered its capacity to imagine, to propose, or to construct alternative realities ... Infrastructural urbanism marks a return to instrumentality and a move away from the representational imperative in architecture ... Material practices (ecology and engineering for example) ... do not work primarily with images or meaning, or even with objects, but with *performance*: energy inputs and outputs, the calibration of force and resistance. They are less concerned with what things look like and more concerned with what they can do.[10]

Standing as a precursor to contemporary conceptions of environmental performance in architecture and related design fields, the shift in emphasis toward material conditions has been exemplified for the last two decades by explorations of landscape and ecological urbanism, a renewed interest in the politics of territory and infrastructure, as well as the omnipresence of mapping as a methodological tool for "design research." And, after almost twenty years, we are currently experiencing the evolution of a third stage, presenting climate change as the *new* "new condition." Ranging from managerial metrics of environmental engineering and maintenance to data accumulation, climate change reveals architecture's desire to engage with the realities of the world—but runs the risk of limiting its unique capacities to narrow negotiations between data and its management. Correspondingly, rather than reducing architecture's agency to the managerial, can we instead project an alternative kind of geographic imagination that builds unfamiliar aesthetic couplings between the representational and the material?[11] Here, instead of negating the representational for the sake of an emphasis on the material, as it was in the nineties, this geographic imagination would aim at a new kind of materialism—one that couples deep engagement with environmental

9

Archizoom Associati, "No-Stop City Residential Parkings: Climatic Universal System," *Domus* 496 (March 1971): 53.

10

Stan Allen, "Infrastructural Urbanism," in *Points + Lines: Diagrams and Projects for the City* (New York: Princeton Architectural Press, 1999), 49–53. Emphasis in original. See also Sanford Kwinter and Michael Feher's introduction to *Zone 1/2: The Contemporary City* (New York: Zone Books, 1986), 10–12.

11

For more on this point, see Neyran Turan, "How Do Geographic Objects Perform?" in *ARPA 03: Performance* (2015), http://www. arpajournal.net/ how-do-geographic-objects-perform.

contingencies with its seemingly opposing counterpart, representation. In an era when humans are described as "geological agents," architecture is a background and a measure against which the world might be read. Like architecture, then, could a new realism represent the world back to itself and instigate potentials beyond?

MORE THAN REALITY, MORE THAN MATERIALITY

> Enjoy Nature! I am glad to say that I have entirely lost that faculty. People tell us that Art makes us love Nature more than we loved her before … My own experience is that the more we study Art, the less we care for Nature. What Art really reveals to us is Nature's lack of design, her curious crudities, her extraordinary monotony, her absolutely unfinished condition … It is fortunate for us, however, that Nature is so imperfect, as otherwise we should have no art at all … As for the infinite variety of Nature, that is a pure myth. It is not to be found in Nature herself. It resides in the imagination, or fancy, or cultivated blindness of the man who looks at her.
> —Oscar Wilde, *Decay of Lying*[12]

To be able to talk about an alternative relationship between the material and the representational, one point of departure would be the subtle yet crucial distinction between realism and reality. Instead of fixed and fastened, what if the reality promised to us by the new condition of climate change is nothing more than an evolving agglomeration of representations in themselves, in the form of statistics, measurements, and simulations? First, consider the thin line that exists between observation and agency when it comes to sensing, surveillance, and subjectivity.[13] Second, thinking of the incomprehensibility of representations of scientific data and the inhumanly large scale of climate change (both in temporal and spatial terms), grasping climate change comes with challenges in terms of making visible its invisibilities. Anthropologist Peter Rudiak-Gould proposes the idea of "constructive visibilism" as a way to go beyond the limitations of "invisibilism" promoted by the physical scientists—who point to the huge gap between the visible reality and climate change as they claim sole authority for judgment—as well as the "visibilism" of activism, which aims to give nonscientists the right to speak about climate change. He writes, "Climate change is neither inherently invisible or inherently visible; it is, like all other objects, *made* visible."[14] Similarly, Birgit Schneider and Thomas Nocke make us aware of the key role that "climate pictures" play not only "in making the future imaginable and making the unimaginable visible" and point to their political dimension:

> Climate pictures are often produced for normative purposes; they are produced to change what they show: possible futures are blueprinted in order to prevent the futures shown with the help of the curves from coming true … Because of their normative character, what is special about images depicting climate change is that they have implicitly also become *political images*. This is likewise true for scientific images of climate change. Climate science is facing considerable pressure from all sides: when scientific graphics produced by climatologists started to gain currency in the field of policy, as climate change became a key

12

Oscar Wilde, *The Decay of Lying* (London: Penguin Group, 2010), 3.

13

See Paul N. Edwards, *A Vast Machine: Computer Models, Climate Data, and the Politics of Global Warming* (Cambridge, MA: MIT Press, 2010). Also see Ryan Bishop, "Smart Dust and Remote Sensing: The Political Subject in Autonomous Systems," *Cultural Politics*, vol. 11, no. 1 (2015): 100–110.

14

Peter Rudiak-Gould, "'We Have Seen It with Our Own Eyes': Why We Disagree about Climate Change Visibility," *Weather Climate & Society* 5 (2013): 128. Emphasis in original.

issue within risk society, they encountered different values and expectations. Climate change is not only a science subject, it is also at the core of socio-political interests.[15]

Moving toward geography, a second point of departure for renewed understandings of the representational and the material is the geological dimension of climate change. As argued by media theorist Jussi Parikka, what if there is no media without geology? From the geological formation and resource extraction of material to its final destination as electronic waste, Parikka depicts media as a geological entity, in that its materiality is derived from the chemical elements that form the earth. He reminds us that "data mining might be a leading hype term for our digital age of the moment, but it is enabled only by the sort of mining that we associate with the ground and its ungrounding. Digital culture starts in the depths and deep times of the planet."[16] Parikka points to the shift of our society's expanding material usage, especially over the last fifty years. Where human society used to use less than twelve basic materials such as wood, brick, iron, copper, gold, silver, and a few plastics, we are now at a point when even a single electronic chip is composed of more than sixty different material elements.[17] Parikka writes:

> Instead of networking, we need to remember the importance of copper or optical fiber for such forms of communication; instead of a blunt discussion of "the digital," we need to pick it apart and remember that also mineral durations are essential … Besides the materials of production, media history is a story of relations between the organic and nonorganic and the waste products emerging from the use and misuse of materials.[18]

If realism is more than reality and materiality is necessarily geological, what role does representation play in this dilemma, especially in the context of climate change? Consider artist and geographer Trevor Paglen's 2009 photograph *DMSP 5B/F4 from Pyramid Lake Indian Reservation*, which depicts Pyramid Lake with the white orbiting lines of the now defunct but still orbiting 1973 military meteorological satellite. This photograph operates on several registers relative to these questions. First, it reveals the political dimension of imaging, and the thin line that exists between military, geological, and meteorological exploration. Second, by juxtaposing the iconic, pyramidal tufa rocks (which give the lake its name) with the planetary ruins of the dead satellite orbits circulating the earth, his photograph points to a subtle distinction between reality and realism (realism understood as more than depicting reality, in this case phenomena visible to the human eye). By expanding the material *reality* of the earth to include the invisible outer space along with its technological waste, Paglen's images depict an unfamiliar coupling of material and representational, making evident our need for an expanded geologic realism.

Perhaps most striking, however, is Paglen's reworking of Timothy H. O'Sullivan's famous 1867 photograph titled *Tufa Domes, Pyramid Lake*, taken in his capacity as official photographer of the famous King Survey, a geological expedition of the 40th parallel north of California, Nevada, and Wyoming. While Sullivan's photograph centers on the tufa domes and main tufa pyramid itself, in Paglen's photograph, the view is centered both on the pyramid and the orbiting satellite tracks, pointing to the doubling of two

15

Birgit Schneider and Thomas Nocke, eds., *Image Politics of Climate Change: Visualizations, Imaginations, Documentations* (Bielefeld, Germany: Transcript-Verlag, 2014), 15. Emphasis added.

16

Jussi Parikka, *The Anthrobscene* (Minneapolis: The University of Minnesota Press, 2014), Kindle edition, 861.

17

Parikka, *The Anthrobscene*, 227–229. Re-quoted from T. E. Graedel, E. M. Harper, N. T. Nassar, and Barbara K. Reck, "On the Materials Basis of Modern Society," *PNAS* (October 2013), early edition: 1. See also Jennifer Gabrys, *Digital Rubbish: A Natural History of Electronics* (Ann Arbor: University of Michigan Press, 2011).

18

Jussi Parikka, *A Geology of Media* (Minneapolis: University of Minnesota Press, 2015), 4, 26.

Trevor Paglen, *DMSP 5B/F4 from Pyramid Lake Indian Reservation* (Military Meteorological Satellite; 1973-054A), C-print, 2009. Courtesy the artist, Metro Pictures, Altman Siegel, and Galerie Thomas Zander; © the artist.

kinds of geological objects as ruins of the earth—one extending to the underground, the other one far above the atmosphere. Here, by referencing an important precedent from the history of photography in relation to geologic expedition, Paglen's photograph portrays a disciplinary specificity as well as a speculative rigor for investigations yet to come. If discussions around climate change and the Anthropocene are simply telling us that the earth no longer exists in the ways that it did before thanks to the indisputable effects of human activity, what would be the architectural equivalent of this disciplinary specificity and speculative rigor in the context of climate change?

SLIGHTLY (UN)FAMILIAR AS MEASURE

> On the one hand, it is perfectly true that if what you want is changes in policy, you are not likely to get them from art. On the other hand, if what you want is a vision of the structures that produce both the policies we have got and the desire for alternatives, art is almost the only place you can find it.
> —Walter Benn Michaels, 2011[19]

19

Walter Benn Michaels, "Interview with Walter Benn Michaels on Photography and Politics," *NONSITE* 2 (June 2011), http://nonsite.org/editorial/interview-with-walter-benn-michaels-on-photography-and-politics.

The idea of the (un)familiar is pertinent here precisely because of its specific legacy within the history of realism in architecture. From the "de-familiarization" project of modernism to the re-appropriation of the familiar during the 1960s and 1970s—through discussions on reality-as-found, typology, and the ordinary, as seen for instance in the work of Mathias Ungers, Alison and Peter Smithson, Aldo Rossi, and Denise Scott Brown and Robert Venturi—various relationships between the familiar and the unfamiliar have defined architecture's framing of the world and its cultural significance.

With this background in mind, the relationship between the familiar and the unfamiliar merits closer attention when one considers its renewed relevance within contemporary architecture and urbanism. On the one hand, one can speculate on the unfamiliar interpretations of what is considered to be familiar, ordinary, or banal architectural qualities of the city. Here, an endless horizon of suburban tract homes, shotgun houses, "typical plans," cloudy curtain walls, office tower atriums, and big-box warehouses and other forms of commonplace or vernacular architectural production are not simply foregrounded but are understood with a renewed rigor. On the other hand, in an attempt to expand our disciplinary imaginary, one can also speculate on the use of familiar architectural strategies on what is considered to be unfamiliar within a disciplinary setting. Here, all that belongs to the environment yet remains invisible comes into focus, gaining a particular relevance—perhaps in the same manner that Schinkel found beauty in the English factories, Gropius in the American grain silos, and Le Corbusier in the ocean liner. Accordingly, the territorial geometries of agricultural and resource extraction fields, agricultural barns and grain silos, expanded infrastructures of resource and matter, and geological layers of the earth draw a particular attention. In both formulations of the (un)familiar, reality is filtered through a new realism and we are enabled to project renewed relationships between the material and the representational. Thought this way, rather than contextualizing the relevance of climate change for architecture as another external reality—or "condition"—to be accommodated, and quickly associating realism with reality, I propose that we contextualize the concept in the lineage of the (un)familiar.

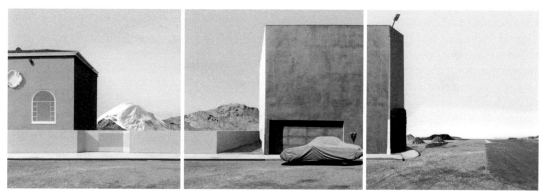

Lauren Marsolier, *Transition* series, part 3, 2012. © the artist.

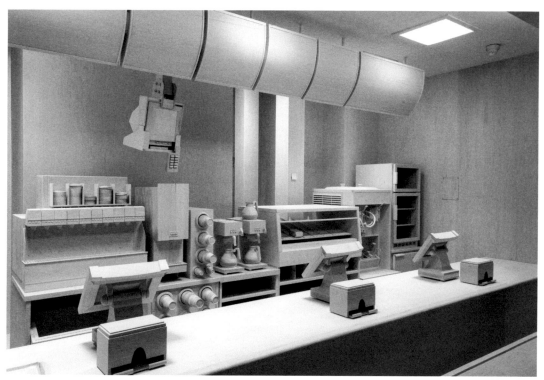

Roxy Paine, *Carcass*, 2013; birch, maple, glass, and fluorescent lighting. Courtesy of the artist and Kavi Gupta Gallery.

The true potential of such an investigation on the idea of the (un)familiar in relation to climate change would be in its capacity to offer unconventional relationships between the real and the abstract, and between the disciplinary and the cultural, rather than resorting to extreme dualities. In other words, while aiming to build a nuanced relationship between reality and realism, we would then be able to forgo the project of hyper-realism (think: righteous scenario planning or environmental engineering of data and performance), hyper-surrealism (think: architectural sci-fi), or hyper-abstraction (think: white noise), and instead project a *slightly (un)familiar* realism as a form of strategic abstraction, i.e., a subtle and unexpected separation from reality.

Rather than probing estrangement for its own sake or negating realism altogether, one can position this slightly unfamiliar realism as an alternative attitude that projects new relationships between the material and the representational. Think of the near-plausibility suggested in photographer Lauren Marsolier's slightly distorted realisms of everyday life that appear in the carefully calibrated digital collages of her *Transition Series*, or of David Reed's paintings inserted anachronistically into the bedroom scene in Hitchcock's *Vertigo*, or of Roxy Paine's meticulous rendering of the banal in his scaled diorama of a fast-food kitchen, produced entirely from birch and maple wood in his *Carcass*. What all of these works share is an abstraction that reinforces reality by pushing back.

Likewise, the potential of a slightly unfamiliar realism for architecture in the context of climate change would be in its carefully calibrated degree of separation from the real, in order to achieve a much deeper and nuanced engagement with reality through the limits and potentials of architectural specificity and its speculative rigor. Realism and abstraction need each other. It is time to consider materialism and representation together as well.

Neyran Turan is an assistant professor of architecture at the University of California, Berkeley and a partner at NEMESTUDIO, a design collaborative based in the San Francisco Bay Area which has recently received the 2016 Architecture League New York Prize for Young Architects. Turan's work draws on the relationship between geography and design to highlight their interaction for new aesthetic and political trajectories within architecture and urbanism. She is founding chief-editor of the Harvard GSD journal *New Geographies*. Her most recent projects include her solo installation titled *STRAIT*, which was on view at SALT (Istanbul) in May–August 2015, and her speculative design project Museum of Lost Volumes (2015), for which she received an Honorable Mention Award in *Blank Space: When Architecture Tells a Story Competition*.

Archives of the Present-Future: On Climate Change and Representational Breakdown

EMILY ELIZA SCOTT

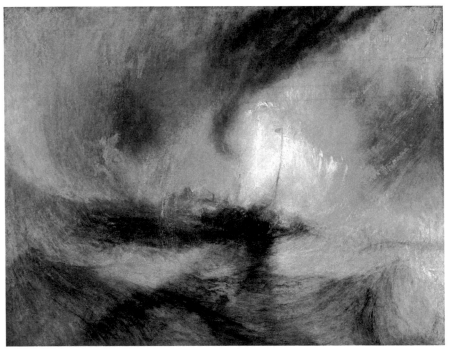

Snow Storm—Steam-Boat off a Harbour's Mouth, Joseph Mallord William Turner, 1842. © Tate, London, 2016.

Climate change is everything, a story and a calamity bigger than any other. It's the whole planet for the whole foreseeable future, the entire atmosphere, all the oceans, the poles; it's weather and crop failure and famine and tropical diseases heading north and desertification and the uncertain fate of a great majority of species on Earth.
—Rebecca Solnit[1]

Apprehending what is significant … may require "escaping the fascination of the picture" by adopting another perspective—a partial or partisan perspective, the perspective of a part. From this …, the whole will not appear as a whole. It will appear with a hole. The perspective from which the hole appears is that of the subject…
—Jodi Dean[2]

Perhaps more than any other phenomenon imaginable, climate change is dizzyingly convoluted, entailing many often correlated and at times seemingly contradictory things happening in multiple places at once, at varying rates and scales, and with myriad types and degrees of consequence. In addition to certain unprecedented material-environmental conditions, it thus poses profound representational dilemmas. This is compounded further by our

[1]
Rebecca Solnit, "Are We Missing the Big Picture on Climate Change?" *New York Times Magazine*, December 2, 2014, http://www.nytimes.com/2014/12/07/magazine/are-we-missing-the-big-picture-on-climate-change.html.

[2]
Jodi Dean, "The Anamorphic Politics of Climate Change," *e-flux journal* 69 (January 2016), http://www.e-flux.com/journal/the-anamorphic-politics-of-climate-change.

3

The philosopher Timothy Morton has famously applied the term "hyperobject" to those contemporary more-than-things, like atomic radiation and global warming, that are, as he characterizes them, "massively distributed in time and space" (1), that we "can't directly see or touch" (12), that have "no center and … no edge" (17), in which "gigantic scales are involved—or rather such knotty relationships between gigantic and intimate scales" that "the social and psychic tools we use to measure them" are utterly confounded (47), and, ultimately, that are "directly responsible for … *the end of the world*" (2). Timothy Morton, *Hyperobjects: Philosophy and Ecology after the End of the World* (Minneapolis: University of Minnesota Press, 2013).

4

Rachel Louise Carson, *Silent Spring* (1962; repr. Boston: Houghton Mifflin, 1992), 238.

5

Emily Apter, "Planetary Dysphoria," *Third Text* 27 (January 2013): 140.

6

See, in particular: John H. Richardson, "When the End of Civilization Is Your Day Job," *Esquire*, July 7, 2015, http://www.esquire.com/news-politics/a36228/ballad-of-the-sad-climatologists-0815, and Meghan Walsh, "It's the End of the World— How Do You Feel?" *OZY*, October 26, 2015, http://www.ozy.com/fast-forward/its-the-end-of-the-world-how-do-you-feel/62757.

immersion in this ever accelerating "everything-ness," the edges of which are challenging, if not impossible, to sense.[3] From what vantage point, then, might we engage climate change—this multiscalar, multitemporal, multidimensional, and multidisciplinary "shadow that is no less ominous because it is formless and obscure," to borrow Rachel Carson's analogy for pesticide contamination from 1962?[4]

The term *dysphoria* (from the Greek: "difficult to bear") denotes a state of extreme unease that can also, in a psychiatric context, mean someone who becomes uncomfortable in her own body or skin. Common symptoms include emotional distress or its seeming opposite, numbed indifference. Indeed, it might be argued that our planetary home, or *oikos*, in its human-caused, heat-trapping state of precariousness, has become a newly foreign medium in which we do not as yet know how to move. In a recent essay titled "Planetary Dysphoria," the literary theorist Emily Apter traces the lineage of a now pervasive and distinctly apocalyptic "aesthetics of planetarity," which she summarizes as "the geo-psychoanalytic state of the world at its most depressed and *unruhig*, awaiting the triumphant revenge of acid, oil, and dust."[5] One might think here, too, of the growing number of publicized cases of depression among leading climate scientists, or what some refer to as the "climate anxiety" or "pre-traumatic stress disorder" experienced by those who most clearly grasp the magnitude of our current atmospheric-environmental crisis, as well as the limited capacity of their own research, thus far, to spur anywhere near a sufficient response.[6]

The *Fram*, caught in sea ice, Arctic Ocean, March 1894. Photograph by Fridtjof Nansen, courtesy of the National Library of Norway.

Under the immense and bewildering weight of climate change, how might we hack out viable positions—or holes within the whole, as the political theorist Jodi Dean puts it—from which to proceed? What kinds of representation (visual and otherwise) are adequate to the task of conveying climate change, and perhaps most importantly, to stemming dysphoric paralysis while triggering critical thinking and action? Using *A People's Archive of Sinking and Melting* (2011–present) by the California-based artist Amy Balkin as one example, this essay will consider those representations that dwell in "the sticky materiality of practical encounters," moving across scales and registers in order to forge not seamless perspectives but ones that are adequately fractured and muddy.[7]

—

The existing visual culture of climate change—brimming with depictions of ravaged landscapes, polar bears atop waning ice, and techno-sublime satellite views—adheres largely to an illustrative mode, despite the incongruousness and even muteness of such a manner in the face of newly complex entanglements between the human and nonhuman, not to mention the kind of attritional "slow violence" associated with climate change.[8] The genre of science fiction may hold a particular potential to wrangle with our present planetary crisis in its oblique relation to the (nonfictional) world, or its invocation of vantage points tied to imagined other times and places. That being said, science fiction's indulgence in the spectacle of disaster has a tendency to turn us into spectators, desensitized and deactivated, by inviting "a dispassionate, aesthetic view of destruction and violence— a technological view," as noted incisively by the cultural critic Susan Sontag in 1965.[9] In any case, the prominent resurgence of dystopian, end-of-the-world imagery in both mass culture and the highly compelling work of some contemporary artists and experimental scholars certainly warrants further attention.[10]

Polar bear in Svalbard, Norway, 2010. Photograph © Rich Kerchner.

7

Anna Lowenhaupt Tsing, *Friction: An Ethnography of Global Connection* (Princeton, NJ: Princeton University Press, 2004), 1.

8

By "newly complex entanglements between the human and nonhuman" I mean the thoroughly hybridized state of the world, in which any notion of a clear distinction between "nature" and "culture" has become untenable, and that is underscored by everything from the existence of "plastiglomerates," a recently discovered geological specimen made of melted plastic fused with rock, to various biotechnological developments, to the presence of synthetic contaminants in even the remotest of locations. A handful of forward-thinking ecologists has adopted the term "ecological novelty" to capture this condition, while "emergent ecologies," "post-nature," and indeed "the Anthropocene" also speak to this reality.

9

Susan Sontag, "The Imagination of Disaster," reprinted in *Against Interpretation and Other Essays* (New York: Dell Publishing Co., 1966), 216.

10

The Collapse of Western Civilization: A View from the Future (New York: Columbia University Press, 2014), by the historians Naomi Oreskes and Erik M. Conway, is one example of a work by scholars that engages the genre of science fiction in order to work through the present. The recent writings and performance-lectures of the sociologist Bronislaw Szerszynski, which invoke extraterrestrial, futuristic, and ghostly personae, are another noteworthy instance.

I want to touch briefly on two dominant visual tropes with a documentary orientation, both of which involve the compression of climate change into a single picture, or its encapsulation "in a data point or disastrous image."[11] First, let us consider that of the iconic, lone polar bear struggling for survival in a rapidly disappearing habitat. As with countless other portraits of the Earth and its beings in peril, here it is a single, tragic victim that occupies center stage. Our own viewpoint is proximate, at the human scale. Imbued with pathos, this picture is at once allegorical and journalistic. In a recent newspaper article on the mainstream press's coverage of a newly erected solar power plant in

11

Dean, "The Ana-
morphic Politics of
Climate Change."

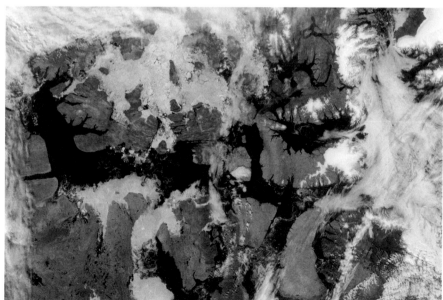

Satellite images—taken on July 17 and August 3, 2012—document the retreat of snow and sea ice in the North-west Passage of Canada's Parry Channel, a link in the long-sought northern route for ship traffic between the Atlantic and Pacific Oceans. Produced with NASA's Terra satellite, by Jesse Allen, Lance system/NASA Earth Observatory.

California, one that has resulted in the rather spectacular death of a number of birds via midair incineration, the writer Rebecca Solnit explores the consequences of such imagery. She cites the media's propensity to feature the loss of individual animals, in this case the charred and broken bodies of solitary birds, insisting that this emphasis on foreground comes at the dire expense of a "big-picture" perspective—one that would, for instance, ask us to think at the scale of whole habitats and the potential loss thereof, or in terms of the respective effects of a fossil-fuel- versus alternative-energy-based economy. "The stories about individual birds can distract us from the slow-motion calamity that will eventually threaten every bird," she succinctly warns.[12] Much more could also be said about the ways in which this type of illustration constructs a dichotomous relation between nature, on the one hand, and humans, with their nature-threatening technology, on the other.

At the opposite scalar extreme is the equally prevalent, techno-scientific representation of the Earth's surface garnered from the quasi-external, top-down vantage point of an orbiting satellite. Today, images taken from outer space reveal numerous signs of environmental distress, from the melting of glaciers and polar ice caps to desertification, air pollution, and more frequent and intense tropical storms.[13] Such pictures are increasingly summoned to support the idea that we have entered an unprecedented epoch, the Anthropocene, wherein humans are interpreted as a newly geological force with the power to alter the planet itself, including its Earth systems. Like the famous photographs snapped aboard Apollo spacecraft in the 1960s and early 1970s that are their precursors, these contemporary satellite images, significantly, do not record national and other geopolitical boundaries, thus framing the planet as a natural body.[14] It is clear, however, that they are the outcome and embodiment of specific geo-techno-politics, namely the decades-long, Cold War space race between the United States and Soviet Union that also gave rise to the (interrelated) development of cybernetics and Earth system science as an ever-more dominant frame for scientific analysis. The gaze of the satellite constructs a particular, highly technologized view of the world, not unlike the computer model through which infinite simulations have been run to assess current and future climate scenarios. Both must be considered relative to the broader visual regime of data through which climate change has largely been studied and communicated, and from which the scales of the global and planetary (which are not precisely the same) have become further solidified.[15]

Despite their differences, both of these image types—the too close and the too distant—presuppose the possibility of documenting objective, single truths and, as such, I want to argue, they are too literal to contend with climate change. The figurative mode crumbles under its weight, fundamentally out of scale and character with the subject at hand. Moreover, both the on-the-ground, embedded view and the global view

12

Solnit, "Are We Missing the Big Picture on Climate Change?"

13

Michael Benson, "Gorgeous Glimpses of Calamity," *New York Times*, August 13, 2013, http://www.nytimes.com/2013/08/18/opinion/sunday/gorgeous-glimpses-of-calamity.html.

14

For detailed discussions of the photographs originating from NASA's Apollo space mission (1963–72) as well as how they have been culturally absorbed and put to work, see: Denis E. Cosgrove, *Apollo's Eye: A Cartographic Genealogy of the Earth in the Western Imagination* (Baltimore: Johns Hopkins University Press, 2001), and Laura Kurgan, *Close-Up at a Distance: Mapping, Technology, Politics* (New York: Zone Books, 2013).

15

The global and planetary arguably carry different connotations with regard to their respective positioning of the human. The global scale/perspective is closely tied to understandings of globalization, especially in the mid-twentieth century forward via the emergence of telecommunication networks, intensified international trade, multinational corporations, uneven geographical development, the aforementioned images of the Earth from outer space, and so forth—all

linked directly to human life. See Paul Edwards's ambitious book on global thinking relative to climate change: *A Vast Machine: Computer Models, Climate Data, and the Politics of Global Warming* (Cambridge, MA: MIT Press, 2010). The planetary, meanwhile, brings the physical and natural world, Earth systems, etc.—the nonhuman, so to speak—into its purview, as well as a broader extraterrestrial context. In spatial terms, we might link the former, in its address of tightly interconnected networks across a surface, to horizontality, and the latter to a more vertical orientation. The scholar Jennifer Wenzel offers a highly useful discussion of the two in relation to each other, including a survey of key thinkers on the matter, from anthropologist Tim Ingold to postcolonial theorist Gayatri Spivak in her essay "Planet vs. Globe," *English Language Notes*, vol. 52, no. 1 (October 2014): 1–30.

16

In their brief examination of the imagery invoked during the 2010 BP oil spill in the Gulf of Mexico, the historian of science Peter Galison and art historian Caroline Jones address three tropes, all of which they deem politically disabling: pictures of oil-soaked animals tended to by emergency workers (suggesting to viewers that the front line is being well cared for by others); satellite and other techno-scientific representations, which they associate with the sublime and aesthetic remove; and the "spill-cam" footage of that which we could see in grainy, seemingly close-up, real-time detail, deep down on the sea floor, but had no power to stop— an experience they liken to "environmental torture." Their own analysis has been formative to my own thinking. Peter Galison and Caroline Jones, "Unknown Quantities," *Artforum*, vol. 49, no. 3 (November 2010): 49–51.

17

I realize that I'm painting a very broad picture here, and, in particular, that there is a wide gap (or even diametric opposition) between the kind of foreground-oriented, close-up images of tragic wildlife in mainstream environmentalist nature photography and those that document or otherwise emerge from social and environmental justice activism. The art historian T. J. Demos makes a strong case that it is precisely documentation of resistance on the ground, using the medium of photography, that has the most potential to counter the far too often apolitical nature of Anthropocene- and climate-change-related visual culture. T. J. Demos, Fotomuseum Winterthur blog (May–June 2015), http://blog. fotomuseum.ch/author/tj-demos.

18

Rob Nixon, *Slow Violence and the Environmentalism of the Poor* (Cambridge, MA: Harvard University Press, 2011), 6.

19

Adrian Lahoud, "Nomos and Cosmos," *e-flux journal 65: SUPERCOMMUNITY* (May 30, 2015), http://supercommunity.e-flux.com/texts/nomos-and-cosmos.

from afar tend to curtail political agency.[16] While the former is typically deployed in activist struggles and therefore commonly associated with the political, unless it somehow opens onto wider contexts, such as the structural conditions leading to particular circumstances, it often remains too narrow in focus.[17] Meanwhile, the satellite view subsumes particularities into totalities, in the process naturalizing, or depoliticizing, states of affairs. In addition to masking its own operations, it fails to document, or even elides, the complex processes and relations through which climate change has come into being, including the socially uneven production and distribution of its many effects. In short, both of these widespread visual tropes blur or conceal crucial contexts, conflicts, and contingencies.

Indeed, one of the key representational challenges that climate change poses is how to maintain a degree of resolution fine enough to capture interrelations that are dispersed across time and space, and are often radically asymmetrical in nature. In his influential book on environmental "slow violence," the postcolonial theorist Rob Nixon addresses climate change's "unevenly universal" burden, as well as the misalignment between "calamities that patiently dispense their devastation" and are "marked above all by displacements— temporal, geographical, rhetorical, and technological," and our spectacle-hungry, "flickering attention spans."[18] In a similar vein, Adrian Lahoud, within the context of his work on climate "forensics," draws attention to the ways that broad planetary scales veil glaring inequities. In particular, he takes on the 2-degrees-Celsius warming limit that has become a de facto target for international climate policy. He explains, "The violent abstraction of a global average negates the uneven scale of climate impact and erases the specificity of people from its calculation. A two-degree average increase globally would allow for a catastrophic 3.5-degrees" in many places.[19] This is, in fact, something that leaders from places like the Maldives,

Maldives Underwater Cabinet Meeting protest event, 2009. Image courtesy of 350.org.

a low-lying island nation at acute risk from rising sea levels, have long argued, as captured in a brilliantly staged protest in the form of an underwater cabinet meeting from which President Mohamed Nasheed and his cabinet ministers sent an "SOS" message to the UN climate change summit in Copenhagen in 2009. In line with climate justice activists, especially those from the Global South, Lahoud calls for "a different scale of calculus," one that would possess "just enough resolution to catch the uneven effects of temperature increase."[20]

In light of these concerns and to reiterate my original question—from what position might we build and mobilize more adequate representations of climate change? In her 2004 book, *Friction: An Ethnography of Global Connection*, the anthropologist Anna Tsing proposes that it is "necessary to begin, again and again, in the middle of things."[21] For, *where*, exactly, does something like the global—or climate change, I would add—reside? Is it ever isolatable to one scene? Global connections, Tsing puts forth, "can only be charged and enacted in the sticky materiality of practical encounters."[22] So it is to friction, "the awkward, unequal, unstable, and creative qualities of interconnection across difference," that our attentions must turn.[23] Extending from this, I, like others, want to advocate for perspectives that are highly situated yet move across registers and scales—both spatial (e.g., the so-called local and global) and temporal (e.g., historical time, evolutionary time, and media time). Distinct from an ocular-oriented zooming in and out, such as that enacted in Charles and Ray Eames's well-known 1977 film, *Powers of Ten*, it is representations that engage multiple, ultimately incommensurable scales and registers (e.g., the molecular, bacterial, geopolitical, geological, and architectural, visible, invisible, material, speculative, and so on) that best suit our present moment.

—

A People's Archive of Sinking and Melting, initiated in 2011 by the California-based artist Amy Balkin, is, as she describes it, "a collection of materials contributed by people living in places that may disappear because of the combined physical, political, and economic impacts of climate change ... Together ... they form an archive of what will have been."[24] The artifacts thus amassed—of modest scale and carrying oblique reference to both their sites and acts of collection (the artist is clear that anything that "happens to be there" is welcome)—include a desiccated mango pit gathered along the shoreline of a small fishing village in Senegal, a pebble from the Ninth Ward of New Orleans, and a rusted nautical hook from Palmer Station on Anvers Island in Antarctica. This participatory archive is rooted in mundane objects that together bear situated, poly-vocal witness to real and projected events. While reminiscent of an eighteenth-century *Wunderkammer* in terms of its eclecticism, this bevy of curiosities is not meant to consolidate one vision or experience of the world, but, on the contrary, to serve as "community-gathered evidence" or "a public record," in the artist's words.[25] It is counter-official documentation that she and the archive's donors are building, both political and elegiac in its tone.

The mostly everyday ephemera comprising the collection have already been cast off as debris—flotsam and jetsam of the contemporary and recent past deemed too insignificant for remembrance, preservation, or even proper disposal, in many cases. Yet, here, within the context of the archive, they are converted into touchstones. Touchstones that are, significantly, things we might theoretically hold in our hands, turn over, inspect closely. Lesser in size and

20
Lahoud, "Nomos and Cosmos."

21
Tsing, *Friction*, 2.

22
Tsing, *Friction*, 1.

23
Tsing, *Friction*, 4.

24
Project website, http://www.sinking andmelting.org.

25
T. J. Demos, "The Law of the Land: An Interview with Amy Balkin," in *Amy Balkin: (IN) VISIBLE MATTER*, ed. Edwin Coomasaru and Tom Snow (International New Media Gallery, 2013), http://www.inmg. org/archive/balkin/ catalogue/demos/ #VrRw5dCCxUQ.

ARCHIVE

Description

A PEOPLE'S ARCHIVE OF SINKING AND MELTING is a growing collection of items contributed from places that may disappear owing to the combined physical, political, and economic impacts of climate change, including glacial melting, sea level rise, coastal erosion, and desertification.

Through common but differentiated collections, contributed materials together form an archive of the future anterior, what will have been. A contribution doesn't have to originate from a location - it can be anything that happens to be there, including detritus, flotsam or jetsam.

As of 2015, the archive contains contributions from Anvers Island (Antarctica), Australia, Cape Verde, Santiago de Cuba, Germany, Greenland, Iceland, Venice (Italy), Kivalina (Alaska), Mexico, Nepal, New Orleans, New York City, Panama, Peru, Republic of Komi (Russia), California, Senegal and Tuvalu.

Links

CONTRIBUTE

OPEN ANTARCTICA

THE PEOPLE'S WEATHER MAP

KVAK.TV

TUVALU PHILATELIC BUREAU

archive

ABOUT

NEW YORK COLLECTION

NEW ORLEANS COLLECTION

CAPE VERDE COLLECTION

KIVALINA COLLECTION

VENICE COLLECTION

Floor Tile Fragment
Concrete, white and black tiling
Sarah Lewison
Cairo, Illinois, USA
Lat/Long: 37.0131° N, 89.1803° W
January 8, 2016

Fragment of tile floor from riverfront building (bar? department store? warehouse?).

Cairo is at the confluence of the Mississippi and Ohio Rivers and has the lowest elevation in Illinois. Surrounded by levees and flood walls, it is always under threat of flooding- and will be more so as the hydrological cycle becomes more erratic.

Notes: The location figures heavily in US racial history. As a sliver of 'free state' jutting into the South, Cairo was historically a safe haven, but contentiously so. In the 'Adventures of Huckleberry Finn" it figures as Jim's last chance of escaping slavery. In the night, the Mississippi current pulls their raft away from the shore and Cairo's city lights are first obscured by an island...and then recede into the distance.

ILLINOIS COLLECTION / USA
01.16/6108
Annex I, II, B

Fisherman's Friend
Printed paper package and mints

Kevin Dunne
Blackrock Co. Louth, Ireland - on the beach
Lat/Long: TBD
October 2, 2014

The location is sinking - always flooding.

Image by Riuza Loko

IRELAND COLLECTION
10.14/6102

Glacier Film
35mm Lomokino film, color, digital transfer
Zoé Leonard
Jökulsárlón, Iceland
Lat/Long: 64° 4' 13" N, 16° 12' 42" W
November 19, 2015

Lomokino films are made on 35mm format film. The camera is hand crank, allowing for a variable frame rate. On average, there are four exposures per 35mm frame.

ICELAND COLLECTION
11.15/6107
Annex I, II, B (First Commitment Period)

Wrench
Nickel-chrome plated alloy steel, marked '9mm Craftsman.'

Walter Brinkerhoff
Rim Fire site, near Rainbow Pools (next to Hwy 120, and Cherry Lake Rd), Stanislaus National Forest, California
Lat/Long: 37°49'21.8"N, 120°00'58.6"W
February 22, 2014

Found near a roadway after the 2013 Rim Fire, the surface of the wrench was discolored by heat from the fire.

CALIFORNIA COLLECTION
05.15/6105

Toy Truck
Wood, metal, plastic

Sarah Kavanagh
Sandymount Strand, Dublin, Ireland
Lat/Long: 53.3236° N, 6.2058° W
September 7, 2014

Rusted steel wire
Steel
AURA Lab
Christian Sø, Brunkulslejerne
Lat/Long: 56.017188, 9.070185
June 7, 2015

This was taken at the edge of a highly acidic lake called Christian Sø in Brunkulslejerne, Herning in Denmark. Many lakes in this area were created by mining, the extraction of brown coal to supply fuel during the Second World War. Mining left a highly unstable and contaminated landscape now littered with trash like this.

+ Land use change, water thermal structure change, historical and indirect emissions

DENMARK COLLECTION
07.15/6106
Annex I, II

Website, *A People's Archive of Sinking and Melting*, Amy Balkin et al., 2011–present. Image courtesy of the artist.

26

For further information on the Kyoto Protocol, see the United Nations Framework Convention on Climate Change (UNFCCC) webpage that includes its full text, in multiple languages, http://unfccc.int/kyoto_protocol/items/2830.php.

27

Dana Kopel, "What Will Have Been: Interviews on *A People's Archive of Sinking and Melting*," the *Brooklyn Rail*, June 5, 2014, http://www.brooklynrail.org/2014/06/art/what-will-have-been-interviews-on-a-peoples-archive-of-sinking-and-melting.

potential nostalgic value than ruins, they serve as intimate, material markers of another person's point of contact. As such, they represent a fundamentally different type of index than a photograph or forensic shred, conveying a more multiple and indirect relation between event and trace, emission and reception.

Under the stewardship of Balkin, these archival objects are not abstracted or calibrated in order to become part of a cohesive sum. Instead they maintain their peculiarity—each tied to a story that we might begin to imagine from the basic details sketched by its collector(s). No single artifact stands in place of all others (as with the symbolic lone polar bear), nor does any attempt to capture a comprehensive view. Rather, like the human subjects who have ushered them into this unlikely fold, each stands as a fragment within a larger, not at all smooth context. Balkin refers to them as "common but differentiated," appropriating language from the Kyoto Protocol of 1997 to acknowledge the "common but differentiated responsibilities" of those developed, industrialized nations that are largely responsible for creating climate change versus those that have contributed almost nothing to the problem yet bear its brunt.[26] In so doing, she steers her gesture into the sphere of international environmental law and, more specifically, environmental justice. Elsewhere, she has characterized the archive's various specimens as "political proxies" for their respective contributors.[27] The

New Orleans Collection, *A People's Archive of Sinking and Melting*, Amy Balkin et al., 2011–present. Image courtesy of the artist.

archive is not political in straightforward or instrumental fashion, however. Instead, Balkin is curious, as she has been in previous projects, about "what the shape of political resistance in the difficult-to-inhabit space of the atmosphere might look like."[28]

In a permanent state of unfinished-ness, *A People's Archive of Sinking and Melting* is oriented toward an already unfolding future yet is situated firmly in the present. As such, it differs sharply from works of contemporary art and design that invoke "alternative futures," often with a strikingly utopian overtones. It occupies an odd and overlapping temporality, what Balkin calls the "future anterior," asking us "to think forward in time to look back."[29] In a recent interview with the art historian T. J. Demos, she addressed this "doubling of temporal perspectives," saying that the archive simul-taneously serves as "a 'time capsule' for a near future when hundreds of millions of people are likely to be displaced" and "a 'public display of evidence' for a present when grave mistakes are being made."[30] The project's aim to construct a forum for collectively articulated testimony and long-term memory is particularly poignant in the midst of our accelerated age of quick-fire obsolescence. Indeed, Balkin's choice to use the medium of the archive—which like the

Snap Hook, Antarctica Collection, *A People's Archive of Sinking and Melting*, Amy Balkin et al., 2011–present. Image courtesy of the artist.

28

Demos, "The Law of the Land." For a discussion of additional projects by Balkin that engage atmospheric politics, see also Theresa Deichert, "Amy Balkin and 'Art Activism': A New Approach to Environmental Activism," in *Amy Balkin: (IN) VISIBLE MATTER*, http://www.inmg. org/archive/balkin/ catalogue/demos/ #.VrRw5dCCxUQ. My own essay, "Field Effects: *Invisible-5*'s Illumi-nation of Peripheral Geographies," *Art Journal* (Winter 2010): 38–47, examines a 2006 critical audio tour about environmen-tal injustice and resistance along the Interstate 5 corri-dor in California developed by Balkin and collaborator Kim Stringfellow.

29, 30

Demos, "The Law of the Land."

31

For a lucid and expansive account of the catastrophic extinction event through which we are living, and that is of our own making, see Elizabeth Kolbert, *The Sixth Extinction: An Unnatural History* (New York: Picador, 2015). Numerous articles have been published in the last several years—too many to count and with new ones appearing almost daily—about the extreme rate of species extinctions currently under way. Some claim that we are on the "brink" of a mass extinction event; others, that we are already in its midst. While there is not an agreed upon start date for this trend, from what I understand, it maps loosely onto the so-called Anthropocene, and more specifically, "The Great Acceleration," which began in the mid-twentieth century.

32

Among many other sources on the Svalbard Global Seed Vault, see the art-research project by Annesofie Norn, Signe Lidén, and Steve Rowell, *The Cold Coast Archive* (2012–ongoing), http://coldcoast archive.org.

33

On the release of seeds from the vault to Syria, within the context of a larger effort to protect seed diversity in the face of war, see Robin Schulman, "Sowing the Seeds of Syria: Farming Group Rescues Plant Species Threatened by War," the *Guardian*, November 4,

2015, http://www. theguardian.com/ environment/2015/ nov/04 syria-seeds-experimental-farm-network-plants-biodiversity. On climate-change-induced drought as a key trigger for the ongoing war in Syria, see, among other sources: Audrey Quinn and illustrator Jackie Roche, "Syria's Climate Conflict," September 4, 2015, http://ycarsof livingdangerously. tumblr.com.

34

Thom van Dooren, "The Last Snail: Loss, Hope, and Care for the Future," in *Land & Animal & Nonanimal*, ed. Anna-Sophie Springer and Etienne Turpin (Berlin: K. Verlag and the Haus der Kulturen der Welt, 2015), 1.

35

Van Dooren, "The Last Snail," 6.

36

Van Dooren, "The Last Snail," 13.

museum, library, and other cultural institutions is aimed at holding the present in perpetuity—raises all sorts of questions about futurity itself, including *by?* and *for whom?*

Balkin's archive likewise relates to archives of a more scientific cast, including the emergency biological "banks" that have proliferated in the last several years in response to growing recognition of an abrupt and sweeping worldwide plummet in biodiversity, which many now consider to be the sixth mass extinction in Earth's history.[31] Take, for instance, the Svalbard Global Seed Vault (SGSV), situated halfway between mainland Norway and the North Pole on the icy island of Spitsbergen. The largest secure seed storage facility in the world, it was established to protect crop diversity as both plant varieties and arable land dwindle due to factors including urbanization, over-farming, pesticide use, and the genetic modification of seeds by corporate giants like Monsanto (one of SGSV's main financiers). Nicknamed the "Doomsday seed bank," it currently houses almost one million samples comprising thousands of seed species within the frigid recesses of an Arctic mountain.[32] While intended to be tapped at an unknown future date only under dire circumstances, the first withdrawal of seeds was in fact authorized in 2015, upon a request by researchers in Aleppo, Syria, whose own facility was destroyed in the ongoing war there—a war spurred in part by a multiyear drought resulting from climate change.[33]

We might also consider the "living collections" of animal species at (or past) the brink of extinction that are maintained by biologists, such as the highly endangered tree snail populations preserved at the University of Hawaii. The environmental humanities scholar Thom van Dooren refers to this laboratory as an "ark," like Noah's—"a place of last refuge" where scientists tend to creatures with remarkable care, in spite of the devastating and unstoppable disappearance, in some cases, of entire species.[34] Van Dooren's interest lies in ways that projects like these, "engaged in practical and concrete acts of care," help hold open the possibility for both hope about the future well-being of the world and a deepened sense of accountability.[35] In the end, he calls for banking practices that—in addition to fulfilling their preservation missions—also *"make visible all those things that they cannot quite hold on to and all those that they cannot hope to ever restore,"* so as "to cultivate some semblance of responsibility for another, whose world we (collectively) have destroyed."[36]

A People's Archive of Sinking and Melting is a representation—if we want to even call it that—which likewise addresses the world as it is breaking through a series of acts that are at once mournful and lively, idiosyncratic yet related. In contrast to a techno-scientific view from above, its objects-acts reflect positions on the ground. Any fantasy of discrete sites and events, however, is replaced by a loose set of alliances drawn between people in places that are undergoing related processes of environmental destruction.

SGSV Inner Vault Doors, Transverse Hall, 30" × 44" color photograph, from the project *Cold Coast Archive*, Steve Rowell, 2011. Image courtesy of the artist.

——

It is strangely fitting that climate change—this biggest of imaginable phenomena, this "thing" which is everything—assumes the form of a motley assembly.[37] An assembly that does not depict or explain. That does not coalesce into a single picture. That sits between an unfolding present and imminent and/or imagined futures. That entails a series of non-monumental acts, each performed with great care, from the plucking of artifacts at their found contexts, to the collectors' recordings of such details, to the methodical maintenance of the archive by the artist. And finally—an assembly that poses a humble yet expansive invitation for us to begin cobbling our various (inevitably partial and imperfect) views of the world, refracted perhaps through a mud-soaked slipper, abandoned seashell, crumpled wrapper, or some such.

37

I arrive at the term "assembly" because of its simultaneous evocations of: a place or event where people gather, often toward a common purpose, to air their various perspectives; of a thing made of multiple components fitted together; and lastly, although somewhat less proximate to my concerns, of *assemblage*— the art form of three-dimensional, or sculptural, collage, which has historically involved the collection and fusing of junk and other found materials, especially by artists in the mid-twentieth century (e.g., Louise Nevelson, Joseph Cornell, and Robert Rauschenberg).

Emily Eliza Scott, currently a postdoc in the architecture department at the Swiss Federal Institute of Technology (ETH Zürich), is an interdisciplinary scholar, former park ranger, and founding member of two long-term, collaborative, art-research projects: World of Matter (2011–ongoing) and the Los Angeles Urban Rangers (2004–ongoing). Her writings have appeared in *Art Journal*, *American Art*, *Third Text*, and *Cultural Geographies* as well as multiple edited volumes and online journals; she is also co-editor, with Kirsten Swenson, of *Critical Landscapes: Art, Space, Politics* (University of California Press, 2015).

Waterdams as Dispossession: Ecology, Security, Colonization

ZEYNEP S. AKINCI AND PELIN TAN

Kavsaktepe Dam, Turkey, 2015. Photograph by Zeynep S. Akıncı.

The effects of climatic transformation on territories are not based only on determinist ecological factors. Rather, architectural elements and the production of infrastructure in general can function as an ecological legitimization for implanting security tools in border zones, for dispossessing people of agricultural lands (which leads to the eviction of villages and replacement of the agrarian economy), and for surreptitiously colonizing territory. This can be seen in the contested border region of Southeastern Turkey, where "security waterdams" form an infrastructural constellation with complex and powerful ramifications for the ecology and political climate of the county.

The security waterdam is a formation unique to Southeast Anatolia—the geographic starting point of upper Mesopotamia—and understanding their true spatial and political nature involves elements of the discourses that surround architecture, security strategies, labor, colonization, and dispossession. The dams themselves are joined by military defense centers (*kalekols*) and the Village Guard System, which was an intense paramilitary organization

1

Haydar Çetinkaya,
"Waterdams and
Hydroelectric
Central Map of
Anatolia," *Kuzey
Ormanlari*, March
22, 2015, www.
kuzeyormanlari.
org/2015/03/30/
anadolunun-il-il-
baraj-ve-hes-
soykirim-haritasi.

2

Mustafa Aydın
and Fulya Ereker,
"Water Scarcity
and Political Wran-
gling: Security
in the Euphrates
and Tigris Basin,"
*Facing Global
Environment
Change: Environ-
mental, Human,
Energy, Food,
Health and Water
Security Concepts*,
ed. Hans Günter
Brauch, Úrsula
Oswald Spring,
John Grin, Czeslaw
Mesjasz, Patricia
Kameri-Mbote,
Navnita Chadha
Behera, Béchir
Chourou, and
Heinz Krumme-
nacher (Berlin:
Springer, 2009),
603–613.

3

In 1924 with the
establishment of
the Sark Islahat
Planı or Orient
Reform Plan,
assimilation or
Turkification of
Kurds found legal
grounding. The
assimilation poli-
cies were applied
via three main
tools: settlement,
education, and
policies to cultur-
ally "Turkify" the
region.

4

Joost Jongerden,
"Dams and Politics
in Turkey: Utilizing
Water, Developing
Conflict," *Middle
East Policy*, vol.
10, no. 1 (Spring
2010): 137–143.

in the 1990s in Southeast Turkey. These infrastructures—that of resource management and that of security—are intimately entangled. The vast ecological remaking (and equally vast human resettlement) that accompanies dam building is well known. These are infrastructural objects that speak to a long history in which environmental and human costs are subordinated to developmentalist statecraft, whether in the case of the Three Gorges Dam in China, the Aswan High Dam in Egypt, or the Tarbela Dam in Pakistan. In the Anatolian dams explored here, however, the dam is mobilized as security infrastructure and weaponized architecture.

DAMS OF SECURITY

It would be naïve to evaluate water policies solely by their ability to supply drinking water, support irrigation, or generate energy. In addition to these more apparent uses, water infrastructure can be deployed for security purposes and as surveillance mechanisms, with many geopolitical repercussions. On the part of the state, colonizing resources or utilizing resources for supposed public security is very common in land expropriation and warfare conditions. Trash, rivers, forests, seeds, sea, land, and food are often tools in managing and controlling resource policy by state governments. Looking at the control and colonization of resource territories reveals a complex map of forces of segregation of land and communities, joint ventures of national and local interests, and minority assimilation in different continents of the world. Waterdams, hydroelectric power plants, and highways are only a few examples of infrastructural projects based on networks of spatial colonization.

Currently, there are 1,487 hydroelectric power plants and water dam projects (including those still in construction or to be built) within Turkey and its waters. In Northern Turkey, along the Black Sea, water is primarily piped from underground, while in the Eastern and Southeastern Anatolia Region, which shares a border with Iran, Iraq, and Syria, water is managed through the construction of large-scale dams.[1] Considering the climate of the two regions, one would expect an inverted approach to water and power. The Black Sea region receives rainfall all year, while Southeast Anatolia is dry for many months, making it a less likely spot to capture much rainwater.

Water irrigation and dams on the Tigris and Euphrates were always objects of political negotiation in regard to sovereignty over shorelines rather than shared resources among Turkey, Iraq, and Syria, which became a more acute dispute during the 1990s.[2] Given this, the construction of dams (especially the ones in Northern Kurdistan) suggest political and military motives beyond energy production, a suspicion affirmed by Turkish settlement policies in the region. Since the 1920s, systematic settlement policies have aimed to eradicate the historical existence of the Kurdish population and to suppress their cultural memory. These settlement policies aimed also to diminish the Kurdish population to the east of the Euphrates River. Some Kurds were exiled west of the Euphrates while Turks were placed east of the Euphrates to increase the population of Turks in the east. The plan suggested settling five hundred thousand people from west to east in a ten-year period.[3]

In an article titled "Dams and Politics in Turkey: Utilizing Water, Developing Conflict," Joost Jondergen examines the domestic and foreign politics of Turkey through dams, regarding dams as objects of contention that have a great potential to instigate violent conflict.[4] Given the low population levels in

Çetintepe Dam, Turkey, 2015. Photograph by Zeynep S. Akıncı.

Southeast Anatolia—the result of massive displacement of rural populations in the 1990s—Jondergen argues that the purpose of these "dams of security" cannot be electricity generation or irrigation. Instead, the only purpose of these dams is to create a wall of water to obstruct PKK guerrillas (Kurdistan Workers Party) from crossing the Iraqi–Turkish border. In fact, the 2007 annual report of Devlet Su Isleri (State Hydraulic Works) calls the dams across the Hezil watercourse on the Habur River (after which the Turkey–Iraq border gate is named) "water-blown dams for security reasons," and in the years since, they have been referred to as such, or simply as "security dams," by the media. In 2012, however, BDP Sırnak Parliamentary Deputy Hasip Kaplan filed a parliamentary question about the matter to the Minister of Forestry and Hydraulic Affairs, Veysel Eroglu. Eroglu replied:

> There is no such dam built under my Ministry under the name of "security dam." The water constructions, which are planned for drinking water, irrigation, energy production, and storm mitigation, are technically called "litigated water embankments (dams)." All the project phases of the mentioned facilities, including the Environmental Impact Assessment report, are built in accordance with legislation.[5]

Not only do statements from government officials clearly contradict one another, statements from different external institutions also differ. In 2010, WikiLeaks leaked a document containing conversations between three employees of the US-based geopolitical intelligence firm Stratfor: Emre Dogru,

5

The Sırnak parliamentary deputy Hasip Kaplan's parliamentary question, number BDP 7/5848, June 1, 2012, http://www2.tbmm.gov.tr/d24/7/7-5848sgc.pdf.

an analyst in Turkey; Yerevan Saeed, who was the Washington Kurdistan Regional Government representative; and the watch officer Michael Wilson. In the conversation, Saeed indicates that he suspects the water could obstruct the entrance to Turkey, and that the PKK has plenty of hiding places like caves and mountains. To this Wilson replies, "if geography provides a natural shelter for guerrillas, simply change the geography."[6]

Whether information from the ministers of Turkey, leaked documents, and corporate disclosures are consistent or not, the watchtowers, *kalekols*, regular police stations, and military protection along the many dams in Northern Kurdistan and the Sırnak province suggest these dams are built for more than electricity generation or irrigation. Moreover, because land surveys for the placement of the dams were done within a short period of three months, errors occurred in the designation of the location of dams, causing incidents like landslides during construction and excavation. Such an accelerated time frame for surveying suggests motivation beyond simply that of water infrastructure.[7]

Under normal circumstances, if the objective is to provide energy or irrigation, the most appropriate place for the dam should have been designated after years of surveying, but with the dams recently built in Sırnak this was not the case; the intent was to complete them quickly. Surveys—both land surveys and polls of local residents—took place in 2007, and construction began in 2008, leaving only a year for research and for the design process. Curiously, none of the participants in local surveys leading up to dam construction indicated irrigation or electricity problems, which makes the effort to finalize this project so hastily suspicious.

It is not easy to determine the entire scope of the effects of these newly constructed dams on endemic species or settlements. But it is clear that Nerweh (Tasdelen) Village, situated between Ballı and Kavsaktepe Dams, and Ziravik (Inceler), which is just below the Uludere Dam, will be affected significantly. As of 2011, 1,371 civilians lived in Tasdelen, which is in Robozik (Uludere), where thirty-four civilians were killed by the Turkish Air Force bombing in 2011.[8] The Ballı and Kavsaktepe dams' flood-territory residents were able to return only eighteen years ago, when a village repatriation program was started after devastating fires. These dams were planned and developed without the consent and knowledge of villagers, yet to submerge grazing land and vineyards will inflict the same violence as forcing people to migrate again. Damage to the natural landscape, moreover, is irreversible. Is this project—which can cause extinction of very important endemic species like the *ters lale* flower that grows only in the Uludere and Hakkari mountains—only a means to "improve" border security and block the passage of guerrillas? Or, is it, even more perniciously, a solution to Turkey's "Kurdish problem"?

"KALEKOL"—BORDER POLICE STATION

Kalekol is an invented word composed of *kale*, meaning "tower," and *karakol*, which means "police station." These tower stations are a recent phenomenon designed for and strategically situated in the Southeast regions of Turkey. Since 2012, 149 *kalekol* have been constructed on the Turkey–Iraq border between the cities of Şırnak and Hakkari, at an elevation of approximately 2,000 meters. In total 402 *kalekol* are planned to be built by the Turkish Housing Development Administration, which has stated that its aim

6

"Turkey to Construct Dams Along Iraq Border to Flood PKK Camps—Daily," *The Global Intelligence File,* released May 5, 2013, https://wikileaks. org/gifiles/ docs/16/1653870_ re-os-turkey-iraq-ct-mil-turkey-to-construct-dams-along-iraq.html.

7

The plan for the construction of these dams is recorded in the annual report of DSI from 2007, and the Sırnak Dam was completed in 2011. See "Planned Second Dam Was Completed at the Border Against the PKK," *Hurriyet,* October 26, 2011, www. hurriyet.com.tr/ gundem/19088520. asp. See also Çetinkaya, "Waterdams and Hydroelectric Central Map of Anatolia."

8

This border village was called Robozik between 1928 and 2012; it was renamed Roboski. See entry in *Index Anatolicus,* http:// nisanyanmap. com/?y=Robozik.

Musatepe Dam, Turkey, 2015. Photograph by Zeynep S. Akıncı

is to protect against the PKK and border smugglers.[9] The design consists of
three leveled thirty-one-inch-thick concrete walls surrounding a central core
with bedrooms, a canteen, and other facilities for the soldiers. Two towers are
placed in the corners and are connected to the main facilities by tunnels. All
gates are made of steel, and the windows are bulletproofed. Drones and other
robotic security devices guard the space. When the peace process between

9

See TOKI website,
www.toki.gov.tr/en.

10

A young boy
(Medeni Yıldırım)
lost his life and
eight people were
injured during the
suppression of the
protest. Lice is
one of the Kurdish
towns that was
harshly tortured
and bombed by
the military during
the 1970s, and it
has been under
high surveillance
and oppression
since then. See
Hayri Demir,
"Güvenlikçi' adı
altında koruculuk"
(security guards
under means of
Koruculuk), *Evren-
sel*, April 10, 2015,
www.evrensel.net/
haber/110064/
guvenlikci-adi-
altinda-koruculuk.

11

Turkish Ministry
of Internal Affairs,
2015/8322 Polis
Güvenlik Noktası
Yapımı Amacıyla
Bazı Tasınmazların
İçisleri Bakanlığı
(Emniyet Genel
Müdürlüğü)
Tarafından Acele
Kamulastırılması
Hakkında Karar
(Legislation of
Extra Appro-
priation for
Building Police
Security Centers),
December 28
2015, www.
resmigazete.gov,
tr/eskiler/2015/
12/20151228.htm.

12

"The legal basis of
the Temporary Vil-
lage Guard System
was the Village Law
numbered 442 that
was legislated in
1924. On March
26, 1985, with the
amendment made
to the Article 74,3
a paramilitary
structure com-
prising temporary
and voluntary
village guards
was established.
During the 1990s
when the clashes
were at their
peak, the number
of these armed
forces was over
ninety thousand
… Procedures and
principles regarding
the employment,
duties, autho-
rization to use
guns, responsi-
bilities, training,
dismissal and other
personnel rights
and benefits of the
village guards were
specified by this
regulation. The
fact that it took
such a long time to
pass this regulation
resulted in ambig-
uous duties of the
village guards and
gave the opportu-
nity for the armed
forces to use this
unit as they please."
Şemsa Özar,
Nesrin Uçarlar, and
Osman Aytar, *From
Past to Present a
Paramilitary Orga-
nization in Turkey.
Village Guard
System* (Istanbul:
DİSA Publications,
2013), 9.

13, 14

Özar, Uçarlar, and
Aytar, *From Past to
Present a Paramil-
itary Organization
in Turkey*, 9.

the PKK and the Turkish government began in 2013, the construction of the *kalekol* accelerated. Public debate ensued in local newspapers: the Kurdish public criticized the construction process while pro-government Turks argued that the tower stations discouraged border smuggling and protected against groups like ISIS. As the debate continued, people began protesting during the construction of *kalekols*; in June 2013, soldiers opened fire on protesters in the village of Hezan.[10] The *kalekol* are said to employ contemporary surveillance tools and advanced military architecture and are said to be constructed at the southeastern borderline of Turkey to protect the public. However, most of them are built in or near Kurdish towns inland from the actual border. According to the 2015 decision of the Council of Ministers of the Turkish government, "urgent expropriation" could be used around the lands of a Kurdish town in order to build the stations.[11] The "enemy" who will attack the *kalekol* is still an empty signifier. Significantly, most are constructed within or near dams in the region.

"KORUCU"—VILLAGE GUARD SYSTEM

In 2014 and 2015, many newspapers reported that guards in Turkey's Village Guard System would be trained and placed as security guards in the dams.[12] The Village Guard System had essentially functioned as a paramilitary organization in Kurdish villages in the 1990s in Turkey. After the peace process, the role of the guards in the region had to be replaced. The village guards, "as their name suggests, were technically charged with the protection of their own villages. However, they were sent on operations that lasted for days, from time to time used as a human barrier between the soldiers and the PKK and were even included in cross-border operations."[13] This system is often compared to similar forces in Guatemala, the Philippines, and elsewhere. According to a report by the Diyarbakir Institute for Political and Social Research, "The village guards were under the command of the village headman administratively and the Commander of the Gendarmerie Squad in professional matters. For fifteen years, this paramilitary system was directed without being subject to any law until the Regulations on Village Guard System, which was published on the Official Gazette on July 1, 2000."[14] During the peace process, the Village Guard System was not completely dismantled; the current government protocol is to train members and place them as security guards in and around security dams. Their use, the authors of this paper argue, is as a surveillance tool that can function as a buffer between local villages around the dams or expropriated land of the villages that have been evicted. The labor replacement also shows how the precarity of the current village guards is mobilized in a constellation of the warfare and colonization along the dams and *kalekols*. Thus, water dams constitute a resource infrastructure as territorial weapon abetted by different architectural systems and labor allocation and is made possible by land expropriation policy.

Can peace and reconciliation move forward when villagers are forced to migrate (yet again) after their homes are flooded in the name of security? What will the cost be to the landscape and all its inhabitants? And how might these transformations to the climate itself be part of a security regime? Ecologist and activist Galip Geylani claims that the Beyyurdu and Aslandağ dams, which have been in construction since 2008 (including *kalekols* on top), are already causing the disappearance of species, forests, and fertile lands in Aktülün (Bézelé) and Beyyurdu (Bédaiwé) towns.[15] He also claims that the constellations of dams and *kalekols* function as tools of segregation between the lands of Şemdinli and Hakkari (Zagros and Behdinan), which border Iraq.

A similar argument is cogently articulated in the recent book *The Conflict Shoreline* by Eyal Weizman and Fazal Sheikh, as well as in the research of Paulo Tavares.[16] By analyzing cases at the threshold of the desert, Weizman demonstrates how conflict and climatic changes are connected—how they affect each other, and how conflict makes use of climate. Weizman asserts, "Existing tensions have been aggravated by climate-related shifts in the aridity line, and, in turn, continued conflict in these areas has caused the destruction of vegetation and agriculture, accelerating the process of desertification."[17] Although the methodology is based on photographic and cartographic analysis, and thus differs from the research cited in this text, the study indicates a grounded argument that "the climate has always been a project for colonial powers, which have continuously acted to engineer it."[18]

When resource infrastructure is used as a means of covert colonization and weaponization of land, the result is forced migration and dispossession, damaging communities and their local climate. Due to the dam construction, many villages were evicted and the lands of the villagers were expropriated in Anatolia. Expropriating fertile land causes forced movement of local communities, who will have to find new sources of employment while the land goes untilled.

The extent of the destruction will remain unknown until the dams begin to hold water, but the people, animals, and plants living in the area—which is to say an entire ecology of Southeast Anatolia—will be affected. These future climatic territorial effects are already revealing the connections between eviction, colonization, and security. The relation between warfare, colonization, and the weaponization of the land is not a new process; it is one of the basic practices in the era of the Anthropocene. In *The Shock of the Anthropocene*, Christophe Bonneuil and Jean-Baptiste Fressoz write that the construction of dams effects the disappearance of villages, which causes the disappearance of local production, handcrafts, the peasant network, and the relationship to nature.[19] Thus, entire future-specific ecological habitats are jeopardized. In Southeast Anatolia it is the future of natural, social, and political ecology that is at stake.

15

Nedim Türfent "Semdinli'nin essiz dogasına baraj hançeri," *Diha*, October 29, 2015, http://diclenewsagency.link/tr/news/content/view/481194.

16

Eyal Weizman and Fazal Sheikh, *The Conflict Shoreline: Colonialism as Climate Change in the Negev Desert* (Gottingen: Steidl in association with Cabinet Books, 2015). Paulo Tavares, "The Geological Imperative: On the Political Ecology of the Amazonia's Deep History," in *Architecture in the Anthropocene: Encounters Among Design, Deep Time, Science, and Philosophy*, ed. Etienne Turpin (Ann Arbor: Open Humanities Press, 2013).

17, 18

Weizman and Sheikh, *The Conflict Shoreline*, 12.

19

Christophe Bonneuil and Jean-Baptiste Fressoz, *The Shock of the Anthropocene* (New York: Verso, 2016), 503–504.

Zeynep S. Akıncı holds a BA in urban design and landscape architecture from Bilkent University, Ankara. She is currently a graduate student in the architecture program at Mardin Artuklu University, Mardin, Turkey.

Pelin Tan is an associate professor in the architecture faculty at Mardin Artuklu University. She is also a fellow in art, culture, and technology at the School of Architecture and Planning at the Massachusetts Institute of Technology and a visiting associate professor at the School of Design at Hong Kong Polytechnic.

Evolutionary Infrastructures

MARION WEISS AND MICHAEL A. MANFREDI

Infrastructural systems are enduring forms of urban evolution, multiplying as cities grow and requiring expanding swaths of territory to accommodate increasing numbers of mono-functional requirements. As the frequency of climate events threatens to overwhelm our urban infrastructures, what new, more resilient forms of public nature might emerge if highways, tele-data rights-of-way, railways, subway lines, and flood resistant structures were to amplify their singular expressions to also become settings for new programs? In so doing, might infrastructures be liberated to respond creatively to the weather extremes?

"Evolutionary Infrastructures" was originally published in Marion Weiss and Michael A. Manfredi, *Public Natures* (Princeton Architectural Press, 2015). This essay has been revised and expanded for this publication.

Larger than life but part of it, infrastructure has an immediate presence—it shapes our environment and urban life in vital, authentic, and often messy ways. Infrastructure, of both movement and culture, must evolve to embrace and expand preexisting conditions. Highways, subways, utility lines, and tele-data networks have the capacity to sever or connect communities, define the static or fluid identity of an urban landscape, and unravel or re-stitch the increasingly fragmented fabric of our metropolitan world. The possibility of this new public territory lies in its activation of a range of scales, its opportunities for the simultaneous accommodation of movement and destination, flooding and drought, and the hybrid programmatic potentials it affords.

We look at the physical elements of infrastructure and the often marginalized sites they produce as possible contributors to a meaningful public realm. What if a new paradigm for infrastructure existed? What if the disparate strands of landscape, architecture, engineering, and urbanism could find a more synthetic convergence? We are interested in a new model of practice, one that integrates all fields of design through yet-to-be codified protocols—a synthesis residing at the periphery of disciplinary definitions but perhaps at the center of a wholly new form.

Evolutionary infrastructure is both projective and pragmatic—a prototypical ideal that is intrinsically agile, capable of optimizing ecological and social extremes, and that leverages the stray spatial consequences of preexisting infrastructures. This definition recognizes that urban centers, particularly those settled in close proximity to water, have experienced great transformation over time. As places of trade, these cities depended on waterfront infrastructures to facilitate boat and barge traffic, but over time train lines and highways came to facilitate greater speeds of exchange. This evolution of trade and development has resulted in a patchwork layering of infrastructural systems, creating odd juxtapositions and remnant spaces between ports, city grids, train lines, roadways, and highways. Evolutionary infrastructure does not condemn the artifacts of infrastructure or depend on an idealized blank slate condition, but rather envisions new reciprocities between preexisting infrastructural systems and more ecologically resilient territories suited to contemporary demands and emerging climatic realities.

During periods of rapid urbanization, particularly after World War II, both developed and developing countries built comprehensive networks of roadways and highways to expedite movement within and beyond the core of old cities. Politically fragile communities lacking the strength to protest this signature of progress offered little resistance to such invasive projects. Ecologically fragile waterways and contested landscapes were equally put at risk. The Cross Bronx Expressway in New York divided and devastated its neighborhood, while the Los Angeles Aqueduct system accelerated the creation of deserts to the north and countless impoverished local ecosystems along its way. Once the greatest asset serving the modern urban landscape, infrastructure has now created cities in perpetual crisis, beholden to the seemingly irreconcilable differences between its systems and its objects.

Realizing the limitations of mono-functional infrastructure, evolutionary infrastructure presents a more hybrid, resilient, "thick" infrastructure, in which large-scale regional ambitions do not preclude programmatic variety, spatial richness, and specificity of detail but rather suggest an alchemy of innovative engineering, ecological imperatives, and compelling architecture. The evolutionary model for infrastructure is a galvanizing public/private model that brings the impatience of the entrepreneurial spirit to the broader collective agendas of public agencies.

While the Ponte Vecchio in Florence, Italy offers an exemplary historic model of program-rich infrastructure, a more idiosyncratic merger of river and urban engineering was realized in Ljubljana, Slovenia, with Jože Plecňik's responsive urban section along the Ljubljanica River. At the heart of the city, his signature triple bridge spans the river with a spectacle of redundant crossings and delicacy of scale more common in architectural follies than infrastructure. The steep section cut is fortified with stone walls inscribed with stairs, ramps, and arcades, producing a kind of reverse fortification. Outside the urban core, the walls recline, broadening the width of the river promenade with multilevel walks and water-tolerant trees. The production of this suburban promenade is both subtle and prescient. When the river floods, as it does at increasingly frequent intervals, only an upper-level walk offers passage; when the water level is low, it follows a slender channel, leaving four levels of parallel walkways free for pedestrians. Between these urban and pastoral states, a series of weirs and bridges creates a meter of landmarks along the length of the river. This dynamically changing section accommodates flooding events, merging engineering obligations with ecological agendas, creating a hardworking infrastructure that offers a new mode of urban experience.

Ljubljanica River Promenade, Jože Plecňik, Ljubljana, Slovenia, 1939. Diagram by Weiss/Manfredi.

Top: Fondazione Querini Stampalia, Carlo Scarpa, Venice, Italy. Diagram by Weiss/Manfredi.

Bottom: Louisville Waterfront Park, Hargreaves Associates. Photo by John Gollings for Hargreaves Associates. Diagram by Weiss/Manfredi.

At a radically different scale is Carlo Scarpa's exquisitely scaled Fondazaione Querini Stampalia in Venice, where the ever-sinking city floods with seawater during the "acqua alta" season. Scarpa creates a series of spaces that include galleries and a renovated garden courtyard. Not satisfied with just creating a series of intimately scaled rooms, his design anticipates Venice's inherent fragility in the wake of increasing flooding. He inventively invites water into the formal vocabulary of his project. The creation of subtle changes in the floor plane at the entry loggia allows water to flow in through the exterior gates without destroying the loggia. The narrative of water extends into the garden via a carved stone channel and creates a model of an aqueous architectural aikido—taking the potentially destructive power of water and redirecting its momentum into something of experiential value.

In North America, the proximity of a river is often integral to a city's founding, and the source of haphazard industrial growth. That, coupled with the increasing unpredictability of flooding, limits public access. In Louisville, Kentucky, a 120-acre park on the Ohio River by Hargreaves Associates integrates patterns of environmental and urban change with new recreational and cultural programs. The geometries of the existing infrastructure and the mutable river are simultaneously privileged: a rectangular Great Lawn is framed by a linear stretch of elevated highways and slopes gradually down to the river, while walking trails and wetland gardens track the river's perimeter. The park is ingeniously graded to protect the downtown from periodic flooding without the need for view-blocking floodwalls.

These very same qualities—programmatic variety, climatic resilience, and spatial richness—are part of the legacy of infrastructure-scaled, modernist utopian visions and are a reminder that the legacy of modernism is complex, and its social motivations often overlooked. Le Corbusier, with his unrealized designs for Algiers and Rio de Janeiro, identified a continuous sectional hybrid of highway and housing. Hugh Ferriss, in his 1929 *Metropolis of Tomorrow*, described a vivid dream of a multileveled

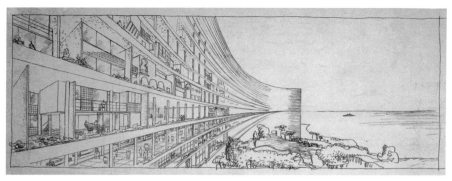

Plan Obus, Le Corbusier, Algiers, 1933. Courtesy of the Fondation Le Corbusier (F.L.C.) / ADAGP, Paris / Artists Rights Society (ARS), New York, 2014.

Manhattan, extended with tendrils of suspension bridges thick with high-rise apartments embedded in the supporting pylons. And, in the decades following the devastation of World War II, the Metabolists, centered in Japan, rendered a vision of elevated cities often sited in the water to sustain growing urban centers. These ambitious proposals, further elaborated in the 1960s in work by Archigram and Paul Rudolph, anticipate a densely inhabited infrastructure, capable of supporting multiple layers of urban life.

Though Ferriss's inhabitable fantasy was never realized, an awkward yet extraordinarily contingent utopia emerged in 1964 in northern Manhattan. Here, the legacy of these aspirations took shape in a piecemeal fashion, where the terminus of the George Washington Bridge translated into the dramatic cut of the Trans-Manhattan Expressway—a massive infrastructural intervention moving high-speed traffic from New Jersey to the highways of the Bronx across one of Manhattan's narrower points. Rather than sever the northern tip of the island with an at-grade highway, the Expressway is depressed to allow the continuation of New York's traditional street grid and to accommodate new programs. Topped by Pier Luigi Nervi's inventive bus station to the west and improbably straddled by four high-rise residential towers to the

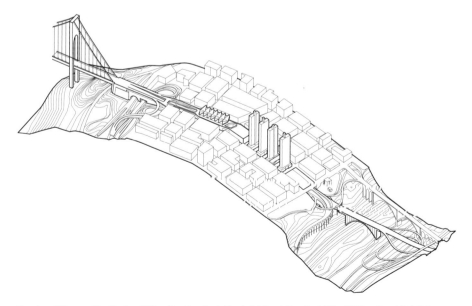

Drawing of George Washington Bridge Bus Terminal, Pier Luigi Nervi, New York City, 1963, with related highway infrastructures. Courtesy of Weiss/Manfredi.

east, this unfinished modernist project crosses the Harlem River and carves through the Bronx with a remnant wake of on- and off-ramps. While this marriage of infrastructure and inhabitation failed to become a contemporary paradigm worth reproducing, it was a bold experiment and prompted the Ford Foundation to commission Rudolph in 1967 to propose an elevated world of highways and housing crossing though Manhattan, complete with extensive parking garages and opportunities for residents to enjoy views of the city. While sectionally rich, the project's avoidance of direct, at-grade engagement with urban life is questionable.

While Manhattan was the focus of both speculative and realized examples of an inhabitable infrastructure, Brooklyn is the site of one of the most enduring successes of an infrastructure hosting speed and slowness, pass-through, and promenade. In the decade after World War II, urban planner Robert Moses and the New York City Planning Commission proposed the creation of the Brooklyn Queens Expressway, an element of progress that would offer efficient routes through New York City's outer boroughs, which included a cut through the historic Brooklyn Heights neighborhood. Led by the Brooklyn Heights Association, the community countered the plan, and won, with a revised proposal by Clarke & Rapuano that cantilevered a two-tiered highway from the urban bluff with a surface street below. Upon seeing this plan, Moses reportedly suggested covering this proposal with a public walk. The project opened in 1959 and included the now famous three-eighths-mile-long promenade, connecting the neighborhood to a shaded urban belvedere with unparalleled vistas of the Manhattan skyline. Over a half century ago, this contingent solution to a specific challenge offered a sustaining sectional paradigm, utilizing topography to eliminate the either/or dichotomy associated with urban fabric and infrastructure.

These heroic infrastructural proposals and seminal projects, seen through the dual lens of pressing ecological imperatives and shifting societal patterns, have renewed our interest in the architectural implication of topography, climate, territory, and urban systems. The late architectural historian Detlef Mertins suggested that these types of utopian models offer relevant hybrid, multivalent, and open-ended strategies to consider in contemporary terms. Against the backdrop of these early inspirational models, we have been challenged to explore more productive relationships between infrastructure, ecology, and public life, in which the logistic obligations of movement and

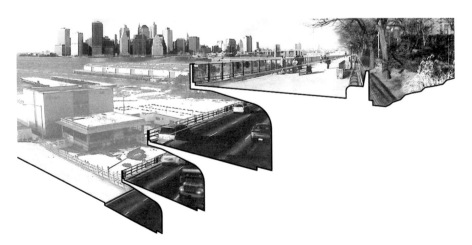

Brooklyn Heights Promenade, Clarke & Rapuano, Brooklyn, New York, 1959. Diagram by Weiss/Manfredi.

systems are modulated to support the "plus" of site-specific investment calibrated to the precise nuances of location, program, and climate demands.

Our preoccupation with a more synthetic potential in these contested
territories has been intensified by our experiences in the waterfront cities
of Seattle, Toronto, and New York, with sites almost completely defined by
transportation and flooding challenges. In each case, we engaged projects
expressly committed to expanding the capacity of infrastructure to host new
programs while connecting communities to waterfronts previously defined
by the imposing footprint of industry and transport.

The Seattle Olympic Sculpture Park, for example, emerged from a proposition to turn three separate parcels of contaminated land, divided by train lines
and highways, into a waterfront public space for the display of art. Segmented in
plan and section, with a forty-foot grade change, the site offered an opportunity
to create three separate parks connected by bridges, or one monolithic cap to
conceal the infrastructure below. We suggested a third alternative—a continuous
Z-shaped landform that alternately concealed and revealed the roadways and
train lines below. Within the boundaries of the site's nine acres, the earthwork is
shaped to create valleys, bridges, ramps, and walkways, beginning at the urban
edge with a pavilion for art, and concluding at the water's edge with a newly
created beach and underwater "habitat benches" for aquatic life. This pedestrian
infrastructure allows long-denied free movement between downtown Seattle and
the waterfront. Beneath the surface plantings of this armature, a new subsurface
infrastructure consisting of two and a half miles of power, water, telephone, and
data lines allows artists to incorporate sophisticated technologies into their work.

This design recognizes that climate has always been one of the primary
determinants for the shaping of architecture and landscape. Climate can refer
either to the weather conditions that prevail seasonally in a region over a long
period, or to the repeating rhythm of changes in a short period, such as day
to night. Seattle's climate frequently shifts from dry to wet, and its fluctuating tides are cyclical effects that are revealed or amplified by the materials

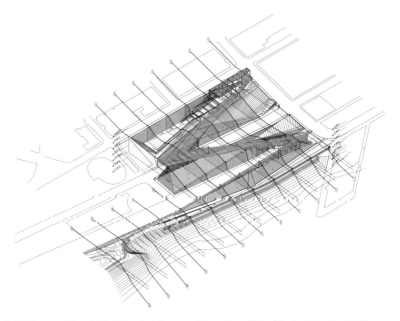

Seattle Art Museum, Olympic Sculpture Park, topographic sections, Weiss/Manfredi, Seattle, 2007.
Courtesy of Weiss/Manfredi.

Seattle Art Museum, Olympic Sculpture Park, Weiss/Manfredi, Seattle, 2007. Photograph by
Benjamin Benschneider.

and orientation of the architecture and landscape. For example, Seattle's
daily variation in light color and dampness from morning, to dusk, to night
is amplified in the pleated stainless steel of the pavilion. The Park also plays a
longer game, expressing the gradual changes of the seasons in its landscapes:
a dense evergreen forest, precincts of meadow grasses, and a deciduous aspen
grove grow and fade in slow annual alternations. At the shoreline garden, the
tidal rhythms of Elliot Bay are given measure through the terraced geometry
of the water's edge while establishing flood-mitigating infrastructure.

Once an ecologically vibrant river delta, the mouth of the Don River in
Toronto was disfigured during the city's industrialization into a concrete

Hunter's Point South Waterfront Park flooding conditions, Queens, New York, 2013. Rendering courtesy of Weiss/Manfredi.

channel, terminating the free flow of water to make room for a port, thereby accelerating the cycles of flooding. The addition of the elevated Gardiner Expressway further disrupted this site. Our project for the Lower Don River sought to transform this waterfront into an ecologically engineered setting for public life through the introduction of a more hydrologically sensitive geometry of arcs and curves, where hard, armored edges gave way to a series of wetlands and walkways. Similarly, in our realized project for Hunter's Point South Waterfront Park, located along New York City's East River, we identified strategies to incorporate the inevitable and often destructive patterns of flooding, as made evident by the impact of Hurricane Sandy just prior to the project's completion. Here, existing concrete bulkheads are alternately utilized for new elevated landscapes and pavilions, or strategically replaced by new wetlands and paths, creating a soft infrastructural edge. A series of new multiuse open spaces creates topographies that accommodate, rather than oppose, rising waters and fluctuating tide swells. In each of these cases, an underutilized and forbidding waterfront is given a second life as a resilient and ecological infrastructural system for public use.

The sensibilities of infrastructure and the project of architecture are latent with reciprocities yet to be imagined. While infrastructure is often incorrectly perceived as hard and inflexible, we see great potential for alternative strategies that structure more lateral, resilient, and pliable systems capable of hosting unpredictable uses and activities, absorbing cycles of flooding and weather extremes and creating cultural value. By bending the loose ends of architecture, landscape, and engineering together, we imagine an alchemy that transcends the limitations of single-use infrastructures, generating a more bountiful, resilient, and inhabitable interpretation of its potential.

WEISS/MANFREDI Architecture/Landscape/Urbanism, founded by Marion Weiss and Michael A. Manfredi, is at the forefront of architectural design practices that are redefining the relationships between architecture, infrastructure, landscape, and art. Their competition-winning projects such as the Olympic Sculpture Park, University of Pennsylvania's Nanotechnology Center, Barnard College's Diana Center, the Women's Memorial at Arlington Cemetery, and the Brooklyn Botanic Garden Visitor Center construct reciprocal relationships between architecture and infrastructure, city and nature. Both partners are actively engaged in academic research and teaching: Marion Weiss is the Graham Chair professor of architecture at the University of Pennsylvania and Michael A. Manfredi is a senior design critic at Harvard University and the Gensler visiting professor at Cornell University. They were recently the Eero Saarinen Visiting Professors at Yale University. Princeton Architectural Press has published two monographs on their work, titled *WEISS/MANFREDI: Surface/Subsurface* and *Site Specific: The Work of WEISS/MANFREDI Architects*. A new book, *Public Natures: Evolutionary Infrastructures*, was published in 2015.

The Resilience Problem: Part 1

JESSE M. KEENAN

1

Andrea Wulf,
*The Invention of
Nature: Alexander
von Humboldt's
New World* (New
York: Knopf,
2015).

2

Wolfgang Beh-
ringer, *A Cultural
History of Climate*
(Cambridge: Polity
Press, 2010).

3

Neil W. Adger,
et al., "Are There
Social Limits to
Adaptation to
Climate Change?"
Climatic Change
93 (2009):
335–354.

4

Pope Francis,
*Encyclical Letter
Laudato Si' of the
Holy Father Francis
on Care for Our
Common Home*
(Rome: Vatican
Press, 2015).

5

See Michael
Northcott and
Peter M. Scott,
*Systematic
Theology and
Climate Change:
Ecumenical
Perspectives* (New
York: Routledge,
2014).

6

C. S. Holling,
"Understanding
the Complexity of
Economic, Eco-
logical, and Social
Systems," *Eco-
systems* 4 (2001):
390–405.

The universal nature of climate change dictates a collision of unanticipated connections between particles, people, and places across an endless array of scales. A diverse set of disciplines and perspectives is required to understand the impact of climate change and to exercise some measure of agency over its causes and ills. Climate Change and the Scales of Environment, a confer- ence convened at Columbia University in December 2015, gathered such an assortment of thinkers. The scientists remarked at the depth of language and meaning that assailed their constrained empiricism. The designers responded to the overwhelming nature of socioecological stimuli from which one could analyze in physical terms. The humanists and historians provided a narra- tive of change and instability that has prevailed throughout human history. It is not just the parameters of uncertainty that animate such conversations around climate change, but also the tension between diverse constructions of the concept of "environment" and the manifested control or agency thereof.

The proceedings highlighted a broader popular struggle to construct an environment conceptualized through an anthropocentric reality tempered by the narrative appeal of equity and justice. The struggle is to make universal any randomized selection of ecologies. The problem with this type of narra- tive is that the characters of pollution and oppression are not so black and white and the divisions between human and natural ecologies are more or less allegorical. This incidental framing relates very closely to the current popu- lar reexamination of Humboldt's "nature."[1] In this sense, one is needlessly repeating a long history of a "man-versus-nature" dichotomy that removes moral culpability and practical agency.[2] While ecology and evolutionary biology have disconnected the human experience in some form or another as a means to bring perspective, the socioecological frameworks for climate change are slow to recognize the tremendous status quo bias latent within the prevailing discourse of socioecological resilience.[3]

Dipesh Chakrabarty has noted the intellectual power of the recent papal encyclical that posited that humans are just one of many equals in a "common home."[4] Have humans had any higher intrinsic value in the manifestation of God's likeness than ants or viruses? It seems odd that theology would be the source of a morally neutralizing agenda that conflicts with the history of paternalistic conservation.[5] Perhaps removing human reality from ecological evolution is useful to home in on the absolute nature of extinction or the col- lapse of civilization as we know it. But one shouldn't forget that the world will be fine with climate change. The aggregate systems of the nonhuman world will follow their associated adaptive cycles along a series of evolutionary stable states continuously interrupted by the cosmos.[6] It is humans (and polar bears) that are at risk. If the built environment is more fundamentally defined in ecological terms, then this dualism between humans and nature may no longer be useful in a world where "old dualisms are being supplanted by transdisciplinary thinking, uneasy synergies, complex networks, and surprising collaborations."[7] Scholars such as Chris Reed and Nina-Marie Lister, Jack Ahern, and Donald Watson and Michele Adams all offer compelling and diverse perspectives that reflect the wide-ranging applications of design thinking within an ontologically and ecologically diverse landscape.[8]

However, despite this plurality of viewpoints, urban ecology scholarship has consistently demonstrated the nonlinear and asymmetric—yet *coupled*—operations of socioeconomic and biophysical processes. This would lead one to believe that there is an opportunity for a designed agency—or agency of design.[9] This may be true at one scale (e.g., toxic waste, individual household consumption, etc.), but at many scales these relationships are subject to the prevailing theory of panarchy, which is reliant in part on the continuous adaptive cycle being periodically broken by "creative destruction" of natural capital by forces unanticipated or unknown to the subject cycle.[10] Therefore, one's agency in an open system is inherently limited by forces that we are unlikely to anticipate or to apply probability to. This point was very clearly made by the climate scientists at the symposium who noted the limitations of using downscaled climate science for anything less than region-specific analysis. In addition, the scalability of designed mechanisms for change is limited by resource allocations. This is perhaps the great conundrum of sustainability of any given system and the sustainability of resources necessary to promote resilience of any given system. Sustainability at one scale comes at the cost of instability at another scale or system.[11]

Therefore, the popular mantra that we designed climate change and that we can design our way out is of limited application. Unfortunately, it is too late to stop climate change, and it is impossible to sustain the resources necessary for any—let's assume desirable—scalable design solutions that promote the status quo posited by socioecological resilience. Therefore, the more appropriate question might be: how can design intervene to lessen the burden of climate change for people and to minimize the deleterious impact to the natural environment (admittedly self-serving)? Unfortunately, the conventional paradigm of conservation and promoting stability in urban ecological systems is in conflict with many of the Western principles of urban design and planning that offer stability, hierarchy, and generalizable and scalable application for mass utility and/or universal aesthetics. If these open, dynamic, and adaptive systems are based on a heterogeneous intent, how does design develop a taxonomy that isn't otherwise linear and homogenous in its application even though its origins are creative, iterative, and pluralistic? This is the central thesis of design and climate change. We didn't design Fuller's spaceship earth—we merely designed the fuel for the ship and the seatbelt to keep us in place.[12]

7
Chris Reed and Nina-Marie Lister, *Projective Ecologies* (New York: Harvard Graduate School of Design, 2014), 17.

8
See Reed and Lister, *Projective Ecologies*; Jack Ahern, "Urban Landscape Sustainability and Resilience: The Promise and Challenges of Integrating Ecology with Urban Planning and Design," *Landscape Ecology*, vol. 28, no. 6 (2013): 1,203–1,212; Jack Ahern, "From Fail-safe to Safe-to-Fail: Sustainability and Resilience in the New Urban World," *Landscape and Urban Planning*, vol. 100, no. 40 (2011): 341–343; and Donald Watson and Michele Adams, *Design for Flooding: Architecture, Landscape and Urban Design for Resilience to Flooding and Climate Change* (Hoboken, NJ: John Wiley & Sons, 2010).

9
Marina Alberti and J. V. M. Marzluff, "Ecological Resilience in Urban Ecosystems: Linking Urban Patterns to Human and Ecological Functions," *Urban Ecosystems* 7 (2004): 241–265.

10
Jianguo Wu and Tong Wu, "Ecological Resilience as a Foundation for Urban Design and Sustainability," in *Ecological Resilience and Urban Design*, ed. Steward T. Pickett, Mary L. Cadenasso, and Brian McGrath (New York: Springer Publishers, 2012), 211–229.

11
See H. E. Daly, *Beyond Growth: the Economics of Sustainable Development* (Boston: Beacon Press, 1996).

12
R. Buckminster Fuller, *Operating Manual for Spaceship Earth* (Zurich: Lars Müller Publishers, 2008).

The metaphor of the seatbelt and stability is at the core of the "Resilience Problem." Resilience is about the elastic function of a system to revert to its pre-stimulus domain of operation.[13,14] By contrast, adaptation is about the capacity to transform to an alternative domain of operation. Resilience has a threshold and beyond that threshold one either adapts or fails. Today in the US, much of the discourse in design and planning is oriented toward the notion of resilience without fully contemplating the nature of adaptation. If only we can build a resilient city or a resilient building, then we can accommodate climate change. This trend fails to acknowledge the limitations of resilience and the limitation of the capacity to sustain resources to promote resilience. As Jianguo Wu and Tong Wu so precisely highlight,

> It is crucial to note that there can also be a negative dimension of having high resilience. A system can sometimes become resilient in a less desirable regime. For instance, urban regions besieged by impoverishment may be stuck in "poverty traps," where a suite of socioeconomic factors have induced a highly robust state of squalor ... The same genre of dynamics can also affect rural regions, urban fringes, and other socio-ecological systems, manifesting in environmental degradation and the depletion of valuable ecosystem services. This is the case in many urban areas of the developing world, and illustrates that resilience can work as *both a vehicle of sustainability and an agent of destitution.*[15]

As the authors cite, if we focus on perpetuating the operations of the status quo, then we run the risk of perpetuating inequality and the less desirable aspects of society and environment. Consumption is in many ways predicated on the status quo. One needs a stable job in order to afford things to consume and a stable place in which to store them. However, when we attempt to make resilient aspects of excessive or perilous consumption—whose costs are borne by the collective—we overlook the central danger of resilience in that it does not allow us to challenge the institutions that have created our vulnerabilities.

In recent months, consumer bath products have been utilizing the word "resilient" in their packaging and marketing. There is a Resilience Bar in Hong Kong, and NPR has daily resilience reporting—or so they say. Nike has a resilience line of shoes, and nearly every car marketed in America is resilient—either to one's lifestyle or to the rugged everyday terrain of America's deteriorating infrastructural landscape. Actors receive awards for their resilient performances, and refugees are lauded for their resilient character. According to Google Trends, since 2004, online searches for the term "resilience" have more than doubled. For some reason, Australians, New Yorkers, and Kenyans can't get enough. The number one search query by a margin of 2:1 to the second most queried phrase is the "definition of resilience."

This highlights the extent to which resilience has morphed into just another buzzword with very little substantive popular meaning.[16] Yet, resilience is the call-to-arms for people around the US as they struggle to recover from disasters and build capacities to accommodate future storms and climate change. However, resilience has precise technical and scientific meanings and applications in psychology, engineering, ecology, anthropology, computer science, and climate science. While resilience is important, we also need to acknowledge the necessity for adaptation. In applied terms, adaptation is about the transformative capacity to shift to alternative modes of consumption

13

Jesse M. Keenan, "Material and Social Construction: A Framework for the Adaptation of Buildings," *Enquiry: Journal of Architectural Research*, vol. 11, no.1 (2014): 18–32.

14

Resilience is a process and not an outcome. One may be resilient along a continuum of the process of resilience. One can only be resilient to known stimuli to which the process of resilience is engaged with.

15

Wu and Wu, "Ecological Resilience as a Foundation for Urban Design and Sustainability," 224. Emphasis added.

16

Eva-Maria Stumpp, "New in Town? On Resilience and 'Resilient Cities,'" *Cities* 23 (2013): 164–166.

and production. If we focus on resilience, then we are setting ourselves up for investing resources into modes of consumption and a way of life that simply isn't sustainable in the future. Adaptation isn't going to be easy, and there will be winners and losers. Some aspects of the built environment need to be resilient, but other aspects will simply need to creatively destruct and adapt.

Politicians and foundations love to frame the world through the lens of resilience because it speaks to the present interests of their respective constituencies. No politician wants to be the one who stands up to advocate for disinvesting in a high-risk neighborhood. However, with climate change, there aren't enough public and private resources to support universal resilience. It is incumbent upon society to acknowledge that we can't do it all. At best, one can promote the resilience of vulnerable populations and help facilitate long-term adaptations. However, this task won't be easy. Resilience at one scale might be maladaptive at another and vice versa. The subjective application of these concepts belies new governance institutions and techniques that adjudicate matters of justice and equity.[17] The struggle for the status quo is not entirely about excess consumption. Elements of culture, identity, and place are as critical to the climate change discourse as economics and resource allocation.[18] So, it is that one comes full circle to the narrative techniques of the humanities and jurisprudence highlighted at the previously mentioned symposium.

Advancing the role of designers and planners in the built environment in the accommodation and mitigation of climate change must be based on principles that acknowledge the dynamism between socioeconomic and biophysical processes. Equilibrium-seeking notions of conservation are no longer adequate as they do not represent our empirical understanding of the built and natural environments under contemporary ecologically framed knowledge. However, design cannot be bound by, nor can it rely blindly on, scientific understandings, projections, and probabilities. The complexity of our world is beyond our capacity to accommodate through the conventional economic decision-making necessary to allocate resources to one iterative design decision over the other. This isn't to say that under many circumstances probabilistic approaches to design and risk cannot be harmonized. Rather, it is to say that designers will never find a panacea in scientific knowledge. Sensitivity to people and place is of equal weighting if mechanisms are in place to reflect upon the nature of excess or perilous consumption. However, weighting is not the same thing as optimization. The built environment generally tends to reject optimization in favor of a qualitatively panarchic and messy state.[19] This seemingly transitive state is where the iteration and mediation of design work best. Designers and planners have an opportunity to translate scientific and social scientific knowledge for application in the construction of the built environment. From translation to taxonomy, this will be many generations in the making. However, a critical first step is to develop consistent meanings for resilience, acknowledge its conceptual and actual limitations, and make productive its counterpoint by promoting a robust capacity to adapt.

17

Ferenc L. Toth, *Fair Weather: Equity Concerns in Climate Change* (London: Earthscan, 2009).

18

Neil W. Adger, et al., "This Must Be the Place: Underrepresentation of Identity and Meaning in Climate Change Decision-Making," *Global Environmental Politics*, vol. 11, no. 2 (2011): 1–25.

19

Joan Iverson Nassauer, "Messy Ecosystems, Orderly Frames," *Landscape Journal*, vol. 14, no. 2 (1995): 161–170.

Jesse M. Keenan is the research director for the Center for Urban Real Estate (CURE), adjunct professor of real estate development at the Graduate School of Architecture, Planning and Preservation, and is affiliated faculty for Extreme Weather and Climate Change at Columbia University.

Next-Century Collaboration between
Design and Climate Science

Kate Orff and Adam Sobel
in conversation

KATE ORFF

I'm happy to sit down with you today to talk about the changing relationships between climate science and design. There are so many questions about how scientists and designers can work together differently in the coming decades to address severe heat, heavy rainfall, storm surge, and flooding, and their ramifications for social life and public space, as these phenomena intensify and affect cities and landscapes across the globe. But working together often involves starting with a very clear understanding of how different disciplines operate, which we designers often don't have. What are our methodologies? Our tools? At what scales do we work? And how do we define the questions and act based on determinations of value? Scientists are trained to put a number on uncertainty in terms of probability (e.g., 1 in 10 versus 1 in 100), but this way of communicating risk is often abstract to the typical person on the street. Designers—with our multiple approaches—can handle it differently. What other ways of communicating climate scenarios do you see, and how do you think designers can help visualize and communicate scientists' assessments and translate their findings for decision-making contexts?

ADAM SOBEL

Before answering this, let me say that I believe communicating the full reality of the oncoming climate change is the most profound and difficult obstacle to dealing effectively with it. I don't think the problem is unique to climate science; it's human nature. Constitutionally, we're not good at visualizing a future that is fundamentally different from the present or past—or at least we aren't able to visualize that future in a way that will motivate us to take action. This is the case even when we know, intellectually, that this future is possible or even probable.

In his book *Thinking Fast and Slow*, psychologist Daniel Kahneman discusses how people think and act in qualitatively different ways depending on whether they're facing something within or without their recent experience. His experiments showed that we think fast and instinctively when dealing with familiar challenges—which is effective most of the time. Whatever we did before will usually work again if we're facing something routine. But when we have to manage future risks that are far outside our experience, we can't rely on instinct—we have to use our conscious intellects to assess the risks and decide what to do. That's thinking slow, and when we do that we don't make such good decisions, even when we have good information.

Kahneman also writes about "availability bias"—our tendency to behave as if the likelihood of something that's recently happened to us happening again is greater than it really is, while if something has not happened recently, or ever, we tend to behave as though its likelihood is smaller than it really is. This tendency holds true even when we actually know what the likelihood is, quantitatively.

KO

Do you see Superstorm Sandy as a wake-up call? That was certainly an extreme event that was deeply felt and had tragic consequences, with so many lives and livelihoods affected here in the New York City region.

AS

Yes, definitely. If you look at the history of substantive, preemptive actions taken to reduce risks of future disasters, they were virtually always carried out in places where similar disasters had recently happened. New York City is a great example—we're seeing tremendous activity here post-Sandy, including your own Living Breakwaters project. Scientific understanding of the risk of a Sandy-type event clearly existed well before the fall of 2012 and was known to government decision-makers. It had been studied, quantified, and understood, but virtually nothing like the investments that are being made now in "resilience" were made pre-Sandy. It took the event to make it real.

Climate change is just about the worst kind of problem in this sense. It is coming on gradually, so even though there is good scientific evidence that it is well under way, most people do not feel it viscerally. And the future global environment is going to be outside the experience of everyone on the planet. So the problem feels remote and far away, it's difficult to visualize, and it's very difficult for people to prioritize it as a political issue among other issues that are ultimately less dire but whose impacts are easier to see in the short term. This holds even for people who understand and accept the scientific predictions. I'm not talking about denial. I'm talking about the difficulty of acting rationally on the basis of information that we believe intellectually to be true but that tells us about a future that's foreign to us.

KO

Can collaborations between scientists and designers, then, advance a new understanding of the issue and forge a different mode of working through it?

AS

This is an important question, not just for designers but for the broader purpose of communicating the reality of climate change. I don't think scientists can solve this problem on their own, because it's not primarily a scientific problem. It is, in part, a scientific communication problem, but on such a large scale that we need talent from outside the scientific community to help us. We need professional communicators. We need more movies, books, and other kinds of storytelling about climate change. There is a growing body of "cli-fi" literature—books that envision futures in which global warming has further advanced. The great fiction writer Amitav Ghosh recently delivered a lecture series at the University of Chicago in which he considers why climate change hasn't been written about much in "serious" fiction.[1] It's important that people like him are starting to recognize this gap in our cultural imaginations. Designers should be part of this project, too—the built environment also has stories to tell, and some of them can be about climate change.

KO

I love hearing that—in urban design we've been working to advance our technical and narrative storytelling skills in order to help people imagine possible futures in ways that go beyond static plans or proposals. I feel, for example, that the climate maps that only show potential urban flooding relative to contour lines—with entire cities underwater, rendered in blue—are actually quite unproductive for any sort of action-oriented discussions. I wonder what other forms of communication we can engage—video, interview, interactive scenario building? The Ghosh reference is a great one. I've also been reading Naomi Oreskes and Erik Conway's book, *The Collapse of Western Civilization*, which pulls in political and social ideology as a backdrop to future conflict.[2] Inspiring and sobering. Along those lines, can you please talk a bit about your Extreme Weather and Climate initiative? To start, I'm particularly interested in how you landed on its name—haven't scientists been desperately trying to shift the conversation away from weather?

AS

Studying extreme weather events is critical for a number of reasons—most obviously because they do a lot of damage. In many cases, we can't anticipate the risks of the most extreme events precisely. For example—what is the probability of a Category 4 tropical cyclone hitting Manhattan or Mumbai, two places where this hasn't happened in known human history? How is climate change affecting that risk? And what kind of damage would the storms inflict if they did happen? As you know, a lot of us at Columbia—in many different parts of the university—are working on questions like this in relation to many types of extreme weather events. The Columbia Initiative on Extreme Weather and Climate aims to make connections between all of these different efforts, stimulate new work in this area, and build relationships with outside partners also interested in extreme events.

Beyond their physical consequences, extreme events also make profound psychological impacts. This makes them very important to the climate change conversation. They generate "teachable moments."

This is part of the justification, I think, for the science of extreme event attribution, in which scientists explicitly calculate the role that human-induced climate change has played in specific weather events. We used to say we couldn't do this, but today it's a rapidly growing field. It's not easy to justify this work on a purely rational basis—if we want to understand climate-related risks, the best way is to study large-scale patterns rather than individual events. But events that have just happened get people's attention like nothing else. And for some purposes it doesn't even matter if attribution studies are conclusive. The mere act of asking the question after an event occurs prompts people to visualize what climate change *could* mean, which may be the more difficult challenge we're facing—more difficult than the science—at this point in history.

I first got the idea for the Extremes initiative, in fact, when Mayor Bloomberg endorsed Barack Obama for president. This was in the immediate wake of Sandy, and Bloomberg stated that whether or not the storm had been affected by climate change, its destruction demonstrated the seriousness of the problem. While he didn't really like either candidate, he felt Obama was better on climate, and experiencing Sandy spurred him off the fence and into making an endorsement. He knew that strong statements about climate change's role in the storm couldn't yet be justified, but he rightly saw Sandy as a signpost to the future of New York City, especially because of its vulnerability to rising sea levels.

Along those lines, I wonder if projects like your Living Breakwaters can play a similar role to extreme weather events—as powerful physical experiences that compel us to visualize different futures. That is, can projects that are being designed and built to address the effects of climate change also enable people to understand its realities in a more visceral way? Do you see Living Breakwaters playing an educational role in this or any other sense?

> KO
>
> Absolutely. It was conceived as a physical project—a chain of rock breakwaters, seeded with oysters—that would initiate a new set of behaviors and a new way of perceiving the landscape of the harbor. High school students seed and monitor the oysters as part of their science curriculum, and Staten Island fishermen and kayakers will steward the structure as it grows—it's designed to foster finfish habitat and recruit additional shellfish to attach themselves to its surfaces. I think we have to begin to integrate social and physical adaptation. We are not adapted to the climate we have now. My goal is that *Living Breakwaters* will help us prepare for the climate of the future—not only by restructuring our physical environment but by reorganizing our ways of intervening, perceiving, and coming together, so that we can build up a more protective landscape over time and channel work in a productive way. What's especially great is that because the project is federally funded and is being implemented by the New York State Governor's Office, we can pilot and test this new way of thinking resilience—as a physical *and* social practice—knowing that it can have an impact at multiple scales. Living Breakwaters is really trying to chart a path of an empowering view of climate response—not only asking how we can reduce and manage risk but also how we can begin to work together in a solution space that's interesting and fun.

AS
How do you account for sea-level rise—or any other aspect of climate change—in the design of Living Breakwaters? Does the project self-adapt, fail gracefully, etc.?

> KO
>
> The project is built of rock, and that rock can be added to as water levels rise. There is also a low onshore dune that can be replenished. We're seeding the in-water structure using five different oyster techniques, with the idea that because the oysters are living organisms, they can accrete and grow and gain complexity over time, further reducing the force of incoming waves and cleaning the water. My point is that extreme weather happens, but rather than being catastrophic, we can modify the built environment to make these events less catastrophic—"failing better," so to speak. In every case, we need to avoid failing catastrophically. We need to learn from the errors of the

past hundred fifty years of levee building, occupying wetlands, and creating single-purpose infrastructure that doesn't provide any other community or ecological benefits, and which actually obscures public perception of environmental risk.

AS

Those are all important points, but let's pursue that last one about perception a little further. How does that play out in Living Breakwaters? What is the interplay in such a project between function and public education, whether it's aimed at climate change or any other environmental issue? Is part of their purpose to make people think, and if so, how?

KO

Well, as the project is growing over time as physical infrastructure, it's also growing a sort of social infrastructure and larger-scale educational project about ecology, water quality, and the vulnerability of our harbor. New York City public school teachers and students participating in the Billion Oyster Project, funded through our HUD grant, will become connected to the shoreline. The big-picture idea is not to "protect people from" water or to keep water out of our city but to bring people closer to it—on an everyday basis, so the risks posed by rising sea levels are better perceived and understood. Although this project was conceived for the New York region, which is facing extreme hurricanes and flooding events, it could be applied to a range of places with different climate risks. I think this is one of the most captivating and challenging climate-related questions we're facing—how climate shocks interact with social systems, and how they will play out relative to the very real constraints of time, budget, and permitting, local planning approvals, serious trade-offs, approvals, and all the other gritty realities of construction. We need some kind of brave new world, or brave new mode of practice, where sociologists, climate scientists, urban designers, and politicians all sit down and try to have the same conversation. We are actually well equipped at Columbia to be the conveners of this conversation.

AS

Yes, we are. We're starting to convene it right here! But going beyond conversation, what about the practical dimension of this kind of collaboration? If tomorrow you could have a climate scientist on your staff as you designed a landscape project, what would you have them do?

KO

Well, I would ask them to first help us define the goals of the project, and the scale of its efficacy and impact, and then to work as an integral member of our design team, modeling and iterating different change scenarios at different scales. One exciting thing we currently have on the boards at SCAPE is a computer modeling tool. We've developed a complex 3-D topographic model that spans land and water, in which we can actively iterate different physical characteristics of different offshore breakwater and onshore dune scenarios in terms of height, length, spacing, location, etc., using the modeling program Grasshopper. We then nest these scenarios into a detailed refraction/diffraction hydrodynamic model—which is used to simulate surface wave propagation—to test its efficacy along the coastline. This is an exciting development that melds a design process with active testing and modification.

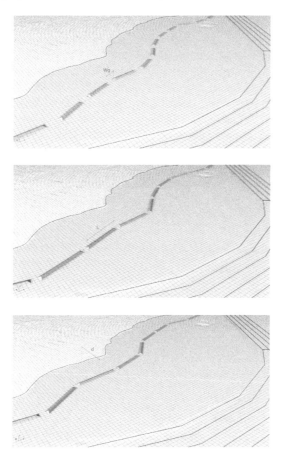

Screenshots from the hydrodynamic model of the Living Breakwaters project. The breakwaters' distance between segments (Wg), single segment length (L), and distance from shore (d) are among a set of parameters that are being calibrated and tested throughout the design process. Courtesy of SCAPE.

I get the sense that scientists and engineers feel that they're expected to provide only the "design-to" criteria: X for earthquakes, Y for the hundred-year flood, Z for sea-level rise, and so on—but that has got to change. We've got to begin working together in an iterative and activist way, and our models have to work as shared design tools as we explore more complex and imaginative futures. How would reforestation of a project's region impact rainfall, for example? Or how would nourishing our bays with sediment, or strategically reducing dredge channels in New York Harbor, reduce flooding, and to what extent? This is a question we worked through for Living Breakwaters with oceanographer Dr. Philip Orton.

Ultimately, the major collaborative challenge we face is that climate science is presented as massive, abstract global data relative to open systems, whereas design and construction require the definition of clear boundaries. At SCAPE we've aimed to advance projects at a landscape scale in order to bridge this difference, but I believe we need to both scale up and scale down as we move forward. When building and constructing resilient landscapes, a project's impetus is so often rooted in what is possible from a regulatory standpoint, or from a capital or funding standpoint. So when we begin to look at "climate" and at "natural systems," many of the values and assumptions and information required for a project are often driven not by design intent but by what can be done—what it will take to implement the project, particularly from a regulatory standpoint. I'd like to continue working with climate scientists to help determine what *should* be done, to expand the perspectives through which climate change adaptation and mitigation projects are viewed and the criteria by which they are evaluated. It's important to highlight new methods for developing projects as well as areas where policy and regulatory change may be required in order to anticipate the next hundred years of adaptation. We need to not only work backwards from what is currently possible but clearly map out the larger systemic changes needed to transform the urban landscape, and to test our hypotheses through a joint science- and design-driven process.

1

Amitav Ghosh, "The Great Derangement: Fiction, History, and Politics in the Age of Global Warming," 2015 Berlin Family Lectures, The University of Chicago, 2015. Videos of all four lectures are available at https://berlinfamilylectures.uchicago.edu/2015-amitav-ghosh. A book that collects them is set to be published by Penguin Random House in 2016.

2

Naomi Oreskes and Erik M. Conway, *The Collapse of Western Civilization: A View from the Future* (New York: Columbia University Press, 2014).

Kate Orff is the director of the Urban Design Program at Columbia GSAPP. She is also a registered landscape architect and the founder of SCAPE, a professional practice based in lower Manhattan.

Adam Sobel is a professor at Columbia University's Lamont-Doherty Earth Observatory and School of Engineering and Applied Sciences. He studies the dynamics of climate and weather, including extreme events such as hurricanes, tornadoes, floods, and droughts and directs the Columbia Initiative on Extreme Weather and Climate. His book about Hurricane Sandy, *Storm Surge*, received the 2014 Choice Award in the popular category from Atmospheric Science Librarians International and the 2016 Louis J. Battan Award from the American Meteorological Society.

A Third Space: Neither Fully Urban nor Fully of the Biosphere

SASKIA SASSEN

There is a specific and critical zone for action and intervention within the larger problem of climate change: a complex assemblage of biospheric and human capabilities that can be thought of as constituting an intermediate space that is neither fully urban nor fully of the biosphere.[1] It contrasts with the more familiar emphasis in critical environmental literature on the rupture in the relation between cities and the biosphere caused by our environmental destructions. This rupture has been described as the *unbiological* consumption by cities of the biosphere. That is, cities today, unlike in past periods, take more from the biosphere than she can regenerate. This is, in fact, mostly the case, especially in very large cities. Yet it is this intermediate space, the interstice of biosphere and city, that can contain multiple articulations of the city and the biosphere, and I want to theorize these articulations as capabilities. Today these are mostly seen as negative capabilities—humankind's potent degradation of the environment, for example—but these capabilities can also be seen as connective tissue that is neither fully urban nor fully of the biosphere, a bridge between these two so different zones. Our challenge is to make these articulations positive, since we are also the ones who made them negative to begin with. The aim is to make this a hybrid working space for experimenting with diverse types of knowledge, from biology to architecture, open and incomplete.

This project is both theoretical and practical: It is predicated on the importance and necessity of using the multiscalar and socioecological properties of cities and recognizing the need to recode these properties as potentials that can be made to work positively. One key aspect of such an effort is to delegate back to the biosphere what she does well. Rather than using man-made chemicals—fertilizers, pesticides, and so on—can we use biospheric elements (for example, bacteria and algae, or, in the case of crops, crop rotation rather than fertilizers and pesticides)? Further, how can we use the knowledge and tools we have developed to strengthen the outcomes of that delegating? The question, then, is how we can extend this type of conception or formula to the case of the city, one of the more complex and incomplete, and thereby unmanageable, systems we have.[2]

Delegating back to the biosphere what she does best is a framing for an analytics that differs from more familiar (and

1

Portions of this essay were originally published as a foreword to Vanessa Keith, *2100, A Dystopian Utopia: The City After Climate Change* (New York: UR Books, 2016). UR Books is the publishing imprint of Terreform. For a fuller treatment of this mode of conceptualizing the relation between city and biosphere, see Saskia Sassen and Natan Dotan, "Delegating, not Returning, to the Biosphere: How to Use the Multi-Scalar and Ecological Properties of Cities," *Global Environmental Change*, vol. 21, no. 3 (2011): 823–834. We humans, and all the animal life in cities, are of course part of the biosphere. My emphasis is on built environments and technical systems when I speak of the city as distinct from the biosphere. Therein lies a hypothesis: that the urban dwelling of human, plant, and animal life can take place in a broad range of settings; some of this range, though a very narrow tranche of it, is evident in the diverse ways in which cities relate to the biosphere—compare, for example, Oslo and Beijing, not only in terms of air quality but also the expanding range of initiatives that entail working with the biosphere that we see in Oslo but not in Beijing.

2

For a full development of this model, see Sassen and Dotan, "Delegating, Not Returning, to the Biosphere"; Saskia Sassen, "Global Finance and Its Institutional Spaces," in *The Oxford Handbook of the Sociology of Finance*, ed. Karin Knorr Cetina and Alex Preda (Oxford: Oxford University Press, 2013); and Saskia Sassen, *Expulsions: Brutality and Complexity in the Global Economy* (Cambridge, MA: Harvard University Press/Belknap, 2014).

romantic) notions of a "return to nature." It can take us well beyond mitigation and adaptation, today's two dominant approaches that, while welcome, are clearly insufficient to address our destructive relationship to the biosphere. Delegating back to the biosphere does not only entail very complex operations; it can also entail very simple ones. But it does entail collaborations across diverse fields of knowledge, including biology, materials science, technology, and engineering. The aim becomes combinations of specialized types of knowledge that can function in an intermediate zone (that is neither fully of the biosphere nor fully urban) rather than directly involving or focusing on the specific settings of people's lives, such as the home, work, transport, or consumption, with all their social and built environment dimensions.

In short, delegating to the biosphere requires a particular kind of intermediation or bridging function. One way of conceiving of such a function is that it is achieved more through *instruments* that can be deployed in diverse domains than, for example, *buildings*. However, I argue that built environments can also be made to function as instruments. That is to say, they become a medium for implementing or incorporating capabilities that are not simply about the building but that have a far larger and heterogeneous role: deploying biospheric capabilities, digital capabilities, and so on, which often inhere in materials or in the aerodynamic or sun-oriented design of a building.

BUILDINGS AS INSTRUMENTS

Architecture, more so than other fields, is marked by its ability to both destroy the biosphere and to work with it in multifaceted ways. What also stands out is its capacity to go well beyond such basics as recycling and gathering rainwater. To repeat, the key here is implementing biospheric capabilities that can transform a building into an instrument for environmental sustainability. Further, working with the biosphere can construct channels that might be of great use for other urban conditions screaming for change.

It is an approach to building that can lead to a whole range of novel biotechnological innovations that advance sustainability in cities and buildings. A growing range of bacteria, algae, and mushrooms are becoming inputs for a variety of applications. For William Myers, biodesign is about "forging relationships with nonhuman life to improve the ecological performance of manufacturing and building." "Evolution," he says, "has shaped a biosphere teeming with miraculous machines. The degree to which we can successfully integrate with them for mutual benefit is limited only by our imaginations."[3]

What makes it work is the marriage between these live elements and forms of technical and scientific knowledge. Many examples of this have become familiar to architects.[4] For instance, living mushrooms can be used to make bricks "that can be assembled and configured into almost any form, and naturally weld into a single object when set together." These "fungal-polyominoes" are the building blocks of what Phillip Ross calls "mycotecture."[5] Algae are another major input for biotech. One example is the Solarleaf façade developed by Arup, the Strategic Science Consult of Germany (SSC), and Colt International. It filters carbon dioxide from the air by using it to grow algae, which in turn can be used as fuel in bioreactors.[6]

Whole building complexes can become sites for incorporating these types of biospheric capabilities. One good example that would not be so difficult to replicate in a variety of diverse buildings is the Omega Center for Sustainable

3

William Meyers, *Biodesign: Nature + Science + Creativity* (New York: Museum of Modern Art, 2012). See also Jake Simons, "Biodesign: Why the Future of Our Cities Is Soft and Hairy," *CNN*, August 29, 2014, http://www.cnn.com/2014/08/27/tech/innovation/biodesign-why-the-future-of-our-cities-is-soft-and-hairy.

4

For a deeper and more detailed account, see Sassen and Dotan, "Delegating, Not Returning, to the Biosphere," 20.

5

"Biofabricate: There's a Bio-Revolution on the Horizon!" *Makezine*, December 18, 2012, http://makezine.com/2014/12/18/biofabricate-theres-a-bio-revolution-on-the-horizon.

6

Patrick Lynch, "For Terreform ONE, Bioengineering Is the Future of Design," *ArchDaily*, January 9, 2016, http://www.archdaily.com/779946/for-terreform-one-bioengineering-is-the-future-of-design.

Living (OCSL) in Rhinebeck, New York, built by Kansas City–based BNIM Architects. This building is carbon neutral and produces 100 percent of its own energy through solar and geothermal systems, and uses photovoltaic power. The green roof collects and cleans rainwater before diverting it to a cistern.[7] These are just a few of the multiple ways in which technical knowledge gets linked to the biosphere, and together they produce a viable solution that is neither fully technical nor fully biospheric.

INSERTING THE THIRD SPACE IN THE CITY

A recent volume by Vanessa Keith explores a variety of concrete cases in very diverse regions of the world that illustrate what I am arguing here, and do so along enormously diverse vectors.[8] And because they focus on extreme conditions in extreme cities, they help make visible key features of this delegating to the biosphere that might be less visible in less extreme situations.

The case of Beijing points both to the need and the possibility of organizing production as more of a loop rather than a line going from extraction to production, consumption, and waste. Keith's work examines how a landfill can be repositioned as a center not only for recycling but also for waste transformation. Carbon will be captured, and the carbon dioxide transformed into carbon fiber, for instance. The project explores the ways in which the city can be made to work more in concert with the biosphere, benefiting from its capacities. Beijing's future is marked by desertification, yellow dust storms, and extreme air pollution. All will require remediation. But the project presented here asks us to reimagine these conditions as opportunities rather than only dangers by designing for and transforming them.

7

Damir Beciri, "Green Architecture—Omega Center for Sustainable Living," RobAid, August 16, 2009, http://www.robaid.com/tech/green-architecture-omega-center-for-sustainable-living.htm.

8

Vanessa Keith, *2100, A Dystopian Utopia: The City After Climate Change* (New York: UR Books, 2016).

Beijing: View from recycling hangars toward hybrid residential. © Vanessa Keith, 2016. This image originally appeared in Vanessa Keith, *2100: A Dystopian Utopia*, UR Books, the publishing imprint of Terreform, New York, 2016.

We can extract a generic lesson from this specific case, and I think it is one with multiple applications. A chemical plant kills and leaves death behind. The biosphere gives us storms, fires, floods, and droughts that are destructive for a particular affected site and that can kill animals, trees, and humans. But we must interrogate each of these different types of destruction. Is the biosphere's capacity to kill similar or different from that of the chemicals and poisons we have made and are now present in more and more areas of the world, leaving death behind—of land, floras, faunas, and humans. When the biosphere kills it is as a part of a larger cycle, not an end point. And that cycle often strengthens life at some point in its trajectory.

A very different version of extreme climate is the case of the Troll Research Station in Antarctica. Here, the project is how to make cities that absorb more of the waste they produce, generate their own clean energy, and produce their own food through urban farming. The urban condition becomes an interconnected system, and urban morphology becomes something shaped by environmental as well as social forces. The buildings are shaped so as to create pinched areas that capture the prevailing wind, with technologies placed along these corridors to generate wind energy. Also factored into the urban morphology are water collection and solar orientation for day lighting and energy generation.

This environmental shaping is further strengthened by the fact of an extreme climate: the urban in this case is a series of interconnected interior spaces, with shared green and recreational spaces within. The project makes visible, in a microsetting, the work of balancing dense urban settlements with natural systems. The Troll project is one illustration of what I think of as an emergent sociobiospheric space/system that belongs neither fully to the city nor fully to the biosphere. It enacts a hybrid in-between space with a positive environmental valence.[9]

9

For a fuller development of this argument, see Saskia Sassen, *Territory, Authority, Rights: From Medieval to Global Assemblages* (Princeton, NJ: Princeton University Press, 2008). Chapters one and nine examine how capabilities produced in a given historical system can switch organizing logics and become constitutive of a novel system. One implication is that systemic change can be furthered by capabilities belonging to the prior regime but on the condition that they switch organizing logics.

Troll, Antarctica: Megastructure interior view overlooking hybrid park / agricultural space. © Vanessa Keith, 2016. This image originally appeared in Vanessa Keith, *2100: A Dystopian Utopia*, UR Books, the publishing imprint of Terreform, New York, 2016.

Johannesburg: Elevated pedestrian / bike pathway with living façade. © Vanessa Keith, 2016. This image originally appeared in Vanessa Keith, *2100: A Dystopian Utopia*, UR Books, the publishing imprint of Terreform, New York, 2016.

But the core concept at work in Troll finds a not-so-evident parallel in a high-density city such as Johannesburg. Here, the effort is how we can manage high-density human settlements without cutting off natural ecosystems. The site has two nature preserves. A series of sections that separate the human from the animal make space for animals to migrate across farmland. These sections become animal bridges and animal tunnels. Further, making buildings that incorporate spaces for bird nesting and encourage the growth of plant life on their surfaces, which in turn absorb CO_2, all contribute to the interconnecting of human, plant, and animal systems.

One aspect of interest to me in this case is that the urban condition can, and often does, include somewhat pastoral settings. Elsewhere I have argued that in today's vast urbanized territories, the rural does not fully disappear, but it is repositioned as an interstice in urban space. I think we should strengthen and perhaps expand these rural interstices. The urban today is increasingly constituted through processes that produce diverse instantiations of space, time, place, and "new natures." Examples on the negative side are new eco-urban conditions such as heat islands and, to scale up, ozone holes. To some extent, this means that the urban also contains some of the transformative possibilities embedded in these same processes. For example, the temporal dimension becomes critical in environmentally sound initiatives. Thus, ecological economics enables us to recognize that what is inefficient or value-losing, according to market criteria with short temporal evaluation frames, can be positive and value-adding when we use criteria driven by the biosphere's orderings, including the longer times of uninterrupted flows and multiscalar shifts. These possibilities need to be distinguished from notions such as a return to nature via rural living or lifestyles.[10]

These projects, as does my own work, emphasize an active engagement with scientific, technological, and architectural knowledge as a way to

10

Two decades ago there was a body of critical analysis on the "return to nature" as a viable option. Thus David Harvey notes that, at best, traditional environmental ecologists can offer some return to an earlier form of urbanization regulated by the metabolic constraints of a bioregional world "as it supposedly existed in the past," a world that for Harvey never really existed. At that time, and perhaps still in much of the world, much of what passed as ecological among social scientists studying cities actually dealt with quality-of-life issues for middle- and high-income people and neglected the needs of the poor. See David Harvey, *Justice, Nature, and the Geography of Difference* (Cambridge, MA: Blackwell, 1996); David Satterthwaite, *Striving for Universal Provision: The Governance of Basic Services in an Urbanizing World* (New York: Routledge, 2014); and David Satterthwaite, *The Earthscan Reader in Sustainable Cities* (London: Earthscan, 1999).

Manila: Elevated plinth with storm and wave energy capture. © Vanessa Keith, 2016. This image originally appeared in Vanessa Keith, *2100: A Dystopian Utopia*, UR Books, the publishing imprint of Terreform, New York, 2016.

11

Nor is the international regime of carbon trading a solution. For a critical analysis, see Mary Kaldor and Joseph Stiglitz, *The Quest for Security: Protection without Protectionism and the Challenge of Global Governance* (New York: Columbia University Press, 2013).

take us beyond the stasis in today's policy debate, with its timid notions of adaptation and mitigation of damage.[11] Manila, for example, is a place that can benefit enormously from such a more radical approach aimed at using particular biospheric capacities. Since this city is subject regularly to strong winds and floods, the aim of Keith's design is to create renewable energy from wave energy as well as storm energy. Secondly, given high levels of pollution, Manila is a good candidate for bioremediation: the use of algae farming to absorb carbon in the air and water, as well as microorganisms in the ocean that will feed on plastic waste. These microorganisms accelerate the process of breaking down the micro-particles of plastics in the water, though not the larger pieces. Importantly, recovered plastic in the ocean can be adapted into fuel to power the city. To complete the cycle, fungi will be used "to clean heavy metals and other toxic chemicals from the water and ground."

One key to this productive encounter of the urban with the biosphere is that it mobilizes the diversity of ecologies and multiscalar capacities of the city. Keith's São Paulo project illuminates the challenges and possibilities. This is a vast city, prone to flooding, overwhelmed by its enormous population and by its unstoppable economic dynamism. The project here is an outpost settlement with a series of elevated sleeping pods "connected by a bridge to a larger 'mothership' pod where shared spaces exist" for people to come and live and work on the project for a period and then rotate out for a new group. Important here are alternatives to fossil-fuel-burning transportation. The bicycle is the key mode of transportation. Tree farming, harvesting of rain forest crops, and growing spirulina and other medicinal crops are major efforts, as is conversion of some of this into medicinal and nutritional extracts. To deal with the pollution, the project uses a "floating river infrastructure as a series of floating modules that combine remediation along with carbon sequestration and coastal protection." All structures are working at engaging the biosphere—the buildings serve to generate renewable energy, including use of surfaces.

São Paulo: Flood state—elevated bikeway looking toward mothership. © Vanessa Keith, 2016. This image originally appeared in Vanessa Keith, *2100: A Dystopian Utopia*, UR Books, the publishing imprint of Terreform, New York, 2016.

CONCLUSION

All these cases include what I think of as the intermediate space where the work of delegating back to the biosphere takes place. It is space with a relative *conceptual* autonomy from power relations. In this regard the use of a concept such as "environmental sustainability" comes with a somewhat more forceful meaning than might be typical in government regulations and international treaties. Here I draw on David A. Sonnenfeld and Arthur P. J. Mol's proposition of a "new world (dis)order."[12] Though power relations are certainly important, they are in a sense orthogonal to that in-between space I am focusing on. Thus, *delegating back to the biosphere* constitutes a mode of theorizing contemporary changes that includes the biosphere as an active partner and turns its back on the endless discussions of policy.

One key starting point is the existence of material and chemical cycles in the biosphere that predate human industry, in the narrow sense of the term, and continue to be responsible for the maintenance of homeostasis on earth. We then need to factor in the limits of this capacity for maintenance given current economic and social logics. Delegating back to the biosphere entails a shift of focus. Every surface and every process needs to be recoded as far as is possible in terms of a deployment of biospheric capacities where we now use chemicals and synthetics that are environmentally destructive (not all chemicals and synthetics are).

Thus, delegating back to the biosphere takes management, making, inventing, discovering: human intervention in the formation of *novel* socio-ecological bridges with positive environmental valence. For instance, the rate of waste production accelerates with urban scale, whereas natural processes for waste removal would tend to decelerate with scale. So delegating waste management back to the biosphere must involve novel socioecological trans-actions that incorporate biosphere-centered methods. One familiar example is using algae and a bioreactor, rather than chemicals, to process wastewater: it is the same process as in the biosphere but accelerated via a human-made

12

David A. Sonnen-feld and Arthur P. J. Mol, eds., Sym-posium on "Social Theory and the Environment in the New World (Dis)Order," *Global Environ-mental Change*, vol. 21, no. 3 (2011).

innovation. That is to say, such processes are often *wrapped* in technology at the moment they are delegated back to the biosphere.

It does imply, perhaps at its most extreme and therefore clearest, that particular kinds of socioecological processes delegated back to the biosphere *must* be managed or accelerated in such a way as to keep pace with the urban material and energy flows that exhibit nonbiological scaling behavior—that is, processes that accelerate with growing urban scale. This contrasts with the biosphere's tendency to decelerate with growing scale. Shrinking or intermediating this unsustainable gap marks the specificity of "delegating back" to the biosphere, in contrast to a simple return to nature.

Saskia Sassen is the Robert S. Lynd Professor of Sociology and a member of the Committee on Global Thought, Columbia University. She is the author of several books and the recipient of diverse awards and mentions, ranging from multiple *doctor honoris causa* to named lectures and various honors lists. Her new book is *Expulsions: Brutality and Complexity in the Global Economy* (Harvard University Press/Belknap, 2014).

An Ecology of Bodies

ROSS EXO ADAMS

1

"Hyperobjects," following a Latourian ontology, offer a new philosophical-ecological framework for thinking phenomena across scales—questions posed by the arrival of Big Data, the Anthropocene and quantum mechanics all demand a trans-scalar, trans-temporal ontology to comprehend the (hyper) objects they illuminate: "things that are massively distributed in time and space." See Timothy Morton, *Hyperobjects: Philosophy and Ecology after the End of the World* (Minneapolis: University of Minnesota Press, 2013), 1. While "flat ontology" can be attributed to Manuel DeLanda, this notion has been more recently reinterpreted by several from the Object-Oriented Ontology/Speculative Realism camp (Levi Bryant, Graham Harman, etc.) to advocate for a nonhierarchical ontology consisting of objects whose ontological consistency extends across spatiotemporal scales. See Manuel DeLanda, *Intensive Science and Virtual Philosophy* (London: Continuum, 2002). In a more muted way,

Jane Bennett's concept of "vibrant materialism" assembles a similar span of issues to build a political ecology that resides as much in human agency as it does in that of non-human and more-than-human materialisms. Again, such a political ecology points to a new ontology of "things": objects, phenomena, and events that resist classical compartmentalization in scale and time. See Jane Bennett, *Vibrant Matter: A Political Ecology of Things* (Durham: Duke University Press, 2010).

2

See, for example, Neil Brenner, "The Limits to Scale? Methodological Reflections on Scalar Structuration," *Progress in Human Geography*, vol. 25, no. 4 (2001): 591–614; or Bruno Latour's short extract from the Geochemistry of the Earth's Surface meeting, published as Bruno Latour, "Some Advantages of the Notion of 'Critical Zone' for Geopolitics," *Procedia: Earth and Planetary Science* 10 (2014), 3–6. Latour's work has, of course, had a great influence on the recent development of Object-Oriented

Ontology, discussions around the Anthropocene, and Radical Ecology, to name but a few effects of this sort that have emerged in the last years.

3

See Neil Brenner, *Implosions/Explosions: Towards a Study of Planetary Urbanization* (Berlin: Jovis, 2014).

It may be that, in the near future, we will come to see how the problems posed by climate change will have opened an entirely new epistemological horizon. Already, such an effect can be registered in the ways many contemporary forms of knowledge have begun approaching the notion of "scale" through radically non-modern ontologies. Emerging across many disciplines are sets of objects, processes, and phenomena visible only insofar as they traverse traditional scalar and temporal categories—"hyperobjects," flat ontologies, and vibrant matter that cannot be made legible unless we abandon our inherited epistemological frameworks and invent new ones.[1] It is a tendency that has altered discourses from geography to philosophy, political theory, and sociology.[2] Architecture, too, has begun to respond to these epistemological cues embedded in problems brought about by climate change (urban ecologies, the "Anthropocene," resource geographies, etc.), all of which call for developing knowledge outside traditional scalar categories precisely because the problems climate change makes visible adhere to none and can be observed across all. The urban, too, is another such category increasingly seen to surpass all traditional scales and spatial confines. From "metropolitan" to "megalopolitan" to "regional" to "planetary" in scale, urbanization now reveals itself traversing deserts, arctic regions, and can even be found in the atmosphere.[3] Indeed, it is possible that we can no longer comprehend the urban as a spatiality isolated to any one particular scale at all but must instead illuminate it in its full, trans-scalar ontology. The urban, it would seem, has no scale.

We architects and urbanists tend to respond to this epistemological shift without fully conceiving the corresponding conceptual tools with which to assess the urban. As a result, we acknowledge the trans/non-scalar nature of the urban by fixating on the very same cues that urbanists have always looked at: the continual expansion of the urban into ever-greater scalar referents. In other words, we see whatever gets bigger. On the one hand, such an expansive perception is confirmed in the increasing scales at which urban design operates. Projects like *Grand Paris* transfer architectural knowledge from what may have been "urban" in scale to what is now ostensibly "regional."[4] On the other hand, architectural knowledge in general has seen more interest in notions of territory, logistics, resource extraction, migration, and even "planning the planet."[5] And if Neil Smith's work in urban geography exposed the process of capitalist urbanization as progressively "jumping scales," it is perhaps this legacy that Neil Brenner and Christian Schmidt have more recently built upon in their work on "planetary urbanization."[6] The work produced in Brenner's Urban Theory Lab in Harvard's GSD is a brilliant and seductive examination of precisely how far such scalar jumps have gone, bringing the urban into hitherto unknown spaces and realms, whose visibility coincides with that of the planet.

4

See Theresa Enright, "Illuminating the Path to Grand Pari(s): Architecture and Urban Transformations in an Era of Neoliberalization," *Antipode*, vol. 46, no. 2 (2014).

5

"Plan the Planet: Jaqueline Tyrwhitt and the Formation of International and Global Architecture" (March 20, 2015), a symposium conducted at the Architectural Association, London.

6

Neil Smith, "Homeless/Global: Scaling Places," in *Mapping the Futures: Local Cultures, Global Change*, ed. John Bird, Barry Curtis, Tim Putnam, and Lisa Tickner (New York: Routledge, 1993): 87–119.

Grand Paris Métropole Douce, LIN Architects, 2008, one of ten proposals within the international consultation initiative launched by former President Nicolas Sarkozy in 2007, Grand Pari(s)—a project to reimagine Greater Paris. Courtesy of LIN + Davide Abbonacci.

7

For more on this
project, see Ross
Exo Adams, "Notes
from the Resilient
City," *Log* 32
(2014): 126–139.

8

Michel Foucault's
well-known notion
of governmental-
ity identifies the
ensemble of mod-
ern power whose
increasingly admin-
istrative character
developed over the
nineteenth century
by tying together
political economy
(as its form of
knowledge),
population (as its
subject-object),
and "mechanisms
of security" (as its
means of interven-
tion). The result
is, as Foucault has
shown, a form of
political power
that, taking "life
under its care,"
expands its realm
of intervention
across "the
whole surface
that lies between
the organic and
the biological,
between body
and population."
This concept, for
Foucault, helps
to explain the rise
of statistics, the
massive investment
in data collection,
and the new tech-
nologies of power
that accompanied
the rise of this
new political
epistemology
from the nineteenth
century onward,
all of which created
a new knowledge
of the state based
on the body and
the population as
at once biological
and political-eco-
nomic categories.
For more on this
notion, see, among
others, Michel
Foucault, *Society
Must Be Defended*
(New York: Pica-
dor, 2003), 253.

Yet if such a tendency has crystalized a certain image of the urban around the spectacle of the "planetary," we inevitably overlook the effects simultaneously produced at other scales by the very same processes. If the urban is continually expanding in scale, what effects, spaces, and technologies are produced at more immediate scales and sites?

Indeed, emerging today is a practice of urbanism whose strategies seem to hinge on their ability to organize circulations and technologies, bodies and ecologies, regions and infrastructures, all *without regard* to scale. It is a form of urbanism that works in far more ambient, subtle, yet radically transformative ways, one whose viability comes insofar as its interventions remains invisible, at most concealed behind other more overt modifications of space. And while it may provide a degree of novelty to focus on the scalar magnitude that such an urbanism proposes, it is the relations it builds *across scales* that makes this practice truly new.

What all this promises is to lay the foundations for a mode of urbanization whose processes are far more complex than the caricatures we tend to conceive of buildings and roads unfurling endlessly across the planet. In fact, it may well be that such caricatures of urbanization are already captured in and put to work for this newer, more elusive form of urbanism as a crucial set of symbolic signifiers that justify its innovations. While this particular breed of urbanism has gone by many names during its gestation, today it calls itself "resilience urbanism." And only today, in direct response to climate change, has it achieved a certain maturity, demonstrating its agenda through a logic that seamlessly coordinates its strategies across multiple scales of space at once. Indeed, if there is a single novel quality that this urbanism displays, it is this. Yet such an attribute is no small thing: in the age of climate emergency, it is precisely this trans-scalar capacity that promises to conceive of spatial design as a strategy increasingly oriented to spatializing law itself.

New York City's *Rebuild by Design* (RBD) is resilience urbanism's most high-profile project to date. Launched in the summer of 2013, following the devastation of Hurricane Sandy, *Rebuild by Design* is an ongoing effort to implement corrective measures across the metropolitan region's coastal zones and the aging infrastructural networks they depend on. The project positions itself as a sober response to the reality of climate change and its promise for more extreme weather to come. While I do not wish to review this project and the ten design proposals that constitute it in detail, I would like to use this project instead to examine how it constructs itself as a set of strategies that, by conceiving themselves as eminently trans-scalar, work in direct relation to a parallel restructuring of law itself.[7] What RBD proposes, in other words, is a radical transformation of governmentality through a coordinated set of spatial technologies.[8] If modern urbanism gained "scientific" consistency by organizing its knowledge and technologies as a hierarchy of discrete scales, resilience urbanism, as evidenced by RBD, is an attempt to invert this logic: it is a program of scalar *coordination*. Its technologies, strategies, and innovations all issue from a common ambition to smooth scalar differences into a single, coordinated space of governance—a regional technology designed for efficient emergency management of the totality of bodies that dwell within it. Indeed, much more than a project of regional-scale design, as it may appear, RBD is a program to urbanize the body across the scale of the earth.

GOVERNMENT BY DESIGN

Initiated by the Rockefeller Foundation in coordination with municipal and national initiatives, RBD has brought together architects, urbanists, engineers, marine ecologists, climate scientists, and economists to propose a new approach to design in the age of extreme weather. The projects assembled by RBD unanimously eschew any radical transformations to the physical spaces of Greater New York City. And while they tend toward regional plans, the strategies of RBD actually operate at a much finer grain. Appearing to promote agency rather than heavy-handed design, what the projects seem to consist in are largely "reclaimed" spaces at the interface of land and water, appropriated by the bodies who inhabit them—shoppers at a local market, families playing on docks, tai chi under the freeway—all of which seem to accentuate the everyday behaviors of individuals. We find urbanism takes on a pedagogical capacity: citizens are now constantly informed about the risks of flood and the local histories of climate catastrophe within the city. Signs, placards, and public institutions distributed throughout the city serve as signposts for a new cult of crisis to establish itself in the metropolitan consciousness. The spaces of RBD are evidently "smart" spaces, with cybernetic technologies coordinating a new sense of public exchange: Citi Bike stands couple real-time hurricane and flood information with public Wi-Fi hotspots and cellphone charging stations; news kiosks distribute "Flood Risk" flyers; and smartphones are equipped with flood-risk apps, all of which will assure the widespread integration of a new public knowledge of climate emergency and its spatio-social management. In fact, it was observed that the process of design itself should play an "essential role" in "educating the civic partners about issues related to resiliency."[9] These strategies, all made evident in renderings, drawings, and diagrams from RBD proposals, and further elaborated in their project descriptions, make the integration of smart city technologies more overt while at the same time departing from any familiar and childish infographics typical of smart city representations. In the final round of RBD, several of the winning proposals suggested that, through the use of distributed Internet and Communications Technology (ICT) infrastructures, new feedback loops of data relevant for government, first responders, insurance companies, and the general population could be created to streamline the management of emergency situations in both immediate (first responders)[10] and long-term time scales (recalibration of risk by insurance companies).[11]

More recently, RBD has begun to aggressively advocate for far-reaching legal reforms that they assert will help to achieve "resilience." In June of 2014, RBD published a report aimed to direct conversations that would take place during a roundtable held later that month. RBD asked its ten design teams to identify areas in which their proposals would encounter legal and policy-based obstacles. The results of this were distilled into three main topics of discussion: how to formalize "civic infrastructures," or how community and civic groups may establish formal relations with government planning; how governmental models could be reformed to "balance human, ecological and economic needs in coastal areas"; and last, how to expand "nature-based solutions" as a form of urban design. The three topics, Collaboration by Design, Governance by Design, and Restoration by Design, respectively, intended to address a realm that is exposed by the design process yet goes beyond what it is able to prescribe: the design of law. "More than design innovation, we will need policy and social innovations to make

9

Rebuild by Design, "Policy by Design: Promoting Resilience in Policy and Practice" (June 2014), http://www.rebuildbydesign.org/wordpress/wp-content/uploads/briefing/RBD_policysession_briefingbook-FINAL.pdf, 11, hereafter cited as "Policy by Design."

10

See OMA's proposal for "communications systems": http://www.rebuildbydesign.org/project/information-systems.

11

See WXY/West8's proposal, in particular pages 9–12 and the discussion on Catastrophe Risk Engineering on pages 106–109: http://www.rebuildbydesign.org/wordpress/wp-content/uploads/briefing/WXY_IP_Briefing_Book.pdf.

sure that the Sandy-affected region and the nation become more resilient as we continue to rebuild."[12]

What came out of this meeting was summarized in a white paper published a year later.[13] Divided into three broad "topic papers," the first two—"collaboration by design" and "governance by design"—address the overall question of how governance should be reformed to accommodate the mandates of resilience. In these papers, RBD participants emphasize the need to coordinate government across all scales. The current model, described as "fragmented," operates through clumsy and often nonexistent mechanisms to coordinate federal, regional, state, municipal, and local authorities when faced with an emergency at the scale of Sandy. Current governance operates as a system whose structural opacity makes the distribution of resources and the communications between levels inefficient, if not dysfunctional, in the face of emergency, all of which was made patently obvious in the handling of Sandy's aftermath. What RBD participants call "Administrative Continuity" is an effort to restructure government across levels, collapsing all levels of governance into a single fluid machine of government tuned to respond to the imminent emergencies that climate change promises to produce. This makes sense: since the climate is not bound to a single scale, neither, too, should the governmental response to its effects. "To address emerging environmental and social challenges, we must operate at the scale of climatic and ecological regions, and in a way that puts communities first."[14]

In this restructuring of government, the model of "administrative continuity" highlights the figure of "the community" as not simply the beneficiary of a reformed governmental structure but rather as a central figure in the governmental space of resilience itself. "Communities must be drivers of resilience," headlines the "Collaboration by Design" paper, a statement that opens questions about how community efforts connect with governmental planning and what "gaps" exist in the span between the two. If it was "the community" that was the victim of poor governance in the recovery efforts following Sandy, it is now "the community" that must be endowed with new powers given by the emerging framework of resilient governance. Just as government must adopt resilience, so, too, the participants of RBD suggest, must communities and civic groups be encouraged to "fully embrace resilience as part of their mission."[15] Community here is reconstituted as a formalized entity *within* resilient governance. Through an "engaged partnership" between government and the local coastal communities of greater New York City, the notion of "community" is transformed into a key site in which a new governmental knowledge relevant to resilience may emerge.

If resilient government must work vertically across scales, it must also do so more "inclusively." The government's expanded field of operation within the category of "community" thus requires a kind of pedagogical campaign through which to elucidate a set of values that "community" must adhere to within a broad framework of resilient governance. The role ascribed to design here takes on new importance, which the participants of RBD highlight. For not only is design a matter of proposing legible strategies for the city, but they must also serve as an educational interface through which to both include communities in the design process and, more importantly, to narrate the values embedded in the program of resilience more broadly. Described as "awareness," design becomes both a medium and a practice through which to penetrate the opacity of community, disciplining it to speak in a common language. "The RBD process developed compelling images and

12

Rebuild by Design, "Policy by Design," 2.

13

Rebuild by Design, "Policy by Design: Promoting Resilience in Policy and Practice, June 2014 Roundtable Discussion Summary" (June 2015), http://www.rebuildbydesign.org/wordpress/wp-content/uploads/2015/07/Updated-Policy-Doc.pdf, cited hereafter as "June 2014 Roundtable."

14

Rebuild by Design, "June 2014 Roundtable," 7–8.

15

Rebuild by Design, "June 2014 Roundtable," 4–5.

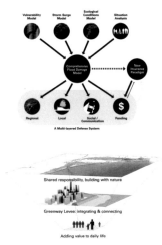

From top: The "community outreach and engagement strategy" of signage to be distributed throughout the city; project strategy diagram; and diagram of multi-scalar thinking. All images by WXY / West 8 team, Rebuild by Design, courtesy of WXY / West 8 / nowhere office.

Design as pedagogy: SCAPE team members held teacher training sessions to incorporate "Oyster Gardening"—a key "nature-based solution"—into public school curricula. Courtesy of SCAPE.

concepts, and these should be utilized for creating awareness and continued engagement … As more stakeholders become aware of the projects and develop greater interest as they continue to go through implementation, a larger community will remain plugged into the Rebuild by Design process."[16] Here, transparency is rife: If this language, carried across in design, images, and spatial diagrams, makes transparent how space is to be reorganized, infrastructures reworked, climate imagined, it uses this language as well to construct and make visible new governmental relations relevant to resilient government.

One of the key forms of design that has been embraced by RBD teams and is actively pursued by the organization as a whole, is so-called nature-based solutions—"sand engines" and oyster beds that can absorb storm surges, breakwaters modified to shield from flooding, water basins optimized to absorb floodwaters more efficiently, and so on. All these proposals constitute ways in which "nature" is incorporated into the design of the urban in ways that selectively cancel out ("naturally") those of its more negative effects.[17] Perhaps more fascinating than these design innovations is the fact that, unlike traditional modes of design, they all seem to require perpetual stewardship. Whether monitored centrally or through a more "bottom-up" engagement with community groups, nature-based solutions present themselves as eco-cybernetic technologies, at once using "nature" and its inherent processes, effects, and qualities in order to normalize the climate. Draped over

16

Rebuild by Design, "June 2014 Round-table," 11.

17

See Adams, "Notes from the Resilient City," 126–139, and Bruce P. Braun, "A New Urban Dispositif? Governing Life in an Age of Climate Change," *Environment and Planning D: Society and Space*, vol. 32, no. 1 (2014): 49–64.

with technologies of monitoring, "nature" becomes at once celebrated in its opposition to old-fashioned, man-made infrastructures, while at the same time it is transformed into a technology known through the copious data it produces. Indeed, nature-based solutions all seem to demand vast assortments of data-collecting technologies, sensors, and computational models, all of which are measured against "performance metrics" to present a real-time image of climate at the interface of land and ocean.[18]

18

Rebuild by Design,
Policy by Design,
10.

In such an eco-cybernetic space, data must be comprehensive, high-quality, and perpetual. Just like resilient governance, the data collected must be consistent across multiple geographies. According to an agenda of "administrative continuity," it must extend a new span of knowledge from the individual "user" to the climatic reaches of extreme weather.[19] Because such a span implies a monumental effort in collecting sufficient data, the RBD participants highlight a concern in the inevitable "gaps" of knowledge that will appear within—the processes, objects, measurements, and metrics that slip outside the datascape of ubiquitous computing and cybernetic sensory technologies. They suggest that a "data clearinghouse" can be opened up in parallel—an entity that aggregates multiple sets of data including community-based open source data, thus also able to identify and fill in gaps in the overall structure of information.[20] For resilience to operate through nature-based solutions, data pertaining to it must be total. Lastly, this data must also be "compelling." It must translate itself not only to be accessible to a broader public, but to seduce it. Its seduction comes in part because it illuminates itself to the inhabitants of the urban in the real-time unfolding of nature, captured in brilliant colors on LCD screens and endless updates communicated through smartphone apps. Less the crisis of climate, what becomes the object of a total visibility in the space of resilient urbanism is a climate of crisis.

19

Rebuild by Design,
Policy by Design,
8, 10.

20

Rebuild by Design,
Policy by Design,
10.

RBD correlates, perhaps more clearly than any other project to date, with what Byung-Chul Han has called the "transparency society."[21] If transparency is a kind of meta-agenda of neoliberalism, as Han has argued, we can see how the production of space proposed in RBD so neatly corresponds to a neoliberal form of government. The constant appeal to "putting communities first" strikes certain tones of a Big Society strategy in which empowerment and "community stewardship" can appear as cheap foils for off-loading governmental responsibility to real communities that are marginalized and precariously exposed to the dangers of climate change. Indeed, the subtle "inclusion" of community as a quasi-formal category of government in effect also converts what is by definition precluded from governmental apparatuses: community, it could be argued, is precisely an ethics that persists in the absence of government. The clearest appearance of such a community manifested itself spontaneously in the wake of Sandy in the neighborhoods and regions that were most affected by the storm, exactly in the *lack of governance* within these areas. Built on top of the movements like Occupy Sandy (an offshoot of the Occupy Wall Street movement), community emerged as a practice that worked through a collectively shared, unchecked debt that circulated throughout spaces in the New York City region.[22] Since the appearance of community understood in this way naturally threatens government, it is no surprise that "community" reappears so prevalently in RBD as the object of reformed governmental care. The semi-inclusion of "community" in resilient governance is also its negation—a move that would constitute what Roberto Esposito has called the *immunization* of community, absorbing and immunizing such communitarian bonds by formalizing this category as a quasi-governmental agency—an army of circumspect

21

Byung-Chul Han,
*The Transparency
Society* (Palo Alto:
Stanford University
Press, 2015).

22

Alan Feuer,
"Occupy Sandy:
A Movement
Moves to Relief,"
New York Times,
November 9, 2012.

Banner from Occupy Sandy, 2011.
Courtesy of West-Park Church.

"stewards" of the resilient city.[23] Ubiquitous computing, sensors, and the constant production and reproduction of data are precisely the mechanisms through which such bonds can be weakened and new ones established. As the participants of RBD emphasize, the use of "nature-based solutions" requires a "fundamental shift" in the status of coastal communities:

> Using natural systems for flood protection requires a fundamentally different way of thinking of coastal communities, and their benefits may be less evident or intuitive than infrastructural approaches. Engagement and education programs must be incorporated into project development, such as in demonstration projects where communities can learn firsthand about the interventions. Helping communities understand how different solutions address different problems or risks also creates local buy-in.[24]

Data, in this sense, complements design as a multifaceted pedagogical interface that actively deterritorializes communitarian relations, reterritorializing them as relations between the body and technology.

URBANIZATION WITHOUT SCALE

If, throughout the twentieth century, the predominant scientific depiction of the body as a set of universal measures allowed it to assist in standardizing the reproduction of urban space, it also, for that reason, made the body visible as the site of abnormalities and incompleteness, requiring evermore effective controls and regulations to intervene on it. With the techno-positivism of cybernetic thought and its introduction into emergent urbanisms from the 1960s onward, an apparent reversal takes place in this relation in which the body, seemingly liberated from its narrow, "top-down," modernist molds and standardizations, presents itself instead as an immeasurably complex, self-regulating organism.[25] Conceived in this way,

23

See Roberto Esposito, *Communitas: The Origins and Destiny of Community*, trans. Timothy Campbell (Stanford: Stanford University Press, 2010), and Esposito, *Bíos: Biopolitics and Philosophy* (Minneapolis: University of Minnesota Press, 2008).

24

Rebuild by Design, 2015, *Policy by Design*, 11.

25

On urbanism and cybernetic thought, see Michael Batty, *Cities and Complexity: Understanding Cities with Cellular Automata, Agent-Based Models, and Fractals* (Cambridge, MA: MIT Press, 2005). The connection between urbanism and cybernetic systems thinking can be attributed to J. W. Forrester's pioneering work: see Jay Wright Forrester, *Urban Dynamics* (Waltham, MA: Pegasus Communications, 1969).

technology will no longer aim to operate on the body through the rationalization of space itself but will attempt instead to achieve this same result by rationalizing behavior, accounting for it as a totality visible through data. It will do so by exposing the body to a far more penetrating visibility than ever before, one that no longer understands it as a biological machine, nor sees it against a background of norms. Rather, this new use of technology constitutes the body as an intricate, responsive social machine, one that voluntarily opens itself to the urban milieu it inhabits. Data, captured in real time, constructs this visibility in catalogs of the most intimate, domestic behaviors, movements, needs, and desires—signatures and patterns, the metadata through which the basis for an entirely new urban knowledge can take shape.[26]

Far from the urbanism of Robert Moses, resilience urbanism appeals to a certain "back to basics" attitude by demonstrating how the totality of existing urban space can be technologically augmented to achieve its goals. A vast array of technologies emerge across scales as the new face of an urbanism of pure disclosure—an urbanism that aims to bring to light all that has remained outside the scope of modernist urbanism. While "population" may have been an indispensable tool for modernist planning and urbanism, today, with high-resolution cybernetic urbanism, such a figure, known through statistical depictions, becomes increasingly redundant. For this reason, it is the body that has become the lens through which a new visibility of life—and of the urban—is brought to light. Through the application of such technologies, the most intimate patterns and everyday movements of bodies in urban space are aggregated as the objects and information from which a new visibility of the city can be constituted, since it is by knowing the habits and activities of every body in real time that life in a theater of crisis can be permanently managed. By articulating a new visibility of the body within a space of climate crisis, it is also the climate and its effects that are made transparent through models and sophisticated risk-forecasts, and re-presented as a permanent and dazzling feature of urban space.

26
See Jennifer Gabrys, "Programming Environments. Environmentality and Citizen Sensing in the Smart City," *Environment and Planning D: Society and Space*, vol. 32, no. 1 (2014): 30–48.

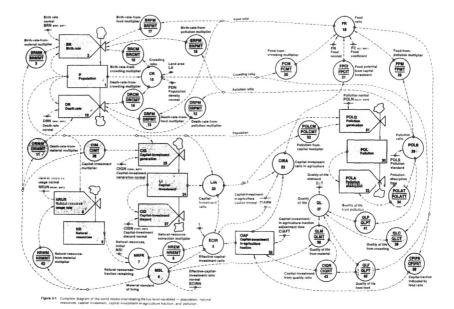

Figure 2-1 Complete diagram of the world model interrelating the five level variables — population, natural resources, capital investment, capital-investment-in-agriculture fraction, and pollution.

A diagram of the complete world model, according to Jay W. Forrester's *World Dynamics*.

Rebuild by Design, picture of winning teams. Courtesy of Rebuild by Design.

If modernist urbanism understood the body through a set of rational metrics, today's unleashes a regime of transparency as an effort to document the body through its *irrationality*. Rather than merely another example of the urban rambling onward through ever greater scalar registers, resilience urbanism constitutes an attempt to treat urban space as a *scaleless* totality, a space understood as a frictionless composition of the rational and the irrational—the rationalized space inherited from twentieth-century urban planning, made sensible through the necessarily irrational behaviors of life that inhabits it. Irrationality here is no longer the negative of rationality: rather, the two are simply qualities of life that can be equally measured, modeled, and predicted. Such a logic allows it to work from the scale of the body to that of the region and, by extension, the world without contradiction. If urban design is to unfold as a spatially mediated campaign of climate awareness, its effect will be to recast the urban as a machine of immunization through the smoothing out of human behavior across all scales into an undifferentiated field of positivities. The urban, no longer comprehensible as something discrete like a "city," perhaps gains more clarity as an ecology. However, if this is the case, the body, though its cybernetic urbanization, also gains a significantly new ontological status, for resilient technologies do not register the body as a discrete entity—the physical correlate to the liberal individual; rather, they work precisely by treating the body as a *dividual* whose presence in urban space is recorded as a catalog of body-effects distributed across space and time and made whole only through the associative metadata relevant to re-presenting the urban as a population-system—an ecology of bodies immersed in a climate of emergency.

Ross Exo Adams is an architect and urbanist whose research looks at the historical and political intersection of circulation and urbanization. He is currently an assistant professor of architecture at Iowa State University.

Cloud Formations: Climate Change and the Figuration of Community

CARSON CHAN

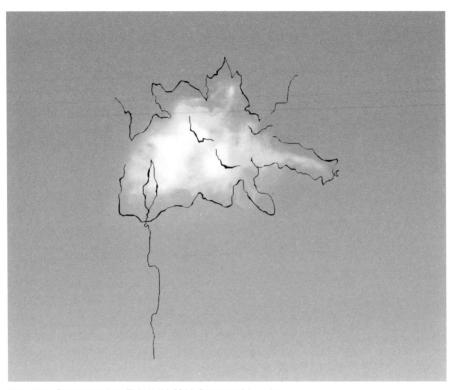

I see a face. Do you see a face, Flaka Haliti, 2014. Courtesy of the artist.

Many of the ideas in this paper were developed in conversations with Sria Chatterjee, Esther da Costa Meyer, and Kimia Shahi. An early version was presented at a conference called "Dealing with Climate Change: Calculus & Catastrophe in the Age of Simulation" (June 25–26, 2015), hosted by Leuphana University, Lüneburg, and convened by Isabell Schrickel and Christoph Engemann. Conference respondent Ulli Vismaier's suggestions helped me expand my scope, and Curt Gambetta's pointers helped focus the argument.

1

"Special 3-Hour Broadcast of the People's Climate March," *Democracy Now!*, September 21, 2014, http://www.democracynow.org/live/peoples_climate_march.

2

Paul James and Manfred B. Steger, "Levels of Subjective Globalization: Ideologies, Imaginaries, Ontologies," *Perspective on Global Development and Technology* 12 (2013): 19.

On Sunday, September 21, 2014, two days before world leaders met at the United Nations for an emergency climate summit, more than four hundred thousand people gathered in New York City for the People's Climate March. Together they moved through Central Park West like a colossal cloud. Advancing past the Museum of Natural History, flowing around Columbus Circle, and funneling through Times Square, the group would stop at different locations, sometimes standing for a moment of silence to remember the victims of climate change, sometimes to sound a "climate burglar alarm"— a collective warning against the fossil fuel industry and politicians intent, as the protesters saw it, "on stealing our future." The march was convened by a coalition of organizations and was intended to bring together a community concerned with the impact of anthropogenic climate change. Beyond its sheer size, the most distinguishing aspect of this community was its diversity of purpose. Since the green movement emerged in the 1970s, protests have shifted focus: from the environment to the climate, from the immediate to the totalizing, from the ethics of saving to the ethics of protecting. Where environmentalists once led the protests, today we see a broader base of support. Indigenous leaders and social justice organizers now join in leading the

3

In his landmark text, "The Climate of History: Four Theses," Dipesh Chakrabarty argues that climate disasters cannot be "reduced to a story of capital" and suggests that structures such as capitalism or even the nation-state will collapse in the face of this new "universal that arises from a shared sense of catastrophe." The category that arises from the dismantling of "us and them" is the human species. Dipesh Chakrabarty, "The Climate of History: Four Theses," *Critical Inquiry*, vol. 35, no. 2 (Winter 2009): 197–222.

For more on these statistics, see Benjamin H. Strauss, "Rapid Accumulation of Committed Sea-Level Rise from Global Warming," *Proceedings of the National Academy of Sciences of the United States of America* 110.34 (2013): 13,699–13,700.

4

Sigfried Giedion proposes a similar model for calibrating oneself to the shifting "constituents" of history. In *Space, Time and Architecture*, Giedion describes "contemporary history" akin to reading the configuration of stars, or the interpretation of constantly moving parts. Contemporary history is also the charge to look deeply at all developments of lived reality, not just the established canon. Sigfried Giedion, *Space, Time and Architecture* (Cambridge, MA: Harvard University Press, 1941), 5–13.

5

Accounts of green architecture's failures, from its ethics of deferral through pollution offsetting, to its post-construction failures, are many. Sustainability through green architecture certification can often lead to its opposite. For a recent opinion, see Michael Mehaffy and Nikos A. Salingaros, "Toward Resilient Architectures 2: Why Green Often Isn't," *Metropolis*, April 4, 2013.

6

Climate, the prevailing weather trends of a large area over a long period of time, is measured through the accumulation of semi-empirical parameters from many sources, often from different countries, with diverse methods of assessment. Global datasets are derived from simulations, constrained by instrument observations. These simulations are "reanalyzed" as technology improves. In this way, climatology is a science driven by modeling. See "Introduction," "Chapter 10: Making Data Global," and "Chapter 12: Reanalysis: The Do-Over," in Paul N. Edwards, *A Vast Machine: Computer Models, Climate Data, and the Politics of Global Warming* (Cambridge, MA: MIT Press, 2010).

protest. Forming the nebulous cadre of marchers was a heterogeneous admixture of environmental justice groups, labor unions, and migrant communities; delegations from around the world's island nations mingled with local interest groups; Buddhist, Taoist, Christian, Jewish, and Islamic religious leaders demonstrated with their attendant congregations; students and academics walked with film celebrities. In great numbers were advocates of the Voting Rights Act, the Hip-Hop Caucus (a civil and human rights organization), and those demanding justice for Michael Brown.[1]

It is easy to dismiss these groups for exploiting the Climate March and its media visibility as a vehicle to further their own agendas, but in recent years, large protest movements unifying disparate concerns have manifested around the world with increasing frequency—take, for example, the so-called Arab Spring or the Occupy movement. Compared to protests convened for a singular cause and that aim to give physical form to solidarity around that cause, uprisings that encompass disparate ontological frameworks challenge dissenters and observers to articulate how the group's multiple purposes, identities, subjectivities, and ideologies lockstep in force. Some have rightly observed that movements like Occupy are symptomatic of globalization and have offered that its processes of material extension not only organize our world through the standardization and amplification of production, trade, communication, and inquiry but also the subjective encounter associated with that spread.[2] Processes like globalization and climate change that on the one hand mark out spaces and temporalities beyond human observation can, on the other, be made tangible and be experienced. To that effect, rising ocean levels will displace up to 1.3 billion people, or 21 percent of the world's population, in the next century, which will confront nation-states with unintended and ruinous weather events more powerful than any legislation.[3] If ideology is a seafaring traveler, then ontologies of being—of how we live temporally and spatially—are the ocean, and our imaginary the guiding stars.[4] Configured as such, in order to grasp the reasons for new, internally diverse, social formations of protest from Occupy to the Climate March, we must take stock of how and what we imagine community to be in today's globalized societies.

A radical restructuring of community demands a reconception of the architecture it produces and inhabits. For the past two decades, architectural practices have reacted to environmental concerns with the material solutions of "green architecture"—positions that ultimately defer or blanket the effects of one of the world's most polluting industries rather than offering any practice-wide change.[5] Perhaps a fresh methodological approach is called for. Fundamental to both architecture and the study of climate is the use of models.[6] In both, models give shape to the yet unformed, and climate specifically is known only through modeling. For both architecture and climate, material for modeling or data isn't collected—it is *made*.[7] Models simulate future conditions, and in both architecture and climate science, they allow the material objects at the core of each discipline—buildings and weather over time—to enter into discourse before and without physical manifestation. Through his analysis of Alberti's writing, Mario Carpo has argued that the drawings of a building and the building itself can be seen as "perfectly identical," if the architectural practice is viewed as autographic—made by the author—rather than allographic, denoted by the author but executed by someone else. "In Alberti's theory," the radicality of which is by now fully intuited by architects, *"the design of a building is the original, and the building is its copy."*[8] To the degree that modeling provides an epistemological affinity between architecture and climate science, can architecture help make the urgencies of the environment as immediate as an architectural plan? How is "taking climate action" also an act of architecture? Can climate change produce a new method for architecture?

In *Imagined Communities*, Benedict Anderson points to print media as a technology that has allowed people from far-flung locations to identify themselves as a collective citizenry. Fully mechanized, and achievable at low costs, printing allowed the novel and the newspaper to flourish in Europe in the eighteenth century and "provided the technical means for 're-presenting' the *kind* of imagined community that is the nation."[9] With characters unknown to each other, and parallel storylines taking place, novels and newspapers evoke both a sphere of events in a discrete space (society), as well as an omniscient observer (the reader) for whom the various events unfold. Through media, a world greater than one's immediate surroundings can be imagined.

As Anderson has argued, these narrative forms helped give rise to the imagined bonds that make the nation-state: a collective sentiment that manifests physically as borders and territory, government and laws, language, religion, and citizenship. Anderson published the first edition of *Imagined Communities* in 1983, before the end of the Cold War—when national identity was code for ideological belonging. This was also just before the invention of communication technologies like mobile telephony and a decade before the advent of Internet-based tools like e-mail and mobile computing. Though the nation-state still has a strong role in contemporary self-identification, beginning in the 1960s—with concurrent growth in migration and in communication technologies—the state found increasing resistance to its ideological, economic, discursive, and even political dominance from non-governmental groups. For populations in flux, the nation was no longer descriptive of *all* the people it harbored. In the postwar imagination of social citizenship, migrant groups became agents of destabilization rather than cosmopolitan citizens.[10] As a result, as diaspora scholar Khachig Tölölyan sees it, "the *nation*'s aspiration to normative homogeneity is challenged not just by immigration but also by various forms of cultural practices and knowledge production," with transnational movements of capital, labor, ideas, and information, dissolving the *state*'s long held hegemony.[11] With identity now fractured into multiple registers, what, then, draws us into communities? If communities are not bounded by borders, how are they organized—and condensed—in space? As a response to the contemporary dispersion of communities challenged by oppressive regimes, or changing climate, architectural solutions like border walls appear all the more senseless.[12]

7

Before the World Meteorological Organization standardized methods for measuring weather in the 1940s, data was of varying fidelity. Climate scientists call the process of adjusting data sets to correct for perceived error "reanalysis." How error is interpreted, and how it is corrected, produces new chains of social, legal, and policy effects. In this sense, data is *made* rather than simply gathered. See Edwards, *A Vast Machine*, 109.

8

Mario Carpo, *The Alphabet and the Algorithm* (Cambridge, MA: MIT Press, 2011), 26.

9

Edwards, *A Vast Machine*, 25.

10

Sunil S. Amrith, "Empires, Diasporas and Cultural Circulation," in *Writing Imperial Histories*, ed. Andrew Thompson (Manchester: University of Manchester Press, 2013), 234.

11

Khachig Tölölyan, "Rethinking Diaspora(s): Stateless Power in the Transnational Moment," *Diasporas: A Journal of Transnational Studies*, vol. 5, no. 1 (Spring 1996): 4. Emphasis added.

12

Saskia Sassen has written extensively on what she calls debordering in *Losing Control? Sovereignty in an Age of Globalization: 1995 Storr Lectures* (New York: Columbia University Press, 1996) and "Bordering Capabilities Versus Borders: Implications for National Borders," *Michigan Journal of International Law*, vol. 30, no. 3 (Spring 2009): 567–597. More recently, Wendy Brown emphasized that twenty-first-century border walls are not built to divide nation-states as in previous centuries but to control the movement of non-state transnational actors. See Wendy Brown, *Walled States, Waning Sovereignty* (Brooklyn: Zone Books, 2010).

13

James Clifford, "Diasporas," *Cultural Anthropology*, vol. 9, no. 3 (August 1994): 302.

14

Avtar Brah, *Cartographies of Diaspora* (London: Routledge, 1996), 208.

15

In relation to footnote 4, we can consider Sigfried Giedion's description of Alexander Calder's mobiles as both a formal diagram of moving parts and how these parts in their relative movement can be read as a unified form. "Calder absorbed the modern means of expression, slowly amalgamating them with his own background until, by 1931, he had attained a sensitivity to states of equilibrium that he stressed in his mobiles ... A motor, a draft of air, or a push of the hand, will change the state of equilibrium and the interrelations of his suspended elements, connected in a mobile wire system, forming unpredictable, ever-changing constellations and so imparting to them the aspect of space-time." Sigfried Giedion, *Mechanization Takes Command: A Contribution to Anonymous History* (Minneapolis: University of Minnesota Press, 1948), 477–478.

16

Brah, *Cartographies of Diaspora*, 183.

17

Brah, *Cartographies of Diaspora*, 204.

18

Brah, *Cartographies of Diaspora*, 204.

The question of figuration—which is to say, the question opening this essay, and the question that evaded Anderson's imagined nationhood—is underscored by contemporary diaspora studies. Led by scholars such as Tölölyan, Avtar Brah, and James Clifford, theories of diaspora acknowledge the unfixed nature of identity. Identity derives from where one grew up, where one resides, and the multiplicity of cultures that compose these places. Attempting to distill and define social conditions of travel, plurality, provisionality, and instability, and organized within porous spheres of identity, diaspora studies is a field that both accepts and warns against the slippages between diaspora as a theoretical concept, as a set of discourses, and as an account of distinct, subjective experiences.[13]

In *Cartographies of Diaspora*, Avtar Brah proposes a frame of analysis: diaspora space. Surfaceless and of shifting dimensions, diaspora space situates diaspora, borders, and the politics of location in orbitlike motion.[14] For Brah, diaspora is understood not just as peoples dispersed from their place of origin but rather the *relationality* between diasporic forms—forms that could be given to architectural diagramming.[15] The conditions that prompt a group to leave, and those that frame how they arrive, are important to understanding the concept of diaspora. In this sense, Brah sees the populations from which a group splintered, and to which a group joins, as belonging to the same diasporic community. It is the "economic, political, and cultural specificities linking these components [groups] that the concept of diaspora signifies."[16] As such, diasporic communities are an amalgamation of the various journeys taken, combined into a confluence of narratives that *imaginatively* defines the community. The concept of borders—the arbitrary divisions between cultures, societies, and territories—is embedded within the concept of diaspora. As an abstraction of differences, borders serve to separate populations and territory, producing a politics of location, a polity of space.[17] In diaspora space, location is not a fixed referent but defined almost apophatically—in the negative—as dislocation, or displacement. If it is indexical of anything, it is contradiction. It is "a positionality of dispersal; of simultaneous situatedness within gendered spaces of class, racism, ethnicity, sexuality, age; of movement across shifting cultural, religious, and linguistic boundaries; of journeys across geographical and psychic borders."[18] The architecture envisioned for this space is one that privileges forms of movement and dispersal, destabilizing the regime of solidity and permanence.

Where narratives of the nation-state feature originary absolutes, ethnic primacy, ideological cohesion, and territorial imperatives, diaspora space addresses multivalent, creolized, contradicting ideologies, where spatial transience is equally valid as concurrent modes of being in the contemporary world. It recognizes that a native-born citizen is also an immigrant, a feminist can very well be racist, a gay rights activist may be a polluter, an organic farmer a human-rights violator, and that all these people with varying identities may nevertheless imagine themselves as a community. Diaspora space, which accommodates the diversity rather than unity of communal experience, manifestly dissolves state narratives, proposing instead the dimensions for a space of conviviality. But dissolution of the state's authority does not mean that contemporary global populations are unburdened by higher authority. "Extrastatecraft," coined by Keller Easterling, is the dynamic matrix of space, information, and power that generates self-arbitrated forms of polity outside of, in addition to, and sometimes in partnership with, the state.[19] The many nongovernmental infrastructures generated by diasporic forms themselves produce a regulatory effect.[20]

As the communities, values, and authorities gathered around the nation-state have been dissipating since the 1960s, are there alternate topologies through which global diaspora can imagine a space of belonging? Furthering Anderson's theory that new technologies of communication foster new forms of belonging, we might consider the broadcasting tools of the World Wide Web generative of other communities. In the 1970s, the early days of the development of computer networks, a goal shared by many of the movement's leaders was expressly to create communities of shared interests.[21] Whether it was the military developers of Arpanet who allowed smaller computers around the US to share capacity (1969); the students from Duke University and the University of North Carolina who established Usenet (1979); the scientists and researchers who developed the Computer Science Network (1982) and National Science Foundation Network (1983); Robert Kahn and Vinton Cerf, who reworked Arpanet's protocol to one that would be adopted by almost all other networks, creating the Internet (1983); or Tim Berners-Lee, who created the World Wide Web, the browser software for navigating the Internet; the aim has always been to establish a way to share information through message services, cooperative devices, and collective documents.

Until around 1993, the Internet was the domain of scientists and academics. The imagined community, its dream of instantaneous access to knowledge and connectivity with others, was limited to an intellectual elite. But at this moment, when the popular press begins to expound on the social and political potentials of the "new virtual community" that would replace traditional public spaces, what Patrice Flichy calls the cyber-*imaginaire* starts to take shape. Structurally descended from academic computer networks, culturally impelled by the promise of a digital revolution, and manifested through user manuals, enthusiastic magazines, and television coverage, as well as a growing number of mass market online access providers like AOL, the Internet became as much a new mode of communication as it did a self-fulfilling myth of universal access to free information—if self-fulfilling only in the minds of its adherents.[22] As noted by Flichy, the forward to one of the early manuals, the *Big Dummy's Guide to the Internet* (1993), by co-founder of the Electronic Frontier Foundation, Mitchell Kapor, is an invitation to enter a new society:

> New communities are being built today. You cannot see them, except on a computer screen. You cannot visit them, except through your keyboard. Their highways are wires and optical fibers; their language a series of ones and zeroes. Yet these communities of cyberspace are as real and vibrant as any you could find on a globe or in an atlas. Those

19

Keller Easterling, *Extrastatecraft: The Power of Infrastructure Space* (London: Verso, 2014), 15.

20

Remittance is a case in point. In 2011, migrants sent approximately three times more to developing countries than these countries received in official aid; this is an amount equal to about half of foreign direct investment to the same countries. Universal standards as imposed by nongovernmental organizations like the ISO (International Organization for Standardization) regulate the distance between train tracks, the turning radii of roads, the thickness of credit cards, codes that let cell phones from different manufacturers communicate with one another, and Internet protocols that allow data to pass from server to server across the globe.

21

Patrice Flichy, *The Internet Imaginaire* (Cambridge, MA: MIT Press, 2007), 67.

22

Flichy, *The Internet Imaginaire*, 92.

23

Mitchell Kapor, "Forward," *Big Dummy's Guide to the Internet*, 1993, http://www.cs.indiana.edu/docproject/bdgtti/bdgtti_3.html#SEC5.

24

Manuel Castells, *Networks of Outrage and Hope* (Malden, MA: Polity Press, 2012), 6.

25

Hu continues: "Though the landscaping of these pocket parks and rooftop gardens bears a superficial resemblance to public space, these overly tidy zones are nevertheless administered by banks, insurance companies, and the like. The eventual consequence is that the lived knowledge essential for imagining and discussing public space has begun to atrophy." Tung-Hui Hu, *A Prehistory of the Cloud* (Cambridge, MA: MIT Press, 2015), 147. Architectural metaphors abound in discussions about the Internet, and artist Constant Dullaart's text, "Balconism," proposes building a metaphorical balcony from the privatized space of the Net to gain a measure of exteriority from what is generally thought to be borderless and infinite. "We are all outside on the balcony now. Standing on a platform made out of a tweet into corporate versions of public space. We are not stored in a cloud, opaque or translucent to whomever ... The most important thing is: you must choose to be seen. We are already seen and recorded on the streets and in trains, in emailz, chatz, supermarketz and restaurantz, without a choice. Remaining unseen, by making a clearer choice where to be seen. We are in the brave new now, get ready to choose your balcony, to escape the warm enclosure of the social web, to address, to talk to the people outside your algorithm bubble. U will not get arrested on the balcony, you and yours should have the right to anonymity on the balcony, although this might seem technically complicated. The balcony is a gallery, balustrade, porch and stoop. The balcony is part of the Ecuadorian embassy ... Balconism is a soapbox in the park. The balcony is connected: stand on a balcony and you will see others. The balcony is connecting: you do not have to be afraid on the balcony, we are behind you, we are the masses, you can feel the warmth from the inside, breathing down your neck. Where privacy ceases to feel private, try to make it private. Ch00se your audience, demand to know to whom you speak if not in public, or know when you are talking to an algorithm." Constant Dullaart, "Balconism," *Art Papers* (March/April 2014), http://artpapers.org/feature_articles/feature3_2014_0304.html.

are real people on the other sides of those monitors. And freed from physical limitations, these people are developing new types of cohesive and effective communities—ones which are defined more by common interest and purpose than by an accident of geography, ones on which what really counts is what you say and think and feel, not how you look or talk or how old you are. The oldest of these communities is that of the scientists, which actually predates computers. Scientists have long seen themselves as an international community, where ideas were more important than national origin. It is not surprising that the scientists were the first to adopt the new electronic media as their principal means of day-to-day communication. I look forward to a day in which everybody, not just scientists, can enjoy similar benefits of a global community. But how exactly does community grow out of a computer network? It does so because the network enables new forms of communication.[23]

As with novels and newspapers, the community that emerged from the Internet is also based on communication. The key difference here is that individuals are not only readers but also producers of text, often the texts they themselves want to consume. "Cohesive and effective communities" are made not through an "accident of geography," but like scientists, by "common interest and purpose." "Ideas are more important than national origin," and in this sense, the imagined community of the Internet is a self-generating diaspora space, its members having *emigrated from* the physical realm to *arrive at* the virtual one. For computer researcher Marc Prensky, there were no "natives" of the digital realm until after 1980. On the Internet, the self was of one's own making. One's online activity on forums, Bulletin Board Systems (BBS), chat rooms, and more recent social media platforms like Facebook, constitute a sociality; what we do, and how we do it manifests into a new and distinct persona, unattached to its physical, real-world referent. Some have termed online correspondence *mass self-communication*, foregrounding the potential mass reach of digital messages as well as the autonomy of the individual publisher.[24] Tung-Hui Hu reminds us that to be online, to be in the cloud, is to partially submit to the neoliberal structures that created the Net in the first place and that maintain it today. Not unlike Zuccotti Park, "what gathering spaces emerge within the cloud are closest to the ambiguously named privately owned public open spaces found in an American city's financial district."[25] But unlike mainstream media outlets, digital communication in the early twenty-first century has the potential to offer a new type of ideological independence, one that is couched within the strictures of corporate and state control while simultaneously unavailable as agents for the propagation of these power structures. Operationally, the nation-state has little to do with the communities imagined on the Internet. Until control is wielded over the totality of online and offline activity by a single power, moving from one realm to another disables the full articulation of a community objectively, its cohesion, like that between the constituents of the Climate March, legible only internally, evading complete authority.

Much study has been made of social movements and protest groups, but few contend with the dynamic nature of their contemporary, post-Internet makeup, reverting instead to understanding them through Marxist power structures.[26] In this vein, Manuel Castells calls groups like those of the Climate March and the Occupy demonstrations networked social movements.[27] These movements began on the Internet. Broadly, networked social movements begin when voices online respond to perceived offline injustices, and if the cause finds substantial purchase, it then migrates to public urban space. In urban space, often in symbolic places controlled by the dominant elite, network society's collective concerns are performed both to strengthen the ties within the group and to display their solidarity of concern to other publics. For Castells, this moment of physical togetherness is fundamental to controlling the fear that comes with resistance while demonstrating to others that they will be among many if they decide to join the group.[28] As explained in *Communication Power*, Castells's formulation is based on the unequal distribution of power, allowing influential social actors to pressure others, bending them to their will, interests, and values. Power is wrought from four processes: the enacting in, making of, instating of, and the combining of social networks. Though this applies well to traditional defiance against hegemonic powers like governments and corporations, the significance given to the possession of power relies on a hegemonic interest in it, concealing other processes that differentiate the vast communities of social actors in question.

Recounting the multitude of interests motivated to march down Manhattan on September 21, 2014, a cloud moving across a grid—spatially transient, multivalent, creolized, and held by internal contradictions—it becomes apparent that theories of diaspora are much more analytically adept at viewing the social and political dynamics of such populations than the well-rehearsed narratives of power distribution. Just as Hubert Damisch has shown that clouds in early modern landscape painting were used to mask perspectival contradictions, bearing them out in the process, reimagining community's shapelessness could sharpen our image of its underlying structures. "In ever changing cloud formations," writes Damisch, "people see what they wish to see: configurations of their desires, images from their theatre of life, signs of their culture."[29]

Like cellphone reception, clouds are ubiquitous but unevenly distributed throughout the sky, and if they are to contribute to a reimagining of community, their unequal mediation of sun, rain, and shade, say, must be accounted for. We must learn how to read clouds.[30] Less than three weeks after the People's Climate March, sociologist Richard Sennett wrote an editorial in the *Guardian* about "why climate change should signal the end of the city-state."[31] In it, he mentions Max Weber, Franz Fanon, and Henri Lefebvre and validates the historic need for city-based conceptions of community and governance. If cities were populated by citizens who felt accounted for and were engaged through inclusive democratic processes, then they would function, and thrive, as autonomous systems. For Sennett, global warming changed this assumption. After the devastation of Hurricane Sandy in 2012,

26

See Yochai Benkler, *The Wealth of Networks* (New Haven: Yale University Press, 2006), particularly "Chapter 10: Social Ties," which elaborates on the thickened ties that bind "virtual communities." Benkler ultimately sees the network as a product, and members of these communities as "users." See also Anna Feigenbaum, Fabian Frenzel, and Patrick McCurdy, *Protest Camps* (London: Zed Books, 2013), which was published after the start of the Occupy movement and the so-called Arab Spring in 2011. *Protest Camps* recognizes the central role digital media plays in contemporary protest groups. Ultimately, the authors see digital media as a tool for resistance within a traditional David vs. Goliath, us-against-them-style antagonism rather than as a model for structuring the community itself.

27

Castells, *Networks of Outrage and Hope*, 3.

28

Castells, *Networks of Outrage and Hope*, 10.

29

Hubert Damisch, *A Theory of /Cloud/: Toward a History of Painting* (Stanford: Stanford University Press, 2002), 185.

30

In *The Marvelous Clouds*, John Durham Peters remarks on how satellite technology has demanded a new literacy of cloud formations. Whether clouds hold meaning has been an unsettled matter in Western culture. Pilots will agree with Aristotle, according to Peters, that they do hold meaning, that if it's cloudy, it will rain. At the same time, they change and vanish quickly, evading scrutiny and discourse. "With space exploration and satellites it has become routine to see clouds, as Joni Mitchell says, from 'both sides now,' and they look different from space than they do from earth, as from the vertigo of a tower looking down." Peters continues: "Meteorologists have discovered what they call 'Hokusai' phenomena, for instance, in satellite images of the atmosphere that resemble his woodcuts of waves ... This is some kind of crucial historical mutation: not only is it legitimate to look for images in the clouds, but it has become urgent. Our survival may depend on knowing how to read the signs in the atmosphere." John Durham Peters, *The Marvelous Clouds* (Chicago: University of Chicago Press, 2015), 260.

31

Richard Sennett, "Why Climate Change Should Signal the End of the City-State," the *Guardian*, October 9, 2014.

many of the coastal mid-Atlantic settlements in the United States were supported with huge federal grants to rebuild damaged houses and to construct storm barriers to protect against future storms. As climate change affects the entire planet, Sennett challenges the Weberian values that maintain legislative sovereignty and calls for us to imagine a global network of cities running in tandem. "Adapting to climate change, in other words, means that coherence of the city's form will alter, due to forces outside human control. Nature is undemocratic. Voting and inclusion cannot change the facts on the ground about how the climate operates." Sennett draws on the incompatibility between civil law and natural law and suggests that the unpredictability and contingency of nature should inform the continued evolution of the city. "We must meet the uncertainty of a physically unsettled world by thinking of the city as itself a more unstable place." In other words, social movements like climate awareness must be imagined together with the physical movement of society's members.

Cities are geographically specific entities. Indeed, as sea levels rise and storm surges alter how planners and designers approach them—determining where people might live and how infrastructure is built—climate change is also transforming our social makeup. How we occupy streets, congregate publicly, and communicate collectively generates new spatial prerogatives that will further determine how communities become manifest. Construed in this way, climate change and its social effects call for a new method in architecture beyond the material responses of offsetting and deferral of liability promoted by green architecture. Sennett imagines a system of legislatively interconnected nodes that, while disparate in their specificity, operate as a unified interest group, collectively supporting the viability of the planet's climate system to support human life. Before this is achieved, if ever, diasporic communities like that which gathered for the Climate March, multitudinous in its internal diversity but moving forward as one, serves well as a model from which to proceed.

Carson Chan is an architecture writer and curator pursuing a PhD in architecture at Princeton University. From 2006 to 2012, he co-directed PROGRAM, a residency and project space for art and architecture in Berlin. Chan co-curated the 4th Marrakech Biennale in 2012, and was Executive Curator of the Biennial of the Americas in Denver in 2013. Also in 2013, he organized "Exhibiting Architecture: A Paradox?", a conference at the Yale School of Architecture, with David Andrew Tasman and Eeva-Liisa Pelkonen. He is a contributing editor of *Kaleidoscope* and editor-at-large of *032c*. His current research tracks the rise of environmentalism and aquarium architecture in the postwar United States.

Corpo

ealities

"It's your human environment that makes the climate," said Mark Twain via Pudd'nhead Wilson. Like many of Twain's aphorisms, it's an insight that only seems to accumulate prescience with time. The human environment can be an intensely personal experience. And our individual desires to create and alter our personal environments, both indoors and out, to make life more comfortable, have collectively resulted in a tangle of far-reaching effects for Earth's climate and ourselves. Exploring large questions about how the human quest for comfort has changed the ways we live, the ways we build our cities, and the ways we govern them can begin at the molecular level. By feeling out the multitude of relations between the body and the environment we can dissolve the built-up layers of abstracted, technoscientific discourse that have come to rule climate change conversation. Returning to the very basic ways in which we experience climate— through temperature shifts and weather events—and attempt to manage it allows us to reapprehend our personal stake in questions and consequences that reach far beyond our immediate selves.

Heather Davis has an intimate engagement with the molecules that move through all our earthly bodies, while Philippe Rahm's fictionalized body spends a winter evening in Tokyo, warmed from the inside out through both sustenance and technology. Reviewing Rahm, Shantel Blakely describes his emotive descriptions of the body as both dwelling and site for acclimatization. Meanwhile, Dehlia Hannah and Cynthia Selin ponder a year without a winter and explore aesthetics as a way to make sense of climatic abstractions. Eva Horn, walking slowly to conserve her "coolth" in a hot place, guides us through the history of mankind's desire for the perfect personal climate and how this quest has transformed humanity. That change is uniquely reflected in our built environment, a notion Jorge Otero-Pailos pushes further in asking whether the atmosphere can be classified as an object for preservation, and if so, what kinds of radical practices it would take to do so. Phu Hoang proposes the radical—and inherently political—idea of bringing the abstractness of the climate closer to the tangibility of the weather in order to view personal meteorological encounters within an altered climatic past, present, and future.

Molecular Intimacy

HEATHER DAVIS

Exhibition view of Terike Haapoja's *Inhale–Exhale* at the Venice Biennale (2013). Courtesy of the artist.

The human-size rectangular glass box emits a slow *whoosh*. A pause. And then again. It sighs. Inside the box are leaves—leaves slowly decomposing. And as they decompose, they exhale. A sound made audible to our ears through the use of carbon dioxide sensors, adapted from forestry management. Ventilation fans on either side of the box open and close, like gills. This is the breath of the earth as it is transforming. Moving through the processes of decay to nourish the soil. As the leaves exhale the carbon they have been storing throughout their lives, it releases into the atmosphere. Molecular structures exchange and transform. As they breathe out, exhale, the leaves become molecular.

This 2008 work, *Inhale–Exhale*, by Finnish artist Terike Haapoja, which was exhibited in the installation "Closed Circuit–Open Duration" at the Nordic Pavilion of the Venice Biennale of Art in 2013, affectively attunes us to the carbon cycle. The numeracy of the climate crisis, the endless barrage of statistics, here gives way to breath—to the sound of breath, to the sound of the breath of the leaves, of the soil. To the exhale we then inhale. Climate statistics pulled inside our lungs. Levels of carbon dioxide and methane and oxygen registered as vibration. "It is not enough," Haapoja says, "to show the workings of carbon in the ecosystem: we need to try to see what does CO_2 mean to us, how does it work its way in our own inner reality, the reality of

Detail of Terike Haapoja's *Inhale–Exhale* at the Venice Biennale (2013). Photograph by Sandra Kantanen, courtesy of the artist.

love, and bodily being, and death."[1] The story of the carbon cycle is the story of things passing in and out of being, of transformation, of composition and decomposition. Haapoja proposes that these cycles are intimate, pulled into our most constitutive and basic elements of being. I would like to take up this proposition. How, in the midst of contemporary political debates about carbon, carbon dioxide, and carbon economies, might a shift in discourse toward affective attunement—toward an intimate engagement with the molecular—augment or intervene in those politics?

Carbon is arguably the most important molecule in an age that has been increasingly framed through the molecular. In the twentieth and twenty-first centuries, molecules have increasingly come to define our bodies and the world around us. The molecule is the primary scientific figure visualizing the inner workings of the world. We understand our sense of self, history, and ethnicity through the fetishization of DNA, just as we alter our bodies through other molecules such as oxytocin, serotonin, estrogen, and testosterone. In other words, we manufacture our subjectivities, especially our gendered identities, on the molecular scale.[2] Similarly, we are asked to think about climate change through molecular composition, including atmospheric levels of carbon dioxide, methane, etc. We increasingly understand ourselves, our identities, and our political realities through the frame of the molecule—so what are the affordances and foreclosures of this framing, especially in relation to contemporary environmental crises and climate justice?

For even as the effects of climate change are being felt and observed in so many communities around the world, scientific data remains abstract for many people. And the carbon molecule is at the heart of this abstraction. The molecular is a historically contingent product of scientific knowledge, with precursors in Ancient Greece—such as Leucippus and Empedocles—through

1

Terike Haapoja, "Closed Circuit– Open Duration in 55. Venice Biennale," http://www. terikehaapoja.net/ closed-circuit- open-duration- exhibition-venice- biennale-nordic- pavilion.

2

On the relation of gendered identity to the molecular, see Beatriz Preciado, *Testo Junkie: Sex, Drugs and Biopolitics in the Pharmaco- pornographic Era* (New York: The Feminist Press, 2013).

3

James Clerk Max-
well, "Molecules,"
in *The Scientific
Papers of James
Clerk Maxwell*, vol.
II, ed. Sir William
Davidson Niven
(Cambridge: Cam-
bridge University
Press, 1890),
361–378.

4

IBM, "IBM Scien-
tists First to Image
the 'Anatomy' of a
Molecule," Zurich,
Switzerland,
August 28, 2009,
https://www-03.
ibm.com/press/us/
en/pressrelease/
28267.wss.

5

Critical Art
Ensemble, *The
Molecular Invasion*
(Brooklyn:
Autonomedia,
2002), 8.

6

Gilles Deleuze and
Félix Guattari, *A
Thousand Plateaus:
Capitalism and
Schizophrenia*,
trans. Brian
Massumi (Minne-
apolis: University
of Minnesota Press,
2007), 275.

to the thought of the Roman Lucretius, to more sustained considerations beginning in the seventeenth century in Northern Europe. In a paper published in *Nature* in 1873, Scottish scientist James Maxwell Clerk claimed that a molecule was "the smallest possible portion of a particular substance. No one has ever seen or handled a single molecule. Molecular science, therefore, is one of those branches of study which deal with things invisible and imperceptible by our senses, and which cannot be subjected to direct experiment."[3] The molecule is the making-abstract of the observable world. The molecular institutes a world beyond our senses, which pushes at the limits of the human sensorium and seems to invite the kind of technological prosthesis that today we take for granted. It wasn't until 2009 that IBM captured the first image of a molecule, which corresponds remarkably well to the diagrams that have been in use since the early twentieth century.[4]

The molecule, as the basis of materiality, as rendering matter knowable and manipulable to the wills and whims of the chemical industry, is ultimately also pure information, pure capital. As the Critical Art Ensemble wrote more than a decade ago, "any form of molecular capital can now be appropriated—it is an open frontier. As with all named and controlled objects, now, genomes, enzymes, biochemical processes, etc., will all be privatized. What was once communal and controlled by traditional authority and common understanding is now usurped by separating its molecular or chemical value from its holistic phenotypic value."[5] This is, in part, what is happening in the current climate debates: the structures of commonality are broken down by the market, by private interests, by national negotiations. The air, the air that we breathe and are so vulnerable to, is rendered molecular, read, contested, and written into legislation through the knowledge of scientific expertise.

The molecule, or molecular, is also a significant figure in contemporary philosophy. Deleuze and Guattari together, and Guattari in his own writings, take the figure of the molecule as central to an anti-capitalist movement. Molecules oppose the category of the molar. "Molar subjects, objects, or form," write Deleuze and Guattari, "we know from the outside and recognize from experience, through science, or by habit."[6] Molecules, on the other hand, articulate the processes of movement—of being in-between, of the interstitial that escapes the confines of the definition of a subject or object—the movement that they call becoming. Deleuze and Guattari write that "all becomings

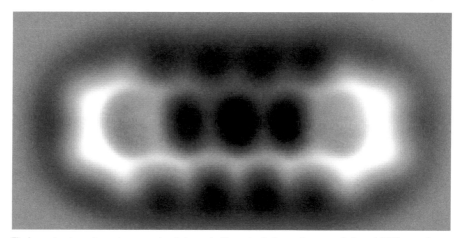

The inner structure of a pentacene molecule imaged with an atomic force microscope in August 2009, revealing its five hexagonal carbon rings and the positions of the hydrogen atoms surrounding them. Image courtesy of IBM Research–Zurich.

As a citizen of _____, "The Nation," I urge you to act now to initiate an extraordinary nomination process to inscribe Earth's Atmosphere on the UNESCO World Heritage List on an emergency basis, consistent with the aims and goals of the World Heritage Convention, and lead a coalition effort to that end,

Recognizing the outstanding universal value of Earth's Atmosphere, and responding to the formidable threats and risks to its integrity from greenhouse gases, including a forecast global temperature rise of 3 to 6 degrees Celsius by 2100,

Finding it in the common interest to protect the Atmosphere for present and future generations, and acknowledging that its preservation is the duty of the international community,

Further recognizing the impacts of climate change on sites of tangible and intangible natural and cultural heritage currently inscribed on the World Heritage List,

The Nation should undertake and faithfully carry out a coalition-led effort for inscription of Earth's Atmosphere on the World Heritage List, consistent with the aims and goals of the World Heritage Convention.

Willing governments should 1) Immediately notify the World Heritage Committee and relevant Advisory Bodies of the decision to present a nomination with the request for processing on an emergency basis, and 2) Register at *d13.publicsmog.org/initiate* to announce the nomination plan publicly via dOCUMENTA (13), the German cultural initiative (Attn: Carolyn Christov-Bakargiev/ Amy Balkin).

UNTERSCHRIFT / SIGNATURE   DATUM / DATE

Amy Balkin, *Public Smog: Earth's Atmosphere as UNESCO World Heritage Preserve*, postcard, 2012. Courtesy of the artist.

are molecular: the animal, flower, or stone one becomes are molecular collectivities, haecceities."[7] This description of the molecular offers what is radical and urgent in our engagement with ecological crisis—that there is no possibility of barricading, containing, or sealing ourselves off. We are radically open, inherently constituted by the molecular outside. We breathe in each other's air, and despite air conditioning and all the attenuating accoutrements of the wealthy, there is no way to shield against our collective molecular becoming.[8] This radical openness to the outside is both what links us to the world and what threatens us. Writer Elias Cannetti, on the occasion of Hermann Broch's

[7]
Deleuze and Guattari, *A Thousand Plateaus*, 275.

[8]
Peter Sloterdijk has written extensively on the vulnerability to the air, and this vulnerability being exploited by state terrorism in the twentieth century in *Terror from the Air* (Cambridge, MA: Semiotext(e), 2009). He writes, "The progressive explication of the atmosphere forces a sustained mindfulness of the air's breathability—above all in the physical sense, and then, more and more, in the metaphoric dimensions of respiration in cultural spaces of motivation and concern ... We begin to understand that man [sic] is not only what he eats, but what he breathes and that in which he is immersed," Sloterdijk, *Terror from the Air*, 84.

Amy Balkin, *Public Smog over Los Angeles in 2004*. Courtesy of the artist.

9

Elias Canetti, "Hermann Broch Speech for His Fiftieth Birthday," Vienna, November 1936, http://www.tanvien.net/Dich_1/Broch_by_Canetti.html.

10

Amy Balkin, "Public Smog," in *Art in the Anthropocene: Encounters Among Aesthetics, Politics, Environments and Epistemologies*, ed. Heather Davis and Etienne Turpin (London: Open Humanities Press, 2015), 344.

fiftieth birthday wrote, in relation to his friend's literature: "It is the defenselessness of breathing, which I would like to talk about in conclusion. One can hardly form too great a notion of it. To nothing is a man so open as to air… Air is the last common property. It belongs to all people collectively… And this last thing, which has belonged to all of us collectively, shall poison all of us collectively…"[9] Molecular-becoming, the carbon cycle, and breath render the body vulnerable while providing the basis of the argument for the necessity of an atmospheric commons. This is at once a political and affective project, a project constituted in and through the filling and emptying of the lungs, again and again. It is a project that ties us to our fellow creatures, as well as to the organic and inorganic.

Amy Balkin's 2004–2012 project *Public Smog* is one artistic proposition for asserting the air as a commons. In this work, she attempted "to submit Earth's atmosphere…for inscription on UNESCO's World Heritage List."[10] This would require that we treat the atmosphere with the same stringent restrictions and regulations that are accorded to public buildings, monuments, and natural preserves, which would effectively limit the amount of greenhouse gases that could be released into the air. The project also consisted of purchasing and retiring emissions in regulated emissions markets, making it impossible for polluters to purchase them. The project effectively takes the privatization of pollution and turns the air into a public resource, creating public space in the sky, or what she calls a "clean air park," while refusing to trade in our collective futures.

In each of these ways, Balkin insists upon the air as a commons and does so by intervening in the United Nation's rights and discourses as well as by manipulating the privatization of the air. The vulnerability to the air, the defenselessness of breathing, is what makes the atmosphere, *a priori*, a commons. But in light of the fact that most of the particles emitted are produced through the private market, and which are then bought and traded in a market economy, the air effectively has become the space of commerce. Balkin reasserts the necessity of the air as a commons through this project, providing a tactic that could, if taken up more broadly, be quite effective in reducing overall emissions.

Susanna Hertrich, *Jacobson's Fabulous Olfactometer (JFO)*, 2014. Courtesy of the artist.

In a different valence, Susanna Hertrich offers a tool for molecular becoming. Her proposition is less about the commons than about the possibility of futurity, and the necessity to become-with animals and techno-objects as a matter of survival. *Jacobson's Fabulous Olfactometer (JFO)*, a sensorial prosthesis, is worn on the head and triggers the "flehmen response" in the wearer when air pollution levels become too high.[11] The flehmen response, which is seen in horses, elk, goats, young elephants, felines, and other animals, is characterized by a raising of the front lip to transfer air containing pheromones and other scents to the vomeronasal or Jacobson's organ, a chemosensory organ located between the roof of the mouth and the palate. The animal uses the response to detect non-volatile organic compounds, like urine and hormones. In the human prosthesis, the movement becomes a warning device. *Jacobson's Fabulous Olfactometer* is a prosthetic becoming-animal. The boundaries of the human sensorium are extended in an adaptation adequate to the molecular threat that we have induced in our environment—capitalism molecularized as the output of fossil fuels, particulate matter, persistent organic pollutants, plastics, flame retardants, and hormones. "Do not imitate a dog," Deleuze and Guattari write, "but make your organism enter into composition with *something else* in such a way that the particles emitted from the aggregate thus composed will be canine as a function of the relation of movement and rest, or of molecular proximity, into which they enter."[12] This becoming-animal, which is figured as the future of human life, an accelerated human evolution driven by existing technology, is meant to help us cope with extreme environments.[13] Hertrich's prosthesis creates a kind of involutionary movement, a transversal space that adapts a response meant, most often, for sexual pleasure, here turned to a world of

11

I first came across Hertrich's work through the exhibition *PROSTHESES: Transhuman Life Forms* at Art Laboratory Berlin (September 26–November 29, 2015).

12

Deleuze and Guattari, *A Thousand Plateaus*, 274.

13

See Susanna Hertrich's website: http://www.susannahertrich.com/research/jfo.shtml.

threat, but that still retains the capacity for play, and excitable communication. It is an "achievement" in the sense that Isabelle Stengers means, "as a collective participation of more-than-human agencies in a changing world."[14]

14

Quoted in Carla
Hustak and
Natasha Myers,
"Involutionary
Momentum:
Affective Ecologies
and the Sciences
of Plant/Insect
Encounters,"
*Differences:
A Journal of
Feminist Cultural
Studies*, vol. 23,
no. 3 (Fall 2012):
83. This article
beautifully lays
out the playfulness
of involutionary
momentum,
contrasted with
Darwinian survival
of the fittest.

The molecule is a figure of contradiction. While it affords the privatization of life itself, it also defies that hubris. Its endless movements, its constant becomings call attention to the ways in which none of us are just us, but rather are composed of everyone else, of everything else, and in this it offers the possibility of an ethics of commonality and of the commons that resists enclosure by both national interests and private enterprise. It makes apparent the ways that we are vulnerable to each other, how we are indebted to each other, and how we are doomed, together.

The molecular offers a framework to re-attune our entangled relations with the world around us, through its perpetual movement and its disregard of the molar categories of the human, animal, leaf, soil, or atmosphere. If we understand our bodies as the temporary stability of a particular form of carbon that inevitably circulates, passing through other bodies, the earth, and the atmosphere, how might this shift our relation to climate change? How might we understand this particular moment as one not just of crisis, but as a point of connection, as a necessary call for a commonality of carbon? And how might we do this without being naïve about the deaths that the chemical revolution has left in its wake, deaths that have happened and those that are foretold?

As Juliana Spahr writes,

How connected we are with everyone.

This space that has just been inside of everyone mixing inside of everyone with nitrogen and oxygen and water vapor and argon and carbon dioxide and suspended dust spores and bacteria mixing inside of everyone with sulfur and sulfuric acid and titanium and nickel and minute silicon particles from pulverized glass and concrete.

How lovely and how doomed this connection of everyone with lungs.[15]

15

Julianna Spahr,
"Poem written
after September
11, 2001," in *This
Connection of
Everyone With
Lungs* (Berkeley:
University of
California Press,
2005), 9–10.

Heather Davis is currently a postdoctoral fellow at the Institute for the Arts and Humanities at Pennsylvania State University, where she researches the ethology of plastic and its links to petrocapitalism. She is the editor of *Art in the Anthropocene: Encounters Among Aesthetics, Politics, Environments, and Epistemologies* (Open Humanities Press, 2015) and *Desire/ Change: Contemporary Feminist Art in Canada* (MAWA & McGill-Queen's UP, forthcoming). Her writing can be found at heathermdavis.com.

"Bacteria and viruses could enter without resistance..." p. 213

CONDUCTION

From Philippe Rahm's Météorologie des sentiments *(Les Petit Matins, 2015),*
excerpted from a forthcoming English translation by Shantel Blakely.

The temperature in the room is glacial. A cloud of steam forms in front of my mouth at each exhalation. I must undress before retiring but, apprehensive about the cold that will surely seem even more vivid to my naked body, I lack the courage to make the first move. Staying at my friend's house, I have been ill for several days, feverish. Snug in bed, under the covers, I continue to be cold. We have turned the heater on the wall to "maximum," but it does nothing.

It is impossible to elevate the air temperature in the house above 16 degrees Celsius [60 degrees Fahrenheit]. Aside from the astronomic, seasonal reason, there is the inclination of the axis of rotation of the earth, which determines that for part of the year, in winter, the Northern Hemisphere will be farthest from the sun. Because of the lengthening of distance between the Sun and the Earth, the latter receives weaker rays. The freezing polar currents descend into more temperate latitudes and cool the ambient air below zero. At the same time, this inclination of the north-south earth axis is responsible for the reduced length of the day. The sun rises later and sets earlier, which shortens the warming period and, by the same token, diminishes the quantity of energy the earth receives during the day. The winter nights become so long that they deplete the earth of whatever feeble warmth it may have acquired in daytime. There is also an intermediate reason, proper to the mode of construction of the building and relative to its poor thermal insulation coefficient. In the absence of supplementary insulation, the walls of the house are exposed concrete (which perform this task poorly). The house is effectively plunged in a bath of freezing air that crosses the small dimension of the walls, bit by bit, to radiate negatively toward the interior and cool everything. The little radiator on the wall, whose electric resistance draws in the air to blow it out reheated, is powerless in the face of the enormous mass of cold that encircles our living space.

CONDUCTION

I had arrived in Japan a week before, at the height of winter. I was badly dressed but nonetheless stayed outdoors for the first few days, venturing into the cold in my cotton undershirt under a fine wool pullover. My body temperature was lowered constantly during the first day and, despite an accelerated combustion of protein in my stomach to compensate for this drop in ambient temperature, I was unable to keep my body from cooling to the threshold where bacteria and viruses could enter without resistance. I nevertheless tried to rebound in the course of the two following days. I adopted the multiple, dissociated, and circumscribed modes of heating that one finds in Japan: drinking warm tea; plunging my body in a hot bath; placing in my pockets, my shoes, against my back, small tissue paper packets containing a material that warmed up when rubbed. I had bought myself a coat but at no great expense and without great results; it was wool, but the absence of a lining and the imprecise mode of closing by three buttons on the front could not really generate an air pocket impermeable to the air and thermally isolated in which my body could be protected from exterior cold. I had also bought a black woolen hat—efficacious but insufficient to compensate for the inefficiency of the rest of my gear. For two days, I intermittently entered shops to heat up, and I took taxis and subways, perpetually seeking the warmest spatial pockets where I could hope to lay down my arms for a few moments in this physiological struggle of my body against the cold. I toured the Kyoto area, visited temples in the snow, crossed dewy forests, walked on icy paths in small shoes. I was accompanied by a friend and two young Japanese women he had met in Tokyo and invited to spend a few days at his home for the New Year. He confessed that he hoped to have an adventure with one of them. I flirted with the other one without asking myself if I really liked her.

—

After a day spent in bed following these two days of tourism, the fever has fallen slightly but I continue to feel the cold inside me; it is impossible to get rid of and I never invested in a better coat. My new Japanese friend invites me to spend a few days at her home in Tokyo. Seeing a means of escaping from this refrigerated mansion, I accept with pleasure. We take the train to Tokyo in the morning. After an afternoon in the city, we arrive at her parents' small, ordinary Tokyo house. Her mother welcomes us kindly in the main space, whose dimensions she encloses around us by pulling two sliding panels. It is cold here too. The wood-and-paper walls seem ridiculously thin and totally inefficacious against the rigors of winter. I notice that there are no radiators except for a single, small, nonfunctioning space heater against the wall.

A short time later we sit on the floor, passing our legs under a low table whose feet are surrounded by a quilt. I have the pleasure of discovering that it is warm underneath: an infrared lamp, protected by a metal grille, is attached to the table's underside and radiates agreeably on our legs. My friend brings out a second infrared lamp, on feet, which she switches on and aims at our faces. Her mother serves us food. We are now in a complex climatic condition: my legs are warm, almost too warm. The left side of my face receives the infrared rays of the other lamp and is heated, while the right side of my face and body, except for my back, remains cold because the air itself in this space is not heated at all.

The only mode of heat in the room is the pair of lamps, which, by direct infrared radiation, heat only exposed skin and clothes, without elevating the temperature of the room air. For this reason, steam rises from everything that was warm: my cup of tea, my bowl of soup, my friend's mouth, reddened by infrared; her mother's mouth, and mine, from which steam passes in front of my eyes at every breath, blurring my vision. The meal appears to me clearly to be a form of complementary heating (which it is, in fact, since it provides nutrients that are burned in the stomach and transformed into energy for maintaining our homeostatic metabolism at 37 degrees Celsius [98.6 degrees Fahrenheit])—one that is distinctly tastier than an electric radiator and releases a wider variety of aromas than a wood fire.

After dinner, I am effectively warmer. I hold in my hands a cup of steaming tea that I lift regularly to my face to drink a mouthful or to feel the radiant heat once again on my jaw. We stay a moment more in this room, then my friend takes me to her room,

CONDUCTION

which she will lend me for the night. It is colder still, but she shows me how to turn on the heated mattress of her bed and how to regulate the temperature. Here, too, the experience will be one of sleeping practically outside, in full winter, but in a bed heated from inside, and happily protected from the wind and snow by the roof and the wall cavities. It is the opposite of modern Western heating, which raises the air temperature of a room to a comfortable level at which one breathes warm air. Here, the air breathed stays icy and it is only against the body, in the body, at specific, circumscribed places, that heat is applied.

I am on my feet, hesitant to undress. It occurs to me that I must still go to the toilet, which is at the bottom of the wooden stair. I leave the room without closing the door, perhaps imagining that the heat of my cup of tea downstairs could rise to my room. At the bottom of the stairs, I open a door on the left and enter the bathroom. The window is open; I close it immediately, shivering. The toilet seat is already down. Poised to undo my trousers and lower them, I dread the moment when my rear end is in the air, where the cold will strike my newly naked skin. But I dread even more the moment when I will place my naked rear end on the toilet seat, which I can only imagine is glacial. I must hurry and do all of this quickly to not give the cold time to rise through my nerves from my skin to my brain. I unbutton my pants, lower them just to my knees, and catch my underwear and lower it onto my pants. Folding my legs, I sit on the seat. It is warm, very warm, heated by electrical resistance integrated in the plastic, and this heat is communicated throughout my entire body, by conduction, to finally reheat me entirely, fully.

Philippe Rahm is a Swiss architect whose office, Philippe Rahm Architectes, is based in Paris. His work, which aspires to extend the domain of architecture to the physiological and meteorological planes, has been widely exhibited and nominated for numerous prizes. Recent projects include the 70-hectare Taichung Gateway Park, awarded by competition in 2011 and currently under construction. His book Météorologie des sentiments *was published in 2015.*

You Are the Weather: Philippe Rahm's *A Sentimental Meteorology*

SHANTEL BLAKELY

1

Philippe Rahm, "Constructed Atmospheres," lecture at Harvard GSD (November 1, 2013), https://www.youtube.com/watch?v=NP6EBTwGcug.

Architects have historically used diagrams representing human occupants of buildings to communicate their approaches to traditional concerns of architecture, such as the proportions of bodies (Leonardo's *Vitruvian Man*, Le Corbusier's *Modulor*) or the space of perception (Herbert Bayer's diagrams of the field of vision). Speaking recently about his work and declaring his preoccupation with body heat, architect Philippe Rahm showed drawings populated with his own representation of the body, a silhouette with a blurry red heart.[1] Rahm has pursued this interest through experimental installations like his "Hormonorium" at Venice in 2002 and in other projects for calibrated air. He has expanded his inquiry to other aspects of metabolism, such as light fixtures whose elements can be adjusted for human or animal vision. But the body in air—his most central and lasting obsession—is the premise of his new book, *Météorologie des sentiments*, a collection of short stories that was recently published in French by Les Petit Matins (an English translation is in the works).

The stories in *Météorologie des sentiments* (hereafter *A Sentimental Meteorology*) are mostly written from a first-person perspective. Rahm makes an impressive show of knowledge and dedication to scientific fact, and the tone of the book is largely set by his detailed descriptions of geographic locales; the physics of cold buildings; or the body's responses to cold, sun, or altitude. But the points of departure for these stories, and for their conflicts and resolutions, are almost always emotions—running the gamut from desire or well-being to dread and fear. The tone of the book hovers somewhere between a memoir and a forensic investigation, as Rahm mines his own experiences to explore different aspects of human experience in general, in a range of environmental and physiological situations.

Perhaps more important from an architectural point of view is that in these stories the mind and body—and the occupied landscape—tend to eclipse architecture as the principal human dwelling place. Architecture may be a collective social response to climate, but physical enclosures are not always sources of shelter and security.

While architecture is nowhere explicitly denounced, the book can be read as a critique of architecture and especially of the Western thermal conventions that are often called "modern." But social relationships are also at the forefront of experience in these stories, which loosely follow a sort of formula. Each vignette finds the protagonist in a situation that offers an unusual vantage point from which habitual experiences are no longer automatic or familiar. From that position we are able to see how, in these situations, personal and social apprehensions of environment can become separated, superimposed, or inadvertently set in conflict with one another. Yet the short titles of these stories imply that each one is a neat description of a property, process, or quality of climate experience, as though the book were a user's manual.

   The protagonist in "Anaerobic," for example, has drifted happily into the sea clutching an inflatable canoe, oblivious to what is about to happen. With the implacability of clockwork the story proceeds to the situation's inevitable crisis.

> If the water is gaseous in the air above me, could it be solid below me? There would then be ice in the depths on which, should I let myself sink in the water, I could put my feet, rest, recover my breath. I sink now, the horizon rises above my field of vision, the blue sky darkens, the shore disappears as I close my eyes to avoid the salt.

The protagonist's innocence about the environment and his position on the verge of discovery are powerfully evoked in "Anaerobic," which revives a literary trope at least as old as *Grimm's Fairy Tales*, which often start with a child heading alone into a hostile wilderness. More ordinary situations, in adult life, ground several of the other stories. There is a backpacker who chooses to spend a night outside ("Cooling"); a soldier participating in a training exercise, who falls asleep and loses track of time ("Obliqueness"); a man on a terrace reading a newspaper in the late afternoon ("Obfuscation"); a motorist driving over the Alps from France to Italy ("Altitude"). Each time, the protagonist has a problem to solve. Will he endure his present condition as a sort of prison? Instead, through informal investigation and using his imagination, he comes to terms with his plight and takes action.
   In many of the situations in the book, the novel climatic condition experienced by the protagonist results from his placement in a liminal social position. But for all Rahm's attention to the individual as distinct from the collective, he handles this distinction with a light touch. This is partly due to his excurses into meteorology and physiology that digress from the protagonist's personal viewpoint. As the reader follows the plot of a story, the narrative may turn from relating a sequence of meaningful events to describing the timing of optical, vascular, cellular systems—or astronomical events, wavelengths, rods and cones, oxygen, degrees of temperature, lux of light, etc. Just as the climax approaches, time slows down. The story then proceeds at a curious, scaleless pace in which the astronomical impinges upon human physiology, and ultimately on thoughts and feelings. The boundary between the subject and environment is blurred.
   Notwithstanding his extended meditations on scientific facts, the narrator's tone is often wistful or warm-heartedly affectionate. Some of the stories involve solitude, but many address a protagonist among friends, classmates, or coworkers. And while all the situations involve a degree of separation from a social group, some are as joyous as others are dark—for instance, in "Greenhouse Effect," in which two lovers find themselves in an ancient garden pavilion whose sequestered indoor air leads them to sense the presence of ghosts; or in "Diffusion," in which a father and son contentedly walk a few blocks to a frozen yogurt shop.
   Even when describing a solitary protagonist, Rahm makes liberal use of the French pronoun *on*, which translates to "we" or to the third-person "one," depending on context. The effect is to conflate the first person with "anyone," which softens the sense of social conflict while the meteorological and physiological descriptions underscore the generality of the protagonist's experience. This ambiguity is especially effective in one of the most delightful moments in the book, a scene in "Radiation." After some time

spent on a scintillating suburban hillside in the hottest hour of a midsummer day, two adolescents waiting for the protagonist's uncle, next to his hot car, give up on trying to cool off.

> At some point, one of us arrives at the conclusion that there is no point in trying to be less hot because we have no means to cool off. In the end our freedom, our liberty to act is the power to choose to be even warmer. A search of the trunk yields a sort of coverlet. We get in the car. The interior air is boiling due to the greenhouse effect that bends the sun's rays to the steel and glass interior, preventing them from escaping the car after having let them in by the glazing and transforming them into heat against the fabric seats and plastic surfaces. The doors are closed, the last open windows put up, the ignition turned on, the heat turned up to maximum. We put on our sweaters, which we had left on the back seat, and bundle ourselves under the coverlet.

In conversation about the book's literary style, Rahm is liable to mention his interest in objectivity and his fondness for the *Nouveau Roman*, particularly the work of Alain Robbe-Grillet. The inspiration is evident if one looks at a story like "Obfuscation," which describes a moment on the terrace of a villa near Lake Neuchâtel. In the opening of his 1957 novel *Jealousy*, which is also set on the terrace of a villa, Robbe-Grillet lays out an extensive, detailed description of the scene, emphasizing the physical setting. Sentences telescope clause after clause in a detailed accounting of geometry, architecture, and furniture arrangement. The reader might discover, deep in one of these sentences, a human motive that has governed a whole string of preceding actions taken by Robbe-Grillet's characters.

Rahm lingers, as Robbe-Grillet does, on the relationships between a terrace and a building, the dimensions of an inhabited space, but the experience of reading Rahm's stories is much more empathetic. We always understand that the point of view belongs to someone, to the narrator.

> The sun lowers, moving in the sky until it disappears behind the vertical line formed by the principal façade. This line suddenly throws a shadow that darkens the terrace's gravel surface except for a very small triangle at the southwest corner. Seated in the sun, on a bench against the house, I read the paper. The paper's white background, striated with characteristic black lines, darkens all of a sudden, attenuating the contrast between the paper and letters and makes reading more difficult, little by little; but by the time I notice it, the sun might have slipped past the southwest corner ten minutes before. I am conscious of being cold, as reading becomes more difficult. Finally there is no more sun

Moreover, mention of weather conditions, however objective their description purports to be, tends to evoke associations of well-being or foreboding. The appearance of the sun or descent into darkness often augurs a shift in emotion for someone in the story. Weather conditions are used with forceful cumulative effect in many paragraphs of another work associated with the *Nouveau Roman*, Albert Camus's *The Stranger*. The narrator feels the sun as a source of encouragement and comfort, or as a volumetric presence—as in "a strip of sand between the sunlight and the sea"—but near the novel's end, the doomed narrator refers to "a dark wind blowing from my future." Arguably his responsiveness to sun, sand, and water are key to the story's capacity to elicit sympathy for its alienated "I" and are therefore crucial to Camus's ability to evoke the inner life of a protagonist who commits a morally unconscionable act. With a similar sense of indexicality, Rahm's weather manifestations are often fringed with affect, however objectively they may be described.

"Meteorology of sentiments"—these words in Rahm's title echo how his stories reflect on the "weather" of emotions, and how closely they are interleaved with shifting, or more or less ideal situations of physical climate. The last story in the collection, "Acclimatization," presents a man at ease, in shirtsleeves, greeting his fiancée. His life has arrived at climatic and psychological resolution at the same time. But while the ostensible dramatic arc of the book is a quest for acclimatization, there is also a progression out of innocence of climate and physiology—or passive subjection to it. The source of free will, free choice, is not so much climate control as self-awareness.

This brings us back to the ambiguity of this book in relation to architecture. The body of each story's narrator may be indoors or outdoors, but the narrative's overarching message is that, in all cases, we "live" the moment through the same basic physiological system. In "Conduction," the protagonist has an unshakeable chill and moves between homes in a country where people warm themselves by unfamiliar means. Normally rendered invisible by habit, these details are painfully evident to a stranger in a strange land. The situation provides the pretext for a welcome review of how architecture is organized to promote insulation, and why the occupants of a building may nonetheless feel cold. But the buildings in this story are inadequate to keep the inhabitants warm. The resolution comes from patience, experimentation, and chance discovery. In contrast to a treatise on a static ideal, *A Sentimental Meteorology* is an exploration of the body in air that aspires to rigorously confront the physiology and emotion of climate, including their contingency.

Shantel Blakely curates and produces the public lectures and conferences program at Harvard Graduate School of Design. Her recent essays in *PLOT*, *Log*, and *AA Files* have probed architecture's relationships to ideologies of experiential practice that consider the human being from distinctly different points of view. She holds an MArch from Princeton and a PhD in the history and theory of architecture from Columbia GSAPP.

Unseasonal Fashion:
A Manifesto

DEHLIA HANNAH AND CYNTHIA SELIN

Warriors of a Downpour City, Anne van Galen, 2014. Courtesy of the designer.

"Imagine we would live in a world with endless rainfall…" muses Anne van Galen, the designer of *Warriors of Downpour City*. The conceptual apparel collection, van Galen's graduate project at Design Academy Eindhoven, responds to a fictional scenario in which postures and textiles have been adapted for rain and "fashion becomes naked, transparent, and layered with thin diluted colors."[1] The molded plastic and silicone pieces serve at once to protect the body and accommodate its continuous movement through relentless fog and damp. Visual disclosure of the body through these gently protective garments, meanwhile, suggests an embrace of exposure to the elements. This project is an exemplary expression of a trend that can be traced from architecture to fashion design in which corporeal exposure to climate is mediated by more and more individualized and adaptable protective enclosures—a trend that should be seen as a reflection of the increasingly unstable condition of the climate itself.

At a historical moment in which the best available scientific predictions portend an almost unimaginably changed—and changing—future, explorations of hypothetical climates prepare us to register these changes perceptually and cognitively. Precisely because they invite us to inhabit fictional worlds, these kinds of projects interrupt our tacit habituation to an increasingly pervasive state of emergency and the unfolding of distant and slow

This essay is supported in part by the US National Science Foundation cooperative agreement #0937591. Any findings, conclusions, or opinions are those of the authors and do not necessarily represent NSF. The authors are also indebted to Jacob Lillemose, curator of *X and Beyond* and postdoctoral fellow at the Copenhagen Centre for Disaster Research, for extensive discussion of the role of fashion in the imagination of disaster.

1

"Anne van Galen Designs Accessories for a 'World with Endless Rainfall,'" *Dezeen*, November 2, 2014, http://www.dezeen.com/2014/11/02/anne-van-galen-warriors-of-downpour-city-silicone-accessories-endless-rainfall-dutch-design-week-2014.

disasters. Like the scenario exercises and disaster rehearsals conducted in the medical, energy, and security industries, aesthetic exercises in reimagining the conditions of everyday life prepare us to chart different paths into possible futures as well as reorganize our sense of the present.

This essay considers the role of speculative fashion design within the context of another such project of anticipatory aesthetics—a collective thought experiment called *A Year Without a Winter*. The project, organized by the authors of this essay, is being staged over the three-year period of 2015 to 2018 through research projects, events, and exhibitions at multiple locations around the world. At once a "futuring" exercise and a creative, historical reenactment of another global climate crisis, *A Year Without a Winter* deploys historical and literary narratives to reframe contemporary imaginaries of climate change. It brings together a broad coalition of artists, scientists, humanists, and policymakers to consider how humans and other species become acclimated to their environments and how, in turn, they alter built landscapes, cultural habits and artifacts, and forms of social organization in order to survive under unfamiliar or inhospitable conditions. Climate must be understood as a lived abstraction, an abstraction that discloses itself through new aesthetic logics and their critical analysis. In tracing a progression from architecture to apparel as sites of climatic mediation, it becomes clear that from the conceptual to commercial sectors, contemporary fashion is a sensitive indicator and rich site for the critical exposition of our increasingly turbulent seasons.

A YEAR WITHOUT A WINTER

April 10, 2015, marks the two hundredth anniversary of the eruption of Mount Tambora on the Indonesian island of Sumbawa. This event set into motion a cascade of environmental, political, and cultural responses whose consequences are still felt today.[2] The largest volcanic eruption in the last ten thousand years, Tambora caused vast destruction locally. It was forceful enough to inject ash and super-heated gases into the upper atmosphere, where stratospheric winds circulated them around the globe. By the following year sulfur dioxide and particulate matter blocked out sunlight and disturbed weather patterns worldwide. Much of the northern hemisphere was plunged into cold and darkness even as the Arctic warmed and tempted explorers north, while both intensified monsoons and drought afflicted the Indian subcontinent and East Asia. The "Year without a Summer," as 1816 was remembered, was the beginning of a three-year climate crisis in which famines and epidemics led to political upheaval, migration, and reforms. It was also a year of profound literary and scientific inspiration. Efforts to understand the causes of this crisis later inspired new approaches to climate management—including today's perhaps Promethean ambitions to cool our rapidly warming planet by intentionally putting sulfur dioxide into the atmosphere, alongside other forms of solar radiation management meant to act as so many planetary parasols.[3]

The cultural legacy of this episode may be traced back to June of 1816, when a group of English literati vacationing on the banks of Lake Geneva found themselves cooped up by the incommodious weather. Lord Byron, John Polidori, Claire Clairmont, Mary Godwin, and her eventual husband Percy Shelley passed the time reading a collection of frightful German ghost stories

2
Gillen D'Arcy Wood, *Tambora: The Eruption that Changed the World* (Princeton, NJ: Princeton University Press, 2014).

3
Jack Stilgoe, Matthew Watson, and Kirsty Kuo, "Public Engagement with Biotechnologies Offers Lessons for the Governance of Geoengineering Research and Beyond," *PLoS Biol*, vol. 11, no. 11 (November 2013).

and, eventually, set for themselves a friendly competition to write the best horror story. "The Dare," as they called it, spawned two great "monsters of modernity:" Mary Shelley's *Frankenstein: Or, the Modern Prometheus* (1818) and John Polidori's *The Vampyre* (1819), an inspiration for Bram Stoker's *Dracula* (1897).[4] According to her own account, Shelley was inspired by the backdrop of dramatic thunderstorms, cold, and darkness and the general atmosphere of terror that these seasonal abnormalities occasioned. The novel, published in 1818 just as the cooling episode was beginning to subside, registered in compelling narrative form a cultural response to central scientific, technological, and political conditions of the day. Its power to shape our encounters with emerging science, technology, and environmental issues remains irresistible.

Two centuries ago, the world endured a year without a summer. We now confront the fearsome prospect of *A Year Without a Winter*—a future in which the elite luxury of escaping to sunny beaches for the holidays is rapidly being transfigured into a nightmare of global seasonal arrhythmia, bleached coral reefs, and sinking paradises. Many scholars, activists, and policymakers claim that climate change has failed to capture contemporary cultural imaginaries in a way that is sufficient to motivate adequate political, economic, and technological responses. Rob Nixon, for example, insists that there is an "urgent imaginative challenge currently facing both the humanities and the sciences, namely how writers and visual artists can embody environmental disasters in literary narratives and images, thereby making imaginatively perceptible and tangible to a broader public what scientists are establishing."[5] New visions and narratives are needed in order to harness the power of our scientific knowledge and intervene constructively in the course of the future.

What such stories and visions might be told or performed today, to capture the public imagination in the context of the all-too-plausible fiction of *A Year Without a Winter*? If the novel was a distinctive aesthetic form of the nineteenth century, what mediums and genres can best encapsulate an awareness of emerging climate crises? An answer just might be found in an outfit.

CLIMATE AS A LIVED ABSTRACTION

In a political atmosphere of carefully policed climate literacy, we are often reminded that no single event—storm, flood, glacier calving—can be directly attributed to climate change. This is not due to any failure of climate science but rather to the nature of climate itself, as indicated by the old adage "climate is what you expect; weather is what you get." Another suggestive anecdote offered is by climate scientists: "If I look at what you're wearing today, I would learn something about the weather. If I were to look inside your closet, I would learn something about the climate you live in."[6] The closet is an archive of one's climatic expectations and aspirations. What might one learn about weather and climate by discovering Jacqueline Bradley's *Boat Dress* in someone's wardrobe? The inflatable dress, part of an installation titled "The Outdoors Type," includes a variety of playfully functional picnic accoutrements. The resourceful feminine archetype evoked here is well prepared to explore and cope gracefully with her environs. She is also prepared for them to change—perhaps quite suddenly.

In order to understand why apparel offers an evocative aesthetic modality through which to comprehend today's environmental crisis, we must consider

4

Kim Hammond, "Monsters of Modernity: Frankenstein and Modern Environmentalism," *Cultural Geographies*, vol. 11, no. 2 (April 2004): 181–98.

5

This description is drawn from an abstract of a talk by Rob Nixon titled "The Anthropocene: Slow Violence and Environmental Time," http://www.princeton.edu/pei/events_archive/viewevent.xml?id=599.

6

As related in private conversations with climate scientists.

Boat Dress, Jacqueline Bradley, part of the installation "The Outdoors Type," 2011. Courtesy of the artist.

more precisely what climate means in contemporary discourse. Climate, along with atmosphere, air, and a constellation of related meteorological concepts, often refers to an ambient background condition perpetually at the edge of perceptual awareness. Like tools that we rely on without noticing their particular properties, climate calls attention to itself phenomenologically only when it misbehaves or manifests itself in particularly dramatic ways—usually in the form of bad weather. If "there will be weather today," then you will probably need to dress for it. As the object of atmospheric science, "climate" denotes the average weather of a particular locale, from a room to the whole world, over a period of time ranging from days to seasons, human lifetimes to millennia. More precisely, climate is a statistical description of the mean and range of variation of temperature, precipitation, wind, barometric pressures, and other factors in a defined spatial and temporal domain—thirty years, by the standards of the World Meteorological Organization. The concept of *global* climate—a notion that emerged only in the last century—must be understood as a mathematical and conceptual abstraction.[7]

A historical answer to the question *what is climate?* offers a very different picture, however, one that places geography and human interests in the foreground. This perspective recalls our attention to the lived experience of climate and the material artifacts we build to control that experience. In their study of the ancient concept of *klima*, James Fleming and Vladimir Jankovic argue that climate has predominantly been conceived of as an *agency* affecting human bodies and affairs. Climate refers to "all the changes in the atmosphere which sensibly affect our organs" and influence "the feelings and mental conditions of men," in the words of Alexander von Humboldt.[8] While atmospheric science treats climate as a set of conditions to be indexed, averaged, and modeled, the language of agency pervades discourses of climate change and its effects on economies, political conflict, food and water security, and, not least importantly, the weather. Considered as an abstraction over space and time, climate lies outside the domain of direct sensory experience. What does it mean, then, to be subject to its agencies, to those temporary states of the atmosphere that the climate models tell us are becoming increasingly turbulent and extreme? How do we live in and with the abstraction of climate, and how do we express our comprehension through quotidian practices, objects, attitudes, and desires?

THE EXPLICATION OF CLIMATE

Attending to climate—to long-term trends and tendencies in how the atmosphere behaves and changes over time—may be done by considering how we modulate its effects on our habitus. Architecture is one obvious site to consider, for example, in its many forms of climate control—heating, cooling, insulation, air circulation, etc. Peter Sloterdijk, for instance, has given us ample conceptual apparatus for conceiving architecture in atmospheric terms. We live, he argues, within so many bubbles within bubbles, which contain and regulate not only airflow but the forms of life that thrive therein. As he shows in *Terror from the Air*, the twentieth century witnessed a proliferation of social and mechanical technologies for manipulating the atmosphere, from air conditioning to flight to chemical warfare. These technologies interrupted our tacit reliance on air for the basic most purposes of breathing, even as its many new affordances were celebrated in the domains of transportation and

7

For further discussion of the emergence of the global concept of climate see Paul N. Edwards, *A Vast Machine: Computer Models, Climate Data, and the Politics of Global Warming* (Cambridge, MA: MIT Press, 2013).

8

James R. Fleming and Vladimir Jankovic, "Revisiting *Klima*," *OSIRIS* 26 (2011): 1–16.

communication. Sloterdijk highlights how moments of atmospheric break-down bring the ambient conditions of life more explicitly into the foreground, a process he calls the *explication* of the environment. Such practices of awareness lie at the basis of any possible experience of climate change.

If we scale this analysis down and individualize it, we can trace a similar mediation of atmosphere through the convergence of architecture with cloth-ing. Sloterdijk recounts a particularly ironic story of Salvador Dalí in which the artist very nearly dies of suffocation when he gets up to deliver a lecture on surrealism and the unconscious dressed in a diver's helmet. Though the hel-met was designed to enable breathing underwater, Dalí neglected to connect the helmet to an air supply, and his wild gesticulations and distraught facial expressions were construed by the audience as an ingenious performance. For Sloterdijk, "The artist's choosing to wear a deep-sea diver suit with an artificial air supply for his performance as the 'Ambassador from the depths' unerringly ties him to the unfolding of atmospherical consciousness, which, as I have attempted to show, is central to the self-explication of culture in the twentieth century."[9] Though understated by Sloterdijk, this story yields a fur-ther insight by pointing our attention to the role of clothing in the explication of a more specifically climatic consciousness.

In an atmosphere of looming environmental crisis, it might seem callous or trite to focus on fashion as such a site of climatic imaginaries. Yet to do so is not to trivialize the matter. On the contrary, whereas the prospect of catastrophic climate change remains on the verge of the unthinkable, clothing remains a domain of daily acknowledgment of and adjustment to our climatic horizons. Like architecture, fashion constitutes a mediating point between the imagination of the local and embodied habituation. We change our clothes more frequently than we change our homes and built environs, however. To continue along Sloterdijk's line of thought, our homes, offices, and cities can be construed as so many spheres—bubbles within and adjacent to other bub-bles. In microcosm, the clothing that we outfit ourselves in may accordingly be thought of as more intimate bubbles blown daily from the options within our wardrobes.

A pair of projects by Archigram—the 1964 *Cushicle* and the 1967 *Suita-loon*—offer a key conceptual transition point in this genealogy of speculative fashion design as a site to explicate climate and thus richly evoke climate adaptation. Inspired by the accoutrements of space travel, these prototypes of "clothing for living in" were designed as inflatable mobile homes, offices, or leisure spaces for one or more occupants. They were also exercises in minimalism, designed to deliver the most basic requirements of shelter and comfort. A point of contrast that marks the boundary between architecture and apparel, these examples of wearable architecture fully enclose their users. In a very minimal sense, one may be considered to be "indoors" when ensconced in such a space and, as such, in a space where climate can be regulated.

Designed well in advance of widespread concerns about climate change, the *Suitaloon* nonetheless gestures toward a question that will be reiterated implicitly in later such garments: how does apparel reflect our needs, desires, and expectations of the climates we inhabit? In the work of Archigram, these needs and desires retain an element of humorous utopianism: "If it wasn't for my Suitaloon I would have to buy a house," wrote designer Michael Webb.[10]

Lucy Orta's 1994 collection *Refuge Wear* similarly includes a range of wearable shelters, yet these are designed for conditions of social and envi-ronmental precarity, for homelessness. Hooded tents and body suits enable

9
Peter Sloterdijk, *Terror from the Air* (Los Angeles: Semiotext(e), 2009), 77.

10
"Cushicle and Suitaloon," http://architecturewithoutarchitecture.blogspot.com/p/cushicle-and-suitloon-were-conceptual.html.

Inflatable Suit-Home, David Greene, 1968; suit made by Pat Haines based on the Suitaloon project by Michael Webb. Photograph by Dennis Crompton, © Archigram, courtesy of the Shelley Power Literary Agency.

sleeping on sidewalks in conditions of rain or cold, in hyper-mobile form. Between individualized refugee tents and protective clothing, these garments do not purport to substitute for a comfortable living environment. Rather, they offer the wearer minimal protection in the context of an inhospitable environment—an environment in which political and economic conditions expose bodies to the ravages of an unstable climate and vulnerability to risk is unevenly borne. Orta's *Refuge Wear*, like the *Boat Dress* and the gear worn by van Galen's *Warriors of Downpour City*, reflects a condition of immersion in and direct exposure to hypothetical climates. And like Dali's deep-sea diving suit, they are legible as material expressions of the long-term trends, cyclical behaviors, or unpredictability of the climates that they enable their wearers to endure—and enjoy.

UNSEASONAL FASHION

As fast-moving consumer goods with a profit structure based on high turn-around, clothing is a highly sensitive indicator—and a driver—of how hopes and fears are invested in consumer cultural imaginaries of climate and environmental disaster. A recent study in fashion marketing notes that "fashion retailers and brands can often be caught out by unseasonable weather. The problems seem to be most acute in the autumn as later warm weather means that consumers are reluctant to begin purchasing heavier clothing."[11] The realization of the fictional scenario of *A Year Without a Winter* could occasion major economic losses for certain sectors of the fashion industry and yet be a boon to others. An exemplary site of aspirational consumption, the

11

Tony Hines and Margaret Bruce, *Fashion Marketing: Contemporary Issues* (London: Routledge, 2007), 172.

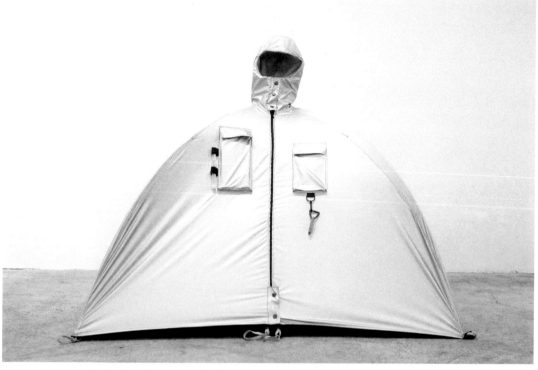

Refuge Wear, Lucy Orta, 1994. Courtesy of the artist, © 2016 Artists Rights Society, New York / ADAGP, Paris.

suggestively named cruise collection appears in late winter with light fabrics and luscious prints to stoke desire for warmer weather and holidays, offering vicarious pleasure even to the non-cruising set. Fashion seasons, industry analysts recognize, will be affected by climate change in form and content, for "although this may be a merchandising issue, it is likely to have an impact on product design and ranging for autumn collections in the future."[12]

The effects of seasonal disruption on fashion design can already be observed on the runways, in art fashion, and even in the streets. A notable "look" seen in the Summer 2010 show of the Canadian designers Dean and Dan Caten, aka Dsquared2, is a veritable montage of survivalist tropes. The model sports an outdoorsy fisherman's hat and rain boots, connoting leisurely self-sufficiency in the wilderness, and camouflage shorts, crossing military and hunting motifs. In his stylish gear pack he carries an umbrella, prepared in case of inclement weather. The whole ensemble is immaculately topped off with a black tie and tuxedo jacket that cannot but imply the survival of the 1 percent—a fine outfit for work and play in the age of disaster capitalism.

Gas masks and other accessories for apocalypse are already familiar on the runways through the work of Alexander McQueen and Rei Kawakubo, for example. High fashion serves at once as a site of critique and a space in which disaster chic normalizes and articulates new possibilities for pleasure within a changing world. A sociological perspective on dress suggests that "body supplements," which may include any article of clothing, "act as alterants of body processes as they serve simultaneously as microphysical environment and as interface between body and the macrophysical environment."[13] Particularly notable in these high-fashion examples is that the means of mediating between body and climate are often detachable, suggesting both that adaptation may take place quickly and that the wearer enjoys flexibility in the environments

12
Hines and Bruce, *Fashion Marketing*, 172.

13
Mary Ellen Roach-Higgins and Joanne B. Eicher, "Dress and Identity," *Clothing and Textiles Research Journal*, vol. 10, no. 4 (June 1992): 4.

Dsquared runway show, summer 2010. Courtesy of firstview.

that she frequents. Umbrellas are detachable appendages, liable to being forgotten, lost, or stolen, even flittering off of their own accord, as though they live lives of their own. This makes the umbrella an unreliable mediator between the body and its social and meteorological milieu.[14] In contrast, protective weather gear that more securely affixes to the body comes closer to architectural stability, on the one hand, and to a second skin, on the other. On both counts, the reliability of a piece of apparel as an environmental mediator is a measure of the relative uncertainty of the wearer and the climate itself.

14

Maria Damkjær, "Awkward Appendages: Comic Umbrellas in Nineteenth-Century Print Culture" (article under review; cited with permission of the author).

Fashion always represents a negotiation of taste and necessity, where the degree to which one's dress reflects taste is a matter of privilege, opportunity, and proclivity for self-fashioning. From the rise of camouflage in the 1960s to the recent proliferation of prints featuring disturbed landscapes and a popular fetish for high-tech outdoor gear, fashion history offers an archive of rapidly shifting attention to social and environmental milieus. The act of putting on clothes each day represents a fine-grain mode of responsiveness to shifting standards of thermal comfort, a register of planned passage through indoor and outdoor climates, and a reflection of our horizons of expectation about what kind of weather the day and season may bring.

Recent fashion trends resonate with a cultural unconscious of a looming environmental disaster. Clothing, like buildings, reflects an understanding of the weather and climate of their users. Because they can be changed much more frequently, wearable interfaces with the environment reflect both the conditions common in that environment and how predictably and frequently change occurs therein. Fashion is uniquely positioned to function as a cultural register of climate change precisely because it captures and is governed by *trends*. Comprehending climate change means recognizing long-term trends in and through short-term variability and cyclical repetition, patterns of change that are part and parcel of fashion design. Moreover, from production to distribution, fashion is *global*.

Dehlia Hannah is a curator and a philosopher of science and aesthetic theory. She is an assistant research professor jointly appointed in the School of Arts, Media, and Engineering and the School for the Future of Innovation in Society at Arizona State University and a guest researcher at Copenhagen University. Her current book project, titled *Performative Experiments*, explores the philosophical implications of artworks that take the form of scientific experiments. She has written and curated art exhibitions about climate change, the Anthropocene, and new media. With Cynthia Selin, she leads *A Year Without a Winter* (2015–2018).

Cynthia Selin is an interdisciplinary scholar focused on the future. She is an assistant professor in the School for the Future of Innovation in Society and the School of Sustainability at Arizona State University and an associate fellow at the University of Oxford's Saïd Business School. She investigates and invents methodologies for clarifying uncertainty and explores more theoretical questions about anticipation in society.

Air Conditioning: Taming the Climate as a Dream of Civilization

EVA HORN

I had no idea the tropics were *that* hot. When I first exited the sliding glass doors at Changi airport, Singapore, I stepped into something I had never quite experienced before. This was not air as I knew it—it was more like a semi-liquid medium, a gel, moist, hot, suffocating. Something you can inhale, but that envelops your body like a mass. Impatiently dragging my luggage through that mass towards the taxi, I started panting, my head and my hands swollen from the heat. The driver opened the door, and I tumbled into the refrigerated inside of the car. For a moment I felt relief as my body escaped the suffocating atmosphere. But I quickly found the stream of cold air on my wet skin to be equally unpleasant. Still damp from the heat, I started to shiver.

The startling chill of the air I found in the car would turn out to be the nature of most of my coming weeks in Singapore. Using the network of subways and subterranean shopping malls, a person can cross the entire city almost completely while avoiding the outside climate. Singaporean temperatures stay around 87 degrees Fahrenheit all year round, with a constant humidity between 80 and 100 percent. If you ever spend time outside, in the streets or lush gardens of the city, a Northern European like me quickly learns what every inhabitant knows: Always seek shade. Don't eat too much during the day. Avoid being outside in the hours around noon. And most importantly, *move slowly*. Most of the time, however, is spent inside, unless you are one of those heat-craving Europeans who have the urge for drinks on rooftops and terraces on balmy Singaporean nights. You will rarely see Asians join the party. Most of the time, whether for work or leisure, is spent in the dry, artificially cooled air of climatized spaces, always rather too cold than too hot. There you can rush and work as hectically as you would in Berlin, London, or New York.

The ascent of a city with such an extreme climate to become one of the most important economic hubs in Asia could not be imagined without a technology that has existed for less than a hundred years. The Singaporean journalist Cherian George called his native city the *Air-Conditioned Nation*.[1] The same holds true for other hot metropoles such as Dubai, Shanghai, Bangkok, and Hong Kong, as well as for the American Sun Belt that in the past decades has rapidly attracted retirees from all over the United States. Air conditioning was not initially developed to enhance human comfort, however, but to facilitate technical procedures in the printing and meat processing industry. At the beginning of the twentieth century, in 1902, Willis Carrier installed the first "apparatus for treating air" in a printing factory to combat humidity; his invention was patented in 1906. Only in the 1920s did air conditioning technology start to be installed in US movie theatres and department stores that suffered from declining customer visits during the sweltering

1
Cherian George, *Singapore: The Air-Conditioned Nation, Essays on the Politics of Comfort and Control, 1990–2000* (Singapore: Landmark Books, 2000).

Proposed conditioned glass biodome for Singapore's Changi Airport, Safdie Architects, currently under construction. Courtesy of Safdie Architects.

summer heat. In the 1950s, air conditioning systems entered private house-holds and cars, and started to spread from the US to the entire world. In Singapore, more than 50 percent of all electricity is consumed by cooling systems; in the US, fewer than 5 percent of newly built houses lack a central air conditioning unit. India and China are quickly expanding markets for cli-mate technology, and even in Europe, where A/C is not yet a default element in private houses, there are virtually no office buildings that do not have it. Stan Cox, who wrote an excellent history of the rise of climate technology in the US, estimates its annual energy consumption for air conditioning in the country at 1,650 kilowatt-hours per person, producing half a billion tons of carbon dioxide.[2] Dutch researchers estimate that, while heating energy consumption will decrease, energy demand for air conditioning will rise by 72 percent during this century.[3]

Air conditioning—the possibility of "fixing" the air's temperature and humidity at one's own comfort level—is one of the oldest dreams of man-kind. It means creating a world without heat or cold, rain or snow, without suffocating humidity or dusty winds. Climate control allows for a life with-out weather, without meteorological contingencies and surprises, extreme weather events, seasonal changes, or locally challenging climate conditions. Air conditioning creates what has long been lauded as a "temperate climate," a climate adjusted to the comfort zone of the human body—a comfort zone that, today, seems to get narrower and narrower. The temperate air that it produces is to be "just right," as in the Goldilocks tale—neither too hot nor too cold, neither too wet nor too dry. Of course, the creation of artifi-cial climate, historically, has not always meant creating "coolth" but rather providing warmth. Creating or finding a temperate climate in the history of mankind has often meant looking for or building a protected space, cut off from the vagaries of the weather outside: find a cave, build a house, wear pro-tective layers to cover the body and preserve its temperature. The philosopher Peter Sloterdijk has called this basic civilizational act "insulation"—a human

2

Stan Cox,
*Losing Our Cool:
Uncomfortable
Truths About Our
Air-Conditioned
World (and Finding
New Ways to Get
Through the Sum-
mer)* (New York:
The New Press,
2012), 133.

3

Morna Isaac and
Detlef van Vuuren,
"Modeling Global
Residential Sector
Energy Demand
for Heating and
Air Conditioning
in the Context of
Climate Change,"
in *PBL: Nether-
lands Environment
Assessment Agency*,
February 2, 2009,
http://www.pbl.nl/
en/publications/
2009/Modeling_
global_residential_
sector_energy_
demand_for_
heating_and_air_
conditioning.

way of being-in-the-world by creating protective "spheres" and thereby isolating human bodies and social environments from the world at large, the natural environment.[4] Clothing creates a warm microclimate around our skin; houses (at least if they are made of stone) are built to preserve heat in the winter and coolth in the summer. In tropical or torrid climates, for centuries, houses have been built in a fashion as to shelter the inhabitants from sun and hot air (in the case of dry and hot zones such as North Africa) or, on the contrary, to enable a constant circulation of air to cool off sweaty bodies (as in the humid parts of the tropics).

Humankind has not limited itself to the creation of such microclimates. Ever since the Neolithic Revolution, people started to transform landscapes by clearing forests or draining swamps in order to create arable land or pastures—thereby also altering local climates. As early as 1784, the German philosopher Johann Gottfried Herder saw man as a climate-altering species. For him, cultural history is the history of humans changing the climate around them:

> As climate is a compound of powers and influences to which both plants and animals contribute, and which every thing that has breath forms as an all-encompassing system, there is no question that man is placed in it as a sovereign of the Earth, to alter it by art. Ever since he stole fire from Heaven, and rendered steel obedient so his hand … ever since he has made not only the beasts but also his fellow men subservient to his will … *he has contributed to the alteration of climate in various ways*. Once Europe was a dank forest; and other regions, at present well cultivated, were the same. They are now exposed to the rays of the Sun; *and the inhabitants themselves have changed with the climate* … We may consider mankind, therefore, as a band of bold though *diminutive giants*, gradually descending from the mountains to subjugate the earth and climates with their feeble arms. How far they are capable of going in this respect futurity will show.[5]

Herder may be one of the first thinkers to understand not only that climate shapes man's living conditions, but that mankind also changes itself through the cultural techniques it employs to alter landscapes and climates. Humankind shapes its life-world by creating atmospheres adapted to its needs or comfort.

Thus, the Anthropocene—that newly-coined epoch in which humans have become a "geological force" transforming the climate and the biosphere on a global scale—did not simply start with the Industrial Revolution.[6] Maybe the Anthropocene actually begins with the start of a civilization that actively intervenes into climates and landscapes in order to create or adapt the atmosphere to their needs—that is, with humankind's transition from hunting and gathering to settlement and agriculture. Climate control, then, is not a product of the twentieth century's society of comfort, but a central element in the project of civilization. This project mainly consists in liberating human society from the contingencies of nature, and particularly of a dimension of nature that is both as elusive and ubiquitous as the air.

"Air"—as an ancient shorthand for "climate"—surrounds and penetrates us, not just our bodies but our minds, souls, and societies. Since ancient times, weather events and seasonal patterns have been the epitome of that which cannot be controlled by humans, but which does mark the imprint of nature upon culture. The seasons, for example, are a circular model of time

4

Peter Sloterdijk, *Sphären III: Schäume* (Frankfurt: Suhrkamp, 2004), 319.

5

Johann Gottfried Herder, *Outlines of a Philosophy of the History of Man*, trans. T. Churchill (London: Hansard, 1800), 176.

6

Paul Crutzen, "The Geology of Mankind," *Nature* 415 (2002): 23. Will Steffen, Jacques Grinevald, Paul Crutzen, and John McNeill, "The Anthropocene: Conceptual and Historical Perspectives," *Philosophical Transactions of the Royal Society A* 369 (2011): 842–867.

that interferes with linear human time and gives it structure. In antiquity, the weather was the stage for the whims of the gods who would fight among each other or against humans by the means of thunder, wind, and inundation. Climate, for both antique geography and medicine, was that which ineluctably shaped bodies and mentalities. Aristotle wrote that

> those who live in a cold climate and in Europe are full of spirit, but wanting in intelligence and skill; and therefore they retain comparative freedom, but have no political organization ... Whereas the natives of Asia are intelligent and inventive, but they are wanting in spirit, and therefore they are always in a state of subjection and slavery. But the Hellenic race, which is situated between them, is likewise intermediate in character, being high-spirited and also intelligent. Hence it continues free, and is the best-governed of any nation...[7]

While Aristotle uses the climate of certain geological zones to explain their political systems, Hippocrates points to the profound physiological and psychological effects of soils, winds, seasons, plants, temperatures, water sources, etc. on the inhabitants of a region. Human health and physical constitution, he argues, can only be understood in the context of climate.[8] From antiquity to the age of Enlightenment, climate was seen as the primary influence on bodies, mentalities, and cultures. In this line of thought, the difference between local atmospheres and livelihoods accounts for the infinite number of different human shapes, colors, social institutions, religious cults, and political systems of power.

Temperature played a particular and vital role in climate theory. Following up on this tradition of thought in the eighteenth century, the legal theorist Montesquieu tried to establish a link between the laws and political institutions of different cultures and the climates in which they are situated.[9] Heat, Montesquieu posited, softens the fibers of the body and thus, he concluded, renders men's bodies slack and lazy, their souls fearful and disorderly. However, heat also entices the imagination and sexual desires. Cold climates, on the other hand, make their inhabitants strong, brave, and virtuous, not to mention relatively disinterested in erotic matters. Montesquieu argues that this explains the differences between social institutions in cold or hot zones. Institutions such as slavery, polygamy, or despotism must be seen as reactions to the slackening effects of hot climates; inventions such as democracy and romantic love are only suitable to the highly disciplined dwellers of Northern spheres, who need an idealized idea of love to be tricked into sexual activities. Peoples who live in the heat therefore need different institutions than those in the cold.

Montesquieu's somewhat bizarre and yet oft-repeated theory about the cultural effects of temperature might look strangely deterministic, from the contemporary standpoint, or even dangerously racist. Is it really the local temperature that determines whether one lives in democracy or under despotism? Clearly not—though in any case, his theory grants humans a certain degree of freedom in positioning themselves in relation to the influence of the climate. While Indians—in terms of their social systems as well as their fatalist religion—yield to the mollifying effect of their native heat, he argues, the Chinese toil against it by establishing a cult of work and discipline. In Montesquieu's Enlightenment style of thought, one always has a choice.

The "temperate climate"—not too cold and not too hot, and thus best suited to both physical and mental work—has been a central facet of all

7

Aristotle, *The Politics and the Constitution of Athens*, book VII, 7, 1,327–1,328, cited here from the edition edited by Stephen Everson (Cambridge: Cambridge University Press, 1996), 175.

8

Hippocrates, "Airs, Waters, Places" [ca. 400 BCE], in *The Medical Works of Hippocrates*, trans. John Chadwick, W. N. Mann, Oxford: Blackwell 1950, 90–111.

9

Charles-Louis de Secondat, Baron de la Brède et de Montesquieu, *The Spirit of the Laws* [1748], book XIV, trans. Thomas Nugent and J. V. Pritchard (London: Bell & Sons, 1914).

attempts to think about the relations between climates and cultures. Eventually, in the twentieth century, truly deterministic models were established by climate researchers such as Ellsworth Huntington, Ellen Churchill Semple, and Willy Hellpach. Huntington, to take an example from 1915, measured the loss of performance of workers in the southern US during the summer months. He concluded that heat unfailingly reduces mental and physical energy and, therefore, that no advanced cultural or scientific achievements could be expected from the inhabitants of hot countries.[10] Such deterministic takes on the relation between climate and culture were often used to legitimize repressive measures for overcoming native "sloth" in colonial contexts; but as the twentieth century progressed, that connection was increasingly used to advertise the intensive use of cooling technology to enhance work efficiency.

Ultimately, for these and other reasons, the idea that climate (or other environmental factors) determines the nature of a people has been discredited entirely. Today, the social world and the natural world are supposed to be separate from one another. To be "modern" means to be independent of such negligible things as air temperature or the degree of humidity. The weather, no matter how often we speak about it, is a background to our social interactions, not a major factor shaping them.

However, this seemingly outdated question about the exchange between climate and culture raised by thinkers from Aristotle to Herder and even the deterministic school of geography also marks something that has been repressed, or at least "cleanly separated" from the modern idea of man as a social and cultural being—the fact that being-in-the-world is also being-in-atmospheres, being-in-a-climate. As Bruno Latour has pointed out, we still feel compelled to cleanly separate "social needs and natural reality, meanings and mechanisms, signs and things."[11] Yet, with the insight at the heart of the idea of the Anthropocene—the fact that man has started to alter and disrupt climate not only on a local but a planetary scale—a separation of climate and culture, nature and technology, environment and humanity is becoming untenable. The old and supposedly defunct tradition of pre-modern climate theory raises the question of how both nature and culture have, in very different forms and degrees of intensity, been shaped by a mutual transformation of climates and civilizations. Climate is the epitome of that which surrounds and impacts human life forms. It is the imprint of nature upon man—yet a nature that, in turn, is massively altered by human technology.

At the core of the modern project to separate and purge the spheres of nature and society from one another—even if, as Latour has stated, this project might have never really been achieved—is a thought of creating spheres in which climate either doesn't play a role anymore or has become a mere option (i.e. a space of pleasant coolth in the hot zones, a comfortably heated room in the cold). Ultimately, the venture of controlling and dominating nature comes into its own in the dreams of creating an atmosphere that is completely adapted to human needs and comfort. This is an old motif of utopian thought: In Thomas More's *Utopia*, the inhabitants "fortify themselves ... against the unhealthiness of their air" not only by their "temperate course of life," but they also "cultivate their soil" and pluck "whole woods ... up by the roots, and in other places [plant] new ones, where there were none before." The transformation of landscapes and vegetation is a way to alter the "unhealthiness of the air" and thereby become healthier than any other population.[12] The more radical utopian Charles Fourier, often hailed as one of the forerunners of socialist thought, dreamt of a thorough reorganization of both

10

Ellsworth Huntington, *Civilization and Climate* (New Haven: Yale University Press, 1915); Ellen Churchill Semple, *Influences of Geographic Environment: On the Basis of Ratzel's System of Anthropo-Geography* (New York: Holt & Co., 1911); Willy Hellpach, *Die geopsychischen Erscheinungen: Die Menschenseele unter dem Einfluß von Wetter und Klima, Boden, und Landschaft* (Leipzig: Engelmann, 1911).

11

Bruno Latour, *We Have Never Been Modern*, trans. Catherine Porter (Cambridge, MA: Harvard University Press, 1993).

12

Thomas More, *Utopia* [1516], translated into English (London: Chiswell, 1684), 131–132.

human society and climate. His first book, *The Theory of the Four Movements* (1808), is a plan for the rearrangement of social as well as sexual relations along the lines of a general theory of the "laws of attraction." These laws, Fourier believed, were as general to all things natural and human as the law of gravity. A rearrangement of society, in order to work with and not against these laws of attraction, would not only bring perfect individual satisfaction but also universal social harmony.

What is striking in Fourier's plan is that this new, sexually liberated and socially harmonious society would come into effect alongside of a fundamental global change of climate. Over the course of the eighty thousand years that such a reorganization of human society would take, Fourier explains, the *aurora borealis* would intensify so much that it would become an additional source of heat, focusing on the poles. This would heat up the polar zones and thereby enable places like Siberia and the north of Canada to enjoy temperatures like the Côte d'Azur.[13] At the same time, Fourier predicts, the tropics would (fortunately) not heat up any further. The entire planet would eventually be bathed in the pleasant weather of the Mediterranean that European pensioners and tourists still deem to be the absolutely "perfect climate." This temperate utopia—which today strikes us as the IPCC's worst nightmare—is, in his logic, a necessary step, since this planetary Mediterranization is necessary in order to provide resources and space for a sexually liberated and thus ever growing global population. A globally harmonized and homogeneous society—and Fourier is actually one of the first to think of a "world society"—needs a globally regulated, temperate world climate. Biopolitics and climate change, for Fourier, must go hand in hand.

Fourier's climatic utopia may, in the days of global warming, sound more like a threat than a promise. However, he brings to the foreground a motif that has been, and remains, central to climate thought—the idea of an ideal climatic "norm," as it were. No matter how differently these philosophers and scientists have regarded the causal link between climate and culture, we find in each the idea that there is such a thing as a singularly "moderate" and thus ideal climate. Aristotle praised the Greek climate as the middle ground between too cold and too hot; Montesquieu seems fixated on an ideal, temperate climate that resembles that of his native Bordeaux; for Kant, it was the general middle European climate zone. While the Scotsman John Arbuthnot praised the effects of Scottish weather as a precondition of high achievements in the arts and in philosophy, the Frenchman Montesquieu pondered the link between the soggy climate of the British Isles and the preponderance of their inhabitants to commit suicide.[14] Every philosopher locates the "ideal climate" exactly in his native country or climate zone. Even the geographer Ellsworth Huntington participated in this "climate-jingoism," as it were, locating the world's best climate in Newport, Rhode Island, a hundred miles from his home at New Haven.

Today, global health has dropped this idea of an ideal climate. The fact that "populations living in different climates have different susceptibilities, due to socio-economic reasons, and different customary behavioral adaptations" is generally acknowledged.[15] Bodies and cultures adapt to the temperatures (and degrees of humidity) they are set in. However, modern medicine also concedes that human health and work performance are related to the temperatures of the environment. The biggest challenge, it seems, is posed by temperature extremes that are unusual in a certain environment, such as heat waves or periods of sudden and unfamiliar cold. Temperature

13

Charles Fourier, *The Social Destiny of Man, or: Theory of the Four Movements*, trans. H. Klapp Jr. (New York: R. M. Dewitt and C. Blanchard 1857), 42–44.

14

John Arbuthnot, *An Essay Concerning the Effects of Air on Human Bodies* (London: J. and R. Tonson, S. Draper, 1733). Montesquieu, *The Spirit of the Laws*, 301–302.

15

Gerd Jenditzky and Birger Tinz, "The Thermal Environment of the Human Being on the Global Scale," in *Heat, Work, and Health: Implications of Climate Change*, special issue of *Global Health Action*, ed. Tord Kjellstorm (2009), 10–21, 10.

stress (and especially "heat stress") also depends on the specific vulnerability of certain groups (the elderly and fragile, or those who don't know how to "acclimatize").[16] Thus temperatures are not just a physiological "given" but rather are culturally coded and subject to acclimatization. While Singaporeans (whether immigrants from Britain or Beijing, or born in the area) know just how to deal physically with the combination of heat and humidity, I did not miss a chance to do everything wrong. I ate too much, I ran hectically through the sunbaked streets in a misguided attempt to quickly find a taxi or a spot of shade, I did not wear a hat, I did my shopping around noon. My first day in the tropics I spent mostly in bed, dizzy and nauseated. Yet, not only did I slowly learn, but somehow my body did too. I wasn't hungry anymore during the daytime, I didn't feel like I was suffocating, and I instinctively started to adapt the relaxed, shade-seeking pace of the locals. Heat and cold, it seems, are codes the body learns to deal with. But they are also freighted with cultural significance. In the tropics, coolth is clearly a sign of status and wealth—hence the over-cooled hotel lobbies and luxury shopping malls. Only the poor dwell outside, in the sweltering heat.

It is this cultural significance of climate that inspires the type of artificial atmospheres that societies dream of, and sometimes even build. While our present times seem to fear nothing more than rising temperatures, the nineteenth century, to the contrary, was obsessed with the idea of a slowly cooling planet. Camille Flammarion, Gabriel Tarde, H.G. Wells, and others devised fictional scenarios of human life slowly dying off in an ever colder global climate. Yet, by this period, architects had discovered how to harness the heat of the sun in closed glass-covered spaces even with chilling outside temperatures. Starting in the seventeenth century with the invention of orangeries to protect Mediterranean plants in Northern Europe during the winter months, glasshouse architecture reached its heyday in the second half of the nineteenth century. The famous "Crystal Palace" built in London for the Great Exhibition

16

Ken Parsons, *Human Thermal Environments: The Effects of Hot, Moderate, and Cold Environments on Human Health, Comfort, and Performance*, 3rd ed. (Boca Raton, FL: CRC Press, 2013), 323–325.

The Crystal Palace built by Joseph Paxton for the Great Exhibition, London, 1851.

of 1851—by Joseph Paxton, who, notably, was a horticulturist and cultivator of
bananas—was the first cast iron and plate glass structure of its size, covering
an immense space of nearly a million square feet and enclosing, together with
the stalls of some fourteen thousand exhibitors, several high elm trees. It was
a ventilated and warm space like a giant marketplace or town square, full of
plants, shops, and people, yet protected from the rain, cold, and wind of British
weather. The Crystal Palace offers the pleasures of being "outside" without the
discomfort of bad weather or seasonal temperatures. The nineteenth century
also saw the birth of covered passages, gigantic imperial greenhouses, elegant
shopping malls—all ways to stroll the streets of a city or to roam the world of
exotic plants without being exposed to meteorological surprises. Little wonder
that the "Familistère," built by the industrialist J.P. Godin on the basis of Fourier's social thought, included a glassed-in central court.

In the twentieth century, these capsules of ever-pleasant climate have morphed into giant covered theme parks, Buckminster Fuller's geodesic domes,
or the *Biosphere II* research project. Glass enclosure offers a world "outside"—streets, trees, marketplaces, cafés, town squares, and various forms
of "wild nature"—yet contained by a sphere that locks out the challenges of
natural climate. These are architectures of a locked-in outside, a stabilized
nature deprived of anything unforeseeable or uncomfortable. Specifically
in the greenhouse architecture at the zenith of colonialism toward the end
of the nineteenth century, the "nature" captured inside the glass sphere is
exotic, allowing the visitor to not only look at exotic plants and flowers but
also to experience the exotic climate of tropical rainforests or deserts. "In
their greenhouses," Sloterdijk remarks, "the Europeans started a series of
successful experiments on the botanical, climatic, and cultural implications of
globalization."[17] Once it is possible to take a short Sunday family jaunt to the
sweaty heat of Malaysia or to endure the aggressive dryness of the Sahara for
some ten minutes, climate becomes an *option*. It ceases to be the inevitable
atmosphere of a given locale, an element of reality that comes part and parcel
with being in a specific place. Glass architecture thus creates "atmotopes," as
Sloterdijk calls them—zones of a carefully manipulated climate, flooded with
natural sunlight, overgrown with plants, and populated with humans (and
sometimes animals). Spaces that seem to be open, transparent, and airy, yet
hermetically sealed from the outside air. They are artificial atmospheres that
experiment on the artificiality of nature itself.

Today, climate is, except for the exotic options in greenhouses and zoos,
well on its way to becoming globally standardized. Landing in Singapore
places you in the same 72 degrees Fahrenheit with 50 percent humidity as
landing in Moscow; working in an office in Montreal has you sit in the same
climate as in Dubai. (This temperature standard is, by the way, adapted to
middle-aged men and generally slightly too cold for women.) It is a culture of
stabilized and standardized climate, which architecture "makes real" by creating smaller or larger pods of identical air all over the planet—a standard set by
European and American ideas of ideal climate. It is also based on a Northern/
Western idea of efficiency and performance, a nine-to-five lifestyle that needs
no siesta or summer break. The modern ideal of open spaces and glass walls
means that architects continue building huge office spaces with glass facades
that require enormous energy to keep the temperatures inside stable. Sometimes a small fraction of this energy consumption is gained by solar panels to
earn these buildings the label of "sustainable." Today, these atmotopes grow
larger and larger. Ironically, even without the sexual and social liberation of

17

Sloterdijk, *Sphären
III: Schäume*, 343.

"Cooled Conservatories," Gardens by the Bay, WilkinsonEyre Architects, Singapore, 2006–2012. Courtesy of WilkinsonEyre.

Fourier's dreams, his vision of a homogenized global climate is about to come true. The irony is that we are getting less and less able to tolerate the very same warming temperatures that we are creating, largely through the massive CO_2 output of air conditioning technology itself. As early as 1992, the British economist Gwyn Prins found that "the physical addiction to air-conditioned air is the most pervasive and least noticed epidemic in modern America."[18]

18

Gwin Prins, "On Condis and Coolth," in *Energy and Buildings*, vol. 18, no. 3–4 (1992): 251–258.

Taking climate advice from an inhabitant of Great Britain might not come naturally to people living in Dubai, Bangkok, or Phoenix. Yet, even youngsters in the tropics are growing more and more intolerant to temperatures above 80 degrees Fahrenheit. What my first and slightly traumatic encounter with tropical heat made me understand was the extent to which climate not only shapes our physical well-being and performance, but also our lifestyles. The alleged "laziness" of Southeast Asians or Africans that infuriated colonial officials, the Mediterranean "siesta" or preponderance for very late meals, all these cultural practices are elements of a climate culture or, as it were, a "climatic intelligence." Dwelling as we do in globally standardized air, we are about to lose this climatic intelligence. We try to run through the streets of Singapore as if we were in Berlin or London. Wherever we go, we tend to count on the temperatures and humidity of a mild Californian day; everything else is an aberration. Our world "outside" has become a climate-controlled interior which we only leave for occasional adventure trips into more extreme climates. The time we spent between the airport door and the cab, the cab and the office, the office and our home gets shorter and shorter. While we're heating the planet by cooling our climate-controlled life-world, it might be worth stepping out into the very wet, very cold, very hot, very dry air that is waiting for us.

Eva Horn is a professor of German literature and cultural theory at the German Department of the University of Vienna, Austria. She has taught at the Universities of Konstanz, Basel, Frankfurt/Oder, NYU, and Columbia University. She is the author of *The Secret War: Treason, Espionage, and Modern Fiction* (Northwestern University Press, 2013), and *Zukunft als Katastrophe* (Fischer, 2014). Her current research revolves around a cultural history of the concepts and representations of climate.

The Atmosphere as a Cultural Object

JORGE OTERO-PAILOS

I've come to think of pollution as the chief product of our civilization, as important or more so than the monuments on which it settles. It's the material that scientists turn to as the evidence of the Anthropocene, a new geological era measured by the permanent mark of man-made airborne pollution on the stone crust of the earth. The mark of buildings on that crust is quite ephemeral by comparison to pollution. That distinction is even more telling in light of the recent Paris climate talks. The best our leaders have been able to do is to propose a cliché borrowed from the field of historic preservation: to restore the atmosphere back to what it looked like in the past.

This is quite a leap. Preservationists know how to restore buildings, cities, and landscapes, but restoring the atmosphere is entirely new territory. All sorts of questions immediately come up. Ontological questions: How can one conceive of the atmosphere as an object worthy of restoration? Since it is made of natural gases and man-made particles, is it more appropriate to call it a natural or a cultural object? Where does this object begin and end? Does it extend continuously from inside our lungs to where airplanes fly? Or does it present itself to us more like a discontinuous series of environments or fragments we can tangibly grasp? Legal questions: Can this object be owned and therefore regulated? Who could own such an object? What rights and responsibilities would that owner have toward the object? Historiographical questions: How far are we to take this atmospheric restoration? The year 2001, 1989, 1492? Cultural questions: What is the cultural significance of each of these dates? Political questions: Who benefits and who bears the cost for this restoration? Technological questions: What restoration technology is appropriate? What if the technological restoration of the atmosphere turns out to damage it further? Could we find a reversible treatment? There are of course many more. My own work in the field, and this essay, are attempts to open up a new set of questions about our relationship to the atmosphere and the central role that pollution plays in our thinking about the atmosphere as our heritage.

This conceptualization of the atmosphere as an object of preservation poses a great civilizational challenge. Since the nineteenth century, the objects that preservationists have cared for have grown in size, from movable artworks to interior rooms, buildings, neighborhoods, landscapes, and even vistas. With each enlargement of the object's scale, we had to devise new institutions capable of carrying out the necessary work—intellectually, politically, culturally, and, of course, technically. Individual collectors could preserve an artwork or even a building, but to preserve a historic neighborhood requires a municipal government, and to preserve areas as large as Yellowstone requires a state robust enough to invent and sustain the National

Park Service. Now we're talking about the entire atmosphere of the planet as one object that needs to be tended to.

We don't currently have an institution in place capable of addressing this task for the atmosphere. We need to build it from the bottom up, starting with the intellectual foundations. To do this, we will need to rethink the discipline of preservation and reimagine what is relevant knowledge for the task before us. The challenge is that most preservation organizations are narrowly focused on improving how to care for the objects that were defined in the nineteenth century. No one is thinking about objects like the atmosphere, which are being defined in our time.

Even at the purely conceptual level, to think of the atmosphere as an "object" worthy of preservation is an intellectual challenge for which we need to call on the greatest minds, irrespective of their discipline of origin. It's important not to talk only among designers, preservationists, and other cultural producers. Beyond a certain level, there are no real boundaries between disciplines. But, regrettably, our institutions are set up to reinforce the idea that knowledge belongs in distinct compartments, or departments, each with its own building. Even in universities there are few opportunities for the great minds of different disciplines to even have a meal together.

In universities, as everywhere else, technology is where the money is. So it is no surprise that world leaders are approaching the preservation of the atmosphere myopically, as if it was strictly a technological problem, to be solved with more new technology, from the smart city to the fuel-efficient car—never mind the lessons of Volkswagen. In so doing, they are imagining the atmosphere as a technical object—a resource to be manipulated and exploited. They seem to have completely written out culture from the solution. This makes no sense to me. Since we are talking about smoke in the end, think of smoking. No technology could make people quit smoking, not nicotine patches, not nicotine gums. It took a cultural revolution, followed by courageous policy to prohibit smoking in buildings, for the majority of people to quit smoking.

A technological upheaval alone won't be enough to preserve the atmosphere. We'll also need a cultural revolution. Not Mao's kind, which was all about destroying heritage, but rather one that grasps culture as intrinsically intergenerational—a shared heritage. To think of the atmosphere as a cultural object means, by definition, to conceive of it as something that belongs to multiple generations. That is the basis of intergenerational equity—leaving the world so that our children may enjoy it as we did, which is not the same as leaving it the way we first found it, which would be impossible, since change is the nature of existence. Since we are talking about the atmosphere, we must also envision an object that belongs to multiple cultures at once, without any single culture laying exclusive claim to it. To move from a strictly technological view of the atmosphere to one that includes a cultural and intercultural understanding of it is very difficult but worth the effort.

The anthropologist Mary Douglas famously found that we typically think of pollution as that which has no place in our culture. Our work should be to bring pollution back *in* to the cultural sphere. Between atmosphere and architecture, we might find a material continuity, pollution. It is through this material continuity that I think we can begin to conceptually extend our traditional idea of the monument to the atmosphere. There is a role for aesthetic work in thinking about climate—reorienting our understanding of the atmosphere toward culture, because it helps to see something in order to understand it.

But there's also much necessary intellectual and scientific work to be done. It is difficult work, because there are few opportunities for collaboration between the visual arts, the humanities, science, and technology. But where there is a will there is a way. In my view, the places best suited to support these encounters are research institutions, not just universities but also museums and the like.

One field that holds much potential in this regard is "experimental preservation"—a theoretically informed practice that tests hypotheses of what preservation can be and do. It is in the nature of this testing to push on the limits of the discipline, to draw relevant knowledge from other disciplines while contributing preservation knowledge back to them. On the surface, experimental preservation can appear to be undermining the very foundations of preservation, but upon closer scrutiny this undermining reveals itself to be a renewal. To push outward is also a way to reaffirm the center, to nourish the unique knowledge that preservation produces and its particular modes of engagement with the world. It is also to remain open to the possibility that preservation knowledge might surface where we least expect it: in biotech, in first societies, in aerospace, in philosophy, in fashion, and so on.

I'm particularly interested in the contributions of contemporary artists to preservation. Ai Weiwei is buying ancient Chinese wooden temples and showing them as his art in an effort to use the power of the art market to prevent their destruction. Olafur Eliasson recently moved an iceberg older than the pyramids to Paris, where it melted in front of world leaders gathered for the climate talks—an expression of the challenge before preservation. There are also fascinating experimental preservation works being produced at the intersection of art, preservation, and history. David Gissen has created some

Ice Watch, Olafur Eliasson and Minik Rosing, Place du Panthéon, Paris, 2015. Photograph by Martin Argyroglo, © Olafur Eliasson 2015.

Top: *Pittsburgh Reconstruction, 1960*, David Gissen, 2006–2010. Image courtesy of the designer.

Bottom: *Air Manifest: Los Angeles 1955, 1965, Instructions for the Reconstitution of Historical Smog*,
Mark Wasiuta, Marcos Sanchez, and Adam Bandler, Studio-X Istanbul, November 1, 2014–January 2, 2015.
Image courtesy of the designers.

Top: *Study of Cirrus Clouds*, John Constable, 1882. © Victoria and Albert Museum.

Bottom: Tomás Saraceno's *Aerocene*, a solar-powered hot-air balloon, White Sands Missile Range, New Mexico. The launch and the symposium "Space Without Rockets" were initiated by Tomás Saraceno with the curators Rob La Frenais and Kerry Doyle for the exhibition "Territory and the Imagination" at the Rubin Center for the Visual Arts. Courtesy of the artist; Pinksummer contemporary art, Genoa; Tanya Bonakdar Gallery, New York; Andersen's Contemporary, Copenhagen; Esther Schipper Berlin. Photograph by Studio Tomás Saraceno © 2015.

incredible visual reconstructions of the atmosphere of Pittsburgh, the Smoky City. Mark Wasiuta, Marcos Sánchez, and Adam Bandler have reconstructed the particulate matter floating in the Los Angeles atmosphere circa 1955 and 1965. We live in a peculiar situation today when certain art practices outside of traditional preservation are at the forefront of a new preservation. It is not the first time in history that art leads the way for emerging disciplines. Think of meteorology in the nineteenth century. For some, John Constable was a great British landscape painter; for others, his paintings are pioneering works of meteorology, made by a scientist who contributed as much to the understanding of the weather as any of his contemporaries.

In our own time, preservation is undergoing a fundamental transformation, emerging as a renewed discipline, with new purposes and means, through experimental practices that are broadening the established boundaries of knowledge. As preservationists, it behooves us to seriously engage those practices, and to recognize their contributions to our common efforts. The way these "outsiders" do preservation may seem unorthodox. But then again, all our conventional treatments of cultural objects, from consolidating stone walls with ethyl silicate to legally designating buildings as landmarks, were at one point in time quite unorthodox things to do. Experimental preservation is radical in the strict sense that it returns to the roots of preservation as an unorthodox treatment of cultural objects. One can pursue orthodox treatments when working on objects that are widely recognized as having cultural value—but in the case of the atmosphere, its status as a cultural object is a question rather than a given.

The unspecialized public tends to assume that cultural objects are out there, and that preservationists come later and treat them. But this is not always the case. For instance, historic districts did not exist before preservationists created them through unorthodox legal treatments. Likewise, National Parks did not exist as such before preservationists invented them, drew artificial lines across vast landscapes, and started forcing ranchers out at gunpoint and employing rangers to keep poachers out. In other words, preservation treatments are constitutive of cultural objects. Preservation in many ways comes before, or at least accompanies the formation of cultural objects.

The question then is what sort of preservation treatments, however unorthodox, can accompany the formation of the atmosphere as a cultural object. My own project, *The Ethics of Dust*, is one such treatment. I paint liquid conservation latex on monuments, which removes pollution—and in so doing, creates an imprint of it as a record. Each cast might seem at first to be about an individual monument. And they are. But taken together the series of casts begins to suggest a material continuity through pollution between all these individual monuments. The pollution itself was once airborne. It belongs to the sky as much as to the monuments. Each cast is a piece of the atmosphere deposited on a particular building. Each piece is also an invitation to do the conceptual work to put them all back together, to reconstitute the atmosphere as a cultural object.

It might seem unorthodox to work from small fragments, strewn about the world, to reconstitute the atmosphere. And it is. But this is precisely how many of the greatest monuments of the world exist. Think of the Parthenon, for example. Where is it? Only a small portion of it is on top of the Acropolis. Much of it is down the hill in a museum. Very important parts of it are in the British Museum. There are original pieces in the Vatican Museum, in the Louvre Museum, in the National Museum of Denmark, in Vienna's

The Ethics of Dust: Trajan's Column, Jorge Otero-Pailos, 2015; commissioned by the Victoria and Albert Museum.
Photograph courtesy of Peter Kelleher and the Victoria and Albert Museum.

Kunsthistorisches Museum, in the Munich Glyptothek, in the Würzburg University Museum, in the University of Heidelberg, and many other places. The Parthenon, from a material standpoint, is a series of fragments strewn about the world. Despite this discontinuous material reality, we think of the Parthenon as something that has unity. But what holds together this cultural object? I would argue that what holds it together is an idea: Western democracy. We cannot imagine Western democracy without its historical anchor in the Parthenon, the symbol of Periclean Greece, at least since the early nineteenth century, when the American and the French revolutions revived democracy as a viable political system.

Just as we can never see the Parthenon all at once, we cannot see the atmosphere in one glance. Even the famous 1972 "Blue Marble" photograph from Apollo 17 shows only half of it, and only an instant of it. Conceived as a cultural object, the atmosphere is not just a space, a sphere, or a moment—1972 or whatever other date politicians might want to restore it to. The atmosphere also has temporal depth; it has a history that can only be borne out, not to mention experienced, in its materiality. It is in this way that the atmosphere is a cultural object, shared across generations, each of which has the right to enjoy and the duty to preserve.

Jorge Otero-Pailos works at the intersection of art, architecture, and preservation. He is associate professor of historic preservation at Columbia University's Graduate School of Architecture, Planning, and Preservation in New York and is the founder and editor of the journal *Future Anterior*.

Can You Believe the Weather We're Having?
The Politics of the Weather Report

PHU HOANG

Early forecasts from the US Weather Bureau were mostly about the current weather in broad regions, which was gained by ground observations and government planes carrying mail.

UNSEASONABLY COLD AND BELOW FREEZING

On February 26, 2015, US Senator James Inhofe famously brought a snowball onto the floor of the Senate as the latest piece of evidence in a long-standing "debate": Since it was "unseasonably cold and below freezing" in Washington, D.C. that day (20 degrees Fahrenheit), it must mean that global warming was, as the title of his 2012 book argues, "the greatest hoax." For throwing the snowball onto the floor, Senator Inhofe (the chair of the Senate Environment and Public Works committee, no less) was widely ridiculed, even by some members of his own political party. Notwithstanding the fact that global climate change will produce hotter summers and colder winters (due to more moisture in the atmosphere), this political stunt visualized a common argument made by those who seek to disprove climate change. Their central case is, essentially, based on separating weather from climate. Weather is immediate and personally "experienced," leading to a belief that it is undeniable. Climate, on the other hand, does not appear to occur "right now"; invisible to the naked eye, it is only made knowable by statistics and computational models. Thus, it seems to be susceptible to the fallibility of the scientific community. But this essential depiction of difference—that weather

The concepts and speculations about micro-weather futures in this article are the result of a design collaboration between Phu Hoang and Rachely Rotem. Together, they codirect the interdisciplinary architecture practice, MODU, which has won numerous international design competitions and awards. MODU was recently awarded a commendation for "21 for 21," an international award that recognizes architects who will "lead the next generation of architects in the twenty-first century." MODU has also been the recipient of architecture and weather research grants from the Robert Rauschenberg Foundation and the New York State Council on the Arts.

is universal and unequivocal, while climate is abstract and debatable—is perhaps the *actual* "greatest hoax ever perpetrated on the American people."[1]

1

James Inhofe, "The Science of Climate Change," Senate floor speech, quoted by Chris Mooney, *The Republican War on Science* (New York: Basic Books, 2006), 84.

The crux of the weather vs. climate "hoax" is not that they are different, because in fact there are important differences between them. But these differences do not support either of the two claims commonly made by climate change skeptics—that global warming is not occurring, or, if it is, that it is not directly linked to human actions. Instead, the differences between weather and climate further support the direct relationship between anthropogenic, or human-influenced, greenhouse gas emissions and global warming in the atmosphere. These differences have important consequences not only for the climate change debate but also for the future of our cities and built environments; thinking about these differences begins with an everyday event occurring in most of our lives—the weather report.

THE WEATHER FORECAST FOR TODAY IS SUNNY

In the US, the weather report is nearly as old as the country itself. The first edition of the *Old Farmer's Almanac*, which included surprisingly accurate weather forecasts, was published in 1792. By the 1840s, the Smithsonian Institution was supplying weather instruments to telegraph companies and had established the country's first weather observation network, which allowed for daily telegraphic weather reports and charts to be shared nationwide. This regular observation soon led the federal government to begin forecasting the weather. In 1870, a Joint Congressional Resolution required the secretary of war to distribute weather reports of approaching storms through the telegraph system. Twenty years later, a new civilian agency formed a national weather service, the US Weather Bureau (housed in the Department of Agriculture), and its weather reports would become part of citizens' daily lives for years to

In 1955, IBM's 701 vacuum tube computers were used to provide daily weather forecasts for the Joint Numerical Weather Prediction Unit. Courtesy of IBM.

come. Recent technological developments in weather forecasting have also stemmed from advancements in communication networks, as was the case with the telegraph system more than a century ago. These developments may radically alter our cities' relationship to the everyday weather report, as well as to the weather itself. The integration of micro-weather forecasting with mobile technologies has made it possible to know upcoming weather with more immediacy and at a higher resolution than previously possible. Using a smart-phone, it is possible to know when it will start to rain in one neighborhood and when that rain cloud will pass to the next neighborhood. Historically, the US National Weather Service provided forecasts based on a grid in which each square measures 7.5 miles on a side. As a result, weather reports were regional and generally the same for an entire city. It is well known, though, that weather can vary considerably over time and distance—in a city like San Francisco, it is entirely normal to have different weather in two neighborhoods. This is where micro-weather forecasting comes into play, as it provides weather forecasts with the granularity of 0.6 miles or less. This isn't simply a convenient precision of information—the high-resolution weather grid can, in fact, be understood as a new kind of urban organization.

Micro-weather forecasts are accessible through a wide range of mobile weather apps: BeWeather, Nooly, and Dark Sky, to name just a few. These apps, all developed by private companies, combine raw data from the National Weather Service with "extra" data from an array of privately man-aged weather sensors, radar, and satellite images. All of this data is processed together using each company's proprietary computational models. For example, the one thousand National Weather Service monitoring stations have been increased sixfold by WeatherBug's networks to provide far more finely grained weather forecasts. Such micro-weather forecasts do not provide regional reports. Rather, they focus on cities and often specific economic development sites (e.g., oil fields). If weather forecasting was originally seen as a blanket of even and generalized coverage, the turn toward nongovern-mental participation recasts the weather as something of an index of special interests, with localized precision playing a more important role (whether for the purposes of capital, development, or security).

This privatization and particularization of the weather forecast has its roots in the citizen science movement, which began in the 1990s and was made possible by the communication networks of the Internet. The Weather Underground, founded at the University of Michigan (and named after the revolutionist faction of the New Left formed at the same school in the 1960s), allowed individuals to link their personal weather stations to the Internet and to form a weather reporting network with some ten thousand monitoring sta-tions. The exponential multiplication of weather monitoring stations has given rise to the idea that weather is universally experienced in its broader forms (a shared language of weather in which a region experiences a heat wave or a blizzard) but nevertheless subjectively different within the same time period in different areas. Micro-weather urban conditions—shade from tall structures, proximity to water, topographical characteristics—all contribute to differen-tial ranges in temperature, humidity, and air pressure. The daily statement of the "weather forecast for today," as conceived of in the worldview of micro-weather, would not be summarized with "sunny," "cloudy," or "raining," but with highly specific conditions tailored to GPS locations.

The impact of microclimatic thinking has been changing the design of building mechanical systems for the past two decades. It is no longer the norm to climatize an entire interior space to a constant temperature; contemporary HVAC design incorporates microclimatic differences based on users' activities in each program. For example, the variation between the indoor temperatures of a public atrium and an office space can be as much as 5 degrees Fahrenheit. The relationship between these generally interior microclimates and the micro-weather that takes place just outside now offers a wide range of potential weather futures, from the socially transformative to the dystopian. Will it be possible for renewable energy building systems (which use solar and wind power) to respond to changes in micro-weather by anticipating and responding before the weather changes? Will we use our smartphones to map a route that accounts for the micro-weather along the way, seeking out better microclimates and thereby relating physical space to its atmospheric experience? In a more dystopian future, will micro-weather reports be used by utility companies to charge higher "surge" pricing in anticipation of inclement weather for a particular neighborhood? In other words, is our urban future one in which businesses can privatize not only the weather report but also the weather? While any of these futures are theoretically possible, it is perhaps more important to understand the role of "futures" in the debate between weather and climate.

WE SURE COULD USE SOME RAIN

As demonstrated by Senator Inhofe, weather occurs *now* and thus can be "trusted," while climate occurs *in the future* or *in the abstract* and so is treated with extreme skepticism. While it is true that the basic difference between weather and climate is time-based—weather occurs over a one-week period at most, while climate is years, decades, or even millennia long—this difference is not what separates them but instead makes their interdependence possible. One way to understand this relationship is through the distinction between meteorology, which is the study of conditions occurring in the atmosphere (temperature, humidity, pressure, density, wind speed, etc.), and climatology, which is the study of everything that affects the atmosphere. It is impossible to believe in the science of meteorology without acknowledging the principles of climatology. Current climate modeling science further interrelates weather and climate by using data sets of past weather to generate predictions for future climate.

With the ever-increasing availability of computer power, climate models are increasingly used as an "instant replay" to analyze extreme weather events. These models are being used to show that global warming is directly connected to human activity. This computational method, described by the climatologist Dr. Heidi Cullen as "weather autopsy," uses data from past extreme weather events to set up two model experiments in which one model includes greenhouse gas emissions and the other does not.[2] The results of both models are compared to understand the risk of future extreme weather events when associated with human influence on climate change. The final step is to run the climate models in a forecasting rather than "autopsy" mode, which inevitably shows a higher risk in the future when human influence is involved.

2
Heidi Cullen, *The Weather of the Future* (New York: Harper, 2011), 53.

Another example of climate as a history of past weather, as well as the interdependent relationship between climate and weather, is demonstrated with recent reanalysis climate modeling. The historian Paul Edwards, in his book *A Vast Machine*, describes reanalysis as asking the basic question, "Is

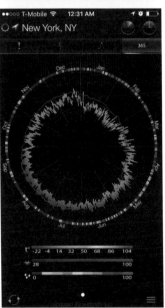

The eWeather HD app provides micro-weather forecasts at a range of time scales, including hourly forecasts per neighborhood and 365 day weather patterns.

it possible to have a 'do-over'?" Up until the later 1980s, climatological stations had their own set of climate data separate from meteorological stations. The concept of the "do-over" relies on rebuilding the entire set of climate data by basing it on past daily weather reports. All of the world's available weather data is input into supercomputers, which provide the computing power to recreate the history of weather at every grid point and altitude in the world. In essence, weather data can be used to look back in time to allow us to see future weather. Several government agencies (NASA, the National Meteorological Center) have undertaken different reanalysis projects in the past decade. While the results have not yet been sufficient to replace traditional climate data—perhaps due to the current limits of computing power—they have been successful in showing that global warming is directly linked to human activity. Reanalysis models have highlighted an increase in the height of the troposphere (from increased atmospheric heating), which, after isolating all possible natural causes, is attributable to human activity. Edwards describes this computational method as "the future of the past."[3] The future, and now the weather, is not what it used to be.

Advancements in computing power are the principal reason why higher-resolution weather models (such as those used in micro-weather forecasts) have become possible. This increase in resolution has also occurred within the climate models themselves, which have less "loss" of information and therefore produce more accurate climate predictions. It is conceivable that the next generation of supercomputers will use the high-resolution data sets from micro-weather forecasts, perhaps combined with the data from citizen scientists' monitoring stations, to predict future climate change with a much higher degree of accuracy. Being able to "see" weather at finer scales has implications not only for the design of the built environment but also for understanding the interdependent relationship between weather and climate, and thus perhaps politicizing the weather report. If this seems counterproductive, it may in fact be a desirable outcome, as that interdependency makes it more difficult

3

Paul Edwards, *A Vast Machine: Computer Models, Climate Data, and the Politics of Global Warming* (Cambridge, MA: MIT Press, 2010), 325.

to use the weather forecast to deny climate change. Cullen has proposed that climate forecasts be included in the daily weather report, providing the public with information about how each day's weather relates both to past weather and future climate change. The saying, "We sure could use some rain" would be contextualized in the weather report with daily climate analysis. In this way, the weather report—especially with the advent of micro-weather apps—would reinforce the connection between the Anthropocene's impact on greenhouse gas emissions and global warming.

CAN YOU BELIEVE THE WEATHER WE'RE HAVING?

Since climate models are increasingly based on past weather forecasts, what is the future for the weather report? If micro-weather apps are already part of our everyday urban lives, what will the future hold for weather forecasts? A current technology developed by the mobile technology company Ericsson offers one possible future. Ericsson's micro-weather technology, called MINI-LINKS, uses microwave links that are already part of mobile phone networks to monitor the local weather at every phone's location. Microwave links in a telecom network operate in the electromagnetic spectrum. When rain passes through the path of microwave links, a degradation of the mobile signal occurs. Low-cost sensors can be added to phones to measure temperature, pressure, and wind as well. These recordings can provide information about local weather, and, multiplied by the millions of phone signals in a city, present the possibility of an extremely high-resolution weather forecasting network. The system is currently being run in a pilot location in Gothenburg, Sweden. Ericsson sees this technology as having the wide-ranging potential to save countless lives from extreme weather events, manage urban waste water systems, and cut business costs by minimizing weather-related transportation delays.

While the future that Ericsson imagines for its MINI-LINK technology is socially and economically important, the technology also

The BeWeather app, using Weather Underground weather data, provides a wide range of micro-weather forecasts, including hourly weather forecasts per neighborhood, minute-by-minute radar maps, webcam feeds from weather monitor stations, and data from personal weather stations.

has the potential to radically change how we experience weather in our cities =and even within the interior environments of buildings, blurring the boundary between them. Since the microwave links and weather sensors of MINI-LINK phones would always be on—outside and inside—millions of smartphones would constantly monitor the weather seamlessly between them. Paired with emerging technologies from Apple and Google that map indoor spaces (Indoor Survey and The Cartographer, respectively), the MINI-LINKS could provide weather information for indoor maps. It would be possible to know the temperature and humidity of all the interior spaces of a city, from offices to restaurants to subway cars. This knowledge would change our perception of the most resolute urban boundary—the border between the "known" outdoor weather and unknown indoor, mechanically climatized spaces. The indoors would have its own history of interior weathers available to guide daily decisions. What is the average indoor temperature in the office of a recent job offer? Which subway line offers the coolest route to get to work during the summer? Which retail stores are the worst offenders, keeping their doors open in the summer and thereby increasing the urban heat island effect?

The resulting "indoor city" recorded by millions of mobile weather stations has fascinating potentials for architects, planners, and engineers to reconceptualize the relationship between the city and the environment. The philosopher Peter Sloterdijk has defined two kinds of atmospheres: the meteorological atmosphere and the interior atmosphere. Sloterdijk believes that people do not react to objects and space but instead to atmospheres.[4] The "indoor city" has the potential to synthesize meteorological conditions with interior atmospheres. It could be possible to "design with weather" such that the decisions made by architects and designers would also account for their meteorological effects—material choices, volumes of space, ceiling heights, etc.

While designing with weather would introduce atmosphere as part of a design methodology, it should not be predicated on yet more techniques of environmental control for architects and engineers. Rather, it should highlight the importance of the variability of interior micro-climates and exterior

4

Peter Sloterdijk, *Terror from the Air* (Los Angeles: Semiotext(e), 2009), 71–95.

The microwave nodes of Ericsson's MINI-LINK technology are integrated into existing cellular networks and provide the possibility of high-resolution micro-weather forecasting.

micro-weather. Instead of the static and constant temperatures instituted by modernism, micro-climates celebrate variability, differentiation, and subjectivity. Not all spaces should be climatized the same way; a building's mechanical systems should be designed to produce mediated environments rather than manufactured weather. The laws of heat transfer that govern the design of building envelopes—namely, that the boundary of a building is where energy exchange occurs—could also inform the organization and morphology of architecture's interior spaces. Mediated exchanges between the different zones of indoor weathers, as well as between indoor and outdoor weathers, could create invisible differences between a building's user groups, programs, or functions. Strategically designed exchanges between indoor weather zones and outdoor weather conditions could be developed using weather data from Ericsson's MINI-LINK networks. The urban landscape of hermetically sealed buildings could be replaced by a gradient of semi-exterior "weather rooms" calibrated for each climate zone. The everyday question "Can you believe the weather we are having?" could be reflected in the design of cities and architecture.

Architecture, rather than striving for environmental separation, can do more to mediate between different weather conditions. The mediation, or negotiation, of architecture with weather could draw on the properties of air as matter. Reyner Banham believed that mediation was part of architecture and described three forms of it: conservative, selective, and generative.[5] Conservative mediation occurs when a surface (like a Trombe wall or a mass floor) provides thermal lag to store heat and radiate it back. Selective mediation, on the other hand, uses adaptation in walls, floors, and roofs to redirect the environment, as with a responsive *brise-soleil* system. Generative mediation occurs when a surface is sealed in order to maximize the efficiency of building mechanical systems. Selective and generative mediation supported Banham's assertion that the environment has an external relationship to architecture, which led to mediation of the environment primarily through high-tech systems. Rather than see the three forms of mediation as discrete, the synthesis of conservative and selective mediation offers a productive way forward to adapt to weather. Instead of seeing this path as either dependent on technology (selective) or too focused on passive strategies (conservative), there is the option of synthesizing active with passive, or high-tech with low-tech, systems. Banham's important contribution of a "well-tempered environment" can perhaps transform into a "multi-tempered environment." Architecture and weather can react to each other in atmospheric exchanges between hybridized active environments and meteorological conditions, made possible by high-resolution micro-weather networks that treat outdoor and indoor weathers as two sides of the same coin.

5
Reyner Banham,
*The Architecture of
the Well-Tempered
Environment* (London: Architectural
Press, 1969).

WEATHER FORECAST FOR TOMORROW: BRIGHT

Senator Inhofe probably did not anticipate that the central argument of his snowball performance could be countered with something as simple as the very weather report that he saw as evidence for his denialism. The promise of micro-weather apps is not only more weather information; pairing them with other technological advances in climate models (high-resolution climate analysis, reanalysis models) and mobile technologies (MINI-LINKS, indoor mapping) allows for transformative social ideas. This could not come a moment too soon, as environmental imbalances are at an unprecedented level (and the complicity of buildings in those imbalances having been thoroughly demonstrated). We are beyond the point at which the methodologies of sustainability—using techno-managerial techniques based on the assumption of environmental balance—offer a viable solution. Resilience thinking, with its assumption that the climate is extremely unbalanced, can mediate future extreme weather events through strategies of mediation and adaptation. The notion of resilience is subtly different from sustainability, which argues that it may still be possible to return the environment to its natural balance. Resilience looks for ways to manage environmental imbalance, including raising buildings above the ground, using natural wetlands as a form of "soft" infrastructure, and returning sand dunes that previously existed to urban, or suburban, development. These strategies, which accept that architecture can no longer be a static environment, selectively allow weather forces into architectural interiors and adapt to their variable effects.

While technology has offered exciting opportunities in this regard, the future depends on the sociocultural practices of architects and designers for long-term, viable directions. It is only with the contribution of the cultural and critical methods of design that the relationship between architecture and weather can truly be rethought. Architecture and its relationship to weather is first and foremost a sociocultural project. Architects, planners, and engineers need to engage with the active design of environments that mediate, rather than separate from, natural forces. What this requires is to see technological advances as social instigators, made possible only by synthesizing high-tech systems with "low-tech" passive design strategies. The crisis of climate change is not simply an environmental project; it is also a cultural project, one that requires all of the resources of the varied design disciplines. The undeniable truth of climate change can learn from weather itself—requiring design strategies that are heterogeneous, differentiated, mediated, and adaptive. All of these challenges and opportunities can be "read" every day in the weather report.

Phu Hoang codirects MODU as an architect with extensive national and international design experience. He is a recipient of the Architectural League Prize and the Core77 Design Award. Before starting his first solo practice in 2006, he served as a director at Bernard Tschumi Architects. He holds a bachelor of science from the Georgia Institute of Technology in Atlanta and a master of architecture from Columbia University in New York. He is currently an adjunct assistant professor teaching advanced design studios in the Graduate School of Architecture, Planning and Preservation at Columbia University.

Encounters with Climate:
A Dossier of Architectural Precedents

In what sites does a concept like the planetary inhere? Where might we locate the nebulousness of climatic thinking in the built world? The answers can be obvious, but more often than not they are surprising and may well be fanciful. Architecture is comprised by myriad imaginaries and multiform realities, always intersecting and messily overlapping. From arctic villages to Noah's Ark, from log cabins to electric cars, from human waste to utopian landscapes, design is inspired by objects both fictional and real, millennia-old and still-unfinished, at the world scale and the microscopic.

The *Avery Review* asked a group of thinkers and designers to each propose a single precedent project—represented by an image and short text describing its significance—that has informed their understanding of "climate." Taken together, the wide-ranging and incisive responses begin to offer something of a cognitive map of how designers might imagine climate anew.

Death Valley Urban Aqueduct, UrbanLab, rendering courtesy of UrbanLab. This text is an excerpt of the introduction to the forthcoming book *UrbanLab*.

FOR AN AESTHETICS OF SUSTAINABILITY

The great artistic and intellectual challenge of the last century was coming to terms with massive technological changes that undermined previously held intellectual certainties and challenged artists to produce work as radical as the new world they saw unfolding around them. As the futurist poet Vladimir Mayakovsky put it, "After seeing electricity, I lost interest in Nature. Not up-to-date enough." But modernism, as a complex cultural construct, is at once dependent on technology and at the same time independent of its precepts and imperatives. Advanced aesthetic work is rarely also technologically innovative.

The final third of the twentieth century saw an accelerating shift from the hard technologies of production to soft technologies of communication, as well as a growing awareness of the complex interdependency of humans and nature. If the grand narrative of the twentieth century was the progress of technology, the challenge for this century will surely be to come to terms with mankind's fraught relationship to nature. A growing body of ecological theory and environmental history is doing just that, and questions of climate change and green technology are now a prominent part of mainstream media debates. In architecture this is evident in the discipline's preoccupation with sustainability; and in urbanism, in strategies to address climate change, sea-level rise, and costal resilience. There is no shortage of innovative work, but it's also true that the primary focus remains technological and performative.

What is missing in this crowded field is a larger cultural and aesthetic project. In the face of massive questions such as resource scarcity; climate change; and the global supply of energy, water, and food; the discipline is relatively powerless to implement effective change. Small victories and modest improvements in efficiency will not turn the tide in the face of unchecked population growth, the modernization of the Global South, and the spread of the megacity worldwide.

Beyond piecemeal intervention, what is required is a change in political will and collective consciousness. And one of the things that creative work in architecture, landscape, and urbanism *can* contribute is to shift the horizon of imagination. This is a unique capacity of the creative disciplines and of architecture in particular. Firms such as Sarah Dunn and Martin Felsen's UrbanLab produce work that addresses hard issues of sustainability at the same time as they map out a new aesthetic territory. It's not a zero-sum game: We have an ethical imperative to respond to climate change with hard, evidence-based research. But creative work that constructs new subjectivities, new ways of being in the world, has a part to play as well. The speculative project of developing an aesthetics of sustainability is not at odds with that ethical imperative, but rather its necessary complement.

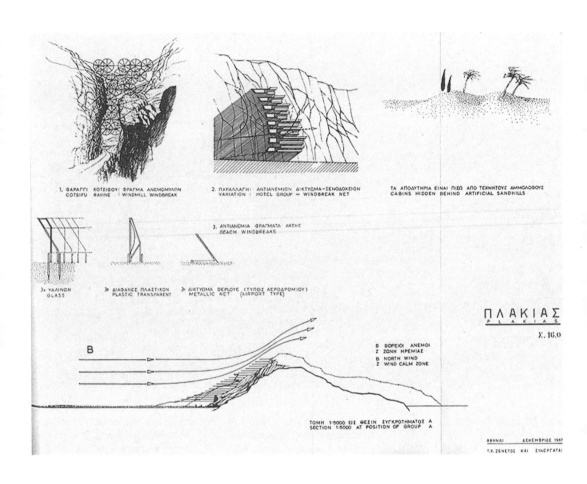

TAKIS ZENETOS'S FUTURISTIC HARDWARE

Climate change is the result of people working against the environment for short-term gains.

The 1967 proposal for a hotel development in Plakias, Crete by Takis Zenetos (1926–1977) changed the way I think about how buildings can occupy the natural world, demonstrating how architecture can work *with* the environment. The proposal—part of Zenetos's radical modernist body of work, which Herman Hertzberger has called "scandalously underrated"—is set in the beach, cliffs, and ravine of Plakias, one of the then-unspoiled wonders of southern Crete. The main body of the complex juts out from the side of the cliff, functioning as a "windbreak net," while the ravine hosts an intricate veil of wind turbines. The beach bungalows are camouflaged beneath sand dunes. Each element of the project plays its own part in making the most of the site's existing wind conditions, together orchestrated into something we might describe today as a large-scale work of land art, a science fiction scenario for inhabiting a strange planet. Neither humble nor discreet, it's an architecture that turns its ecological functions into a natural wonder— a spectacle merging the ancient and the modern, much like a drone flying over the Grand Canyon. The Plakias project manages to be environmentally conscious and perhaps even energy independent while boldly proposing itself as a futuristic piece of hardware, meant to run with the complex operating system that is nature and, subsequently, climate change.

Courtesy of Diana Balmori.

THE ITSUKUSHIMA SHRINE'S CONSTANT CHANGE

In the Seto Inland Sea on the island of Itsukushima, near Hiroshima, stands a shrine, Miyajima, that might be the most beautiful encounter of architecture and landscape I have seen. Originally built in 593 CE to honor a goddess of the sea, the area's complex tides make the shrine, a Heian style wood building, stand on sand at low tide. At high tide, the shrine and *torii* (gate) stand in water, seeming to float on it.

This bold setting in the sea is haunting. Unlike the usual Western approach to waterside projects, in which defensive walls are constructed at the water's edge to separate sea from building, here water and architecture form an intimate association. It's this association that made my work change direction.

It may seem that the shrine is set in defiance of the sea, yet its back and sides are protected by the hilly island. And it is further protected as it is located on an inland sea. Conceptually, Itsukushima is working *with* the natural elements, not against them. This is the direction I have chosen for my work, accepting landscape's constant change.

Architecture strives to be timeless; landscape, to be ever-changing.

The shrine has been rebuilt several times because of typhoons, fires, and seawater's natural processes of decay.

The rising sea levels of climate change will bring about one more rebuilding effort—in my mind, its structure will need to be floated so that it can rise accordingly. In a new way, the constantly changing sea will remain the shrine's timeless element.

Cover of first edition published by the Museum of Modern Art, 1964.

BERNARD RUDOFSKY'S LESSONS TO BE RELEARNED

Bernard Rudofsky's *Architecture Without Architects* is the architecture book from my parents' library I remember best. Modestly subtitled "A Short Introduction to Non-Pedigreed Architecture," the name of the volume concealed his rather grand ambition to challenge Western architectural history and recast the study of the built environment. Historians, he believed, had constructed architecture's chronology as if "dating the birth of music with the advent of the symphony orchestra." Seeing the absurdity of this, Rudofsky set out to reset the clock and the canon.

Through his highly popular book and an exhibition of the same name at the Museum of Modern Art, he exposed alternative examples of architecture—remarkable expressions of human ingenuity—to a large public audience. Functioning as collector, curator, writer, and book and exhibition designer, Rudofsky literally expanded the historical frame of reference, using dramatically cropped, full-bleed, and often aerial photographs of hill towns, bridges, arcades, and subterranean dwellings to draw formal and visual connections that traversed continents and centuries.

Architecture Without Architects relates directly to the current conversation on the environment in that many of the buildings and structures it examines work in concert with natural systems and do not rely on external energy sources. But more importantly, Rudofsky's pluralism—aesthetic, geographic, disciplinary—reflects the kind of thinking we as architects need to employ today. As he wrote in 1964, "the wisdom to be derived [from studying the wide-ranging work of anonymous builders] goes beyond economic and esthetic considerations, for it touches the far tougher and increasingly troublesome problem of how to live and let live, how to keep peace with one's neighbors, both in the parochial and universal sense."

Courtesy of Ruth/Kaniere. Parts of this text appear in Ila Berman, "Fluid Machines," from *GAM.07 ZERO LANDSCAPES: Unfolding Active Agencies of Landscape.*

KUTH/RANIERI'S
FOLDING WATER

Folding Water, designed by Kuth/ Ranieri Architects, is a permeable ventilated seawall intended to wrap the coastal regions of San Francisco most vulnerable to flooding. (The project was initially designed for the City of the Future competition in San Francisco organized by the History Channel, Infinity, and IBM in San Francisco, January 2008, and was later redeveloped as a winning entry for the Rising Tides competition in July 2009.) The architects' strategy, which attempts to mitigate dramatic global and climatic transformations by regulating rising sea levels, is a complex response to the need

for urban containment that both acknowledges, and draws from, the ever-changing fluctuations of the natural environment. Part of a new breed of experimental infrastructural projects, *Folding Water* occupies that amphibious zone between land and water while attempting to capitalize on the opportunities proffered by the dynamic hydrogeography in which it is immersed. Operated by the inflection and bifurcation of the water's surface, the folded wall of water is proposed as an alternative to the static, solid levee wall—its fluidal perforated surface dependent upon natural tidal movement to manage alluvial and biotic flows across the wall's variable depth. Solid and liquid, defensive and permeable, the seawall is

therefore both a barrier and an instrument for fluvial exchange, a flexible system generated from a blending of mobile and static attributes that coalesce to create a new bio-technological ecological infrastructure. The fold enables the waters between ocean and bay to be divided, spaced, and mediated, yet still remain continuous, producing a dynamic flexible joint in a seamless surface—a critical response to our need for both separation from, and connectivity to, the mutable nature that surrounds us.

ARTIFICIAL ISLANDS

Staggering urban sprawl has had an undoubtedly dramatic influence on the environment, as expansive urbanization and the architecture that accompanies it has spread layers of concrete across any available inch of land. Construction, meanwhile, is one of the most pollution-generating industries, creating an intrinsic confrontation between architecture and climate change.

The most sustainable way to build, we know, is not building. Rather than reinforcing excess and extravagance, architecture should aspire to greater humility, to be a practice that may even ameliorate the harm our buildings have caused so far. This requires us to start thinking seriously about global warming and how, as design practitioners, we can alleviate climate change as much as possible.

If building remains this destructive to nature, we are condemned to living with scarce resources sooner rather than later. Instead of looking to build more, or even to build green, we should seek longevity in the existing built environment. We must join forces to ensure these expanding urban enclaves remain habitable and that our designs vibrate harmoniously with (and within) their context, aiming to improve local conditions and to not harm the environment more than the harm of building itself.

Dubai is a city that has grown incredibly fast and spent a fortune raising buildings and creating outrageous artificial environments. It has, in turn, received much deserved criticism over the amount of resources invested in building such hyperbolic architecture. To me, this is an example of what is not to be done as a responsible designer. We must come to a deeper understanding of the complex relationship between our labor and climate change—from the fundamentals of our own practice to the details of what we design.

Where will building a system of artificial islands take us? How can we redeem architecture from its negative impact on climate change and resource exploitation? Planting some trees or turning a rooftop into a green roof is only seeking "ecological redemption"—and is definitely not enough.

FEDERICA M. SOLETTA'S *OCULUS*

Federica M. Soletta's 2013 *Oculus*—an iterative sequence of visual inquiries into what is probably the most significant aperture in the history of architecture (the circular lacuna at the summit of the Pantheon in Rome)—stands, for me, as a paradigmatic instance of generative obliquity at the intersection of art and research. A hybrid sketchbook/lab-report in the form of a scroll 30 feet in width and 24 inches high, *Oculus* offers a highly original window (series of windows?) onto the nexus of weather, thought, and the built environment. Roofs are, of course, "epistemological," in that the history of being sheltered from the elements is inextricable from the history of rational inquiry. By these lights, holes in roofs (and particularly an intentional hole 27 feet in diameter and originally gilded at its lip)

merit close attention in any effort to triangulate building, dwelling, and thinking. It is also the case, however, that holes—which are, after all, conditions of local absence—are difficult to see clearly. In this sense, *Oculus* can be understood as a visual aid: it works as a viewer for a key void. Notable, for those of us who concern ourselves with climatological cyclicities and secular drift, is the way in which the Pantheon's oculus transforms the interior space of this significant edifice into a microcosm of celestial dynamics, registering the seasonal and diurnal movements of the sun as the play of light and shadow on the circumferating walls, even as the entire structure is made to center on an axis of direct columnar access to the rain and snow.

Is there a moral? One can always try: *Those who build glass houses leave home to throw stones. Those who close their temples to the sky eventually need smokestacks.*

THE BOULEVARD DE L'IMPÉRATRICE

The Boulevard de l'Impératrice and the accompanying Ramps d'Argel (later renamed Boulevard République and most recently Boulevard Che Guevara), built in nineteenth-century colonial Algiers by the French architect Charles Frederick Chasseriau, exemplify the value of a complex and multifaceted urban infrastructure. In its form and organizational structure, the boulevard effectively responds and adapts to multiple cultural, economic, technical, and social pressures, allowing it to accommodate new urban forms and uses over time. In section, the project makes up a series of utilitarian ramps that move goods from the level of the port up to the level of the market, creating a thick ground that mediates between city and water. In plan, the project favors a boulevard facing the water, setting the stage for an urban *enfilade* that defines the main public thoroughfare for the city and provides it with a monumental gallantry when seen from the sea. While conceived centuries before the notion of climate change—and under a colonial rule geared toward the extraction of both environmental and human capital—the layered boulevard provides important lessons on how to move beyond the mono-functional legacy of post–WWII urban infrastructure, much of which still guides discourse and proposals for the contemporary city. As climate change continues to gradually alter urban form, particularly in its interface with bodies of water, the spatial synthesis brought forth by the boulevard can provide powerful insight for meeting these challenges.

NICHOLAS DE MONCHAUX'S
REAL ESTATES

Inspired by Matta Clark's *Fake Estates*, Nicholas de Monchaux's 2009 project *Local Code/ Real Estates* is an active form that proposes to use new GIS mapping techniques to identify thousands of publicly owned abandoned sites in major US cities. In New York the total area of this "scattered resource" when seen in aggregate is comparable to Central Park. In San Francisco, de Monchaux found fifteen hundred remnant sites that also happened to correspond to the areas of the city troubled with economic and environmental problems. Rejecting previous models of parks as visible areas of recreation, the project proposes to develop these areas into a connective tissue of green. When land-banked and analyzed for its performance as a climate infrastructure, the network presents "an essential archipelago of opportunity"—values residing in land beyond those associated with the typical real estate ledger. De Monchaux writes, "A focused web, threaded through the grain of existing urban fabric our systematic intervention will be... a distributed immune system for the 21st-century city." While climate change responses often call for belt-tightening, *Real Estates* crafts an expression of *interplay* between spatial variables.

Nicholas de Monchaux's "Local Code," San Francisco Case Study, installed at SFMOMA, 2012; photograph by Matthew Millman.

"PITCH IT WITHIN AND WITHOUT WITH PITCH"

"Even since childhood," one of his biographers writes of George Bush, he had "been a bookworm." There is an unnoted element of irony to this familiar enough sort of observation. For bibliovores it is perhaps worthwhile mentioning that Eric Carle's perennially popular children's picture book, *The Very Hungry Caterpillar*, was originally named *A Week with Willi the Worm*, until Carle's editor suggested that a lepidopteran protagonist would be more appetizing. Not entirely unrelatedly, it was "The Pet Goat," a reading exercise about a girl, her pet goat who "eats too many things," and, of course, a "car robber," that then-President George W. Bush was sharing with students at the Emma E. Booker School in Sarasota, Florida, when, at 9:05 a.m., September 11, 2001, White House Chief of Staff Andrew Card, a former high-ranking General Motors executive, went up to the president and whispered in his ear: "A second plane has hit the second tower. American is under attack."

This brings us to the element of irony, related as it is to the passage: "Make thee an ark of gopher wood; rooms shalt thou make in the ark, and shalt pitch it within and without with pitch" (Genesis 6:14 KJV). What in the world—a world for the imminent climatic (or is it climactic?) destruction of which the righteous Noah was building his ark—is gopher wood? In his *Notes, Critical and Practical, on the Book of Genesis* (1838), George Bush enumerates the linguistic and paleographic complications emerging from the Hebrew hapax legomenon "*atzei gopher*." Bush was not alone in opting for the cypress hypothesis, arguing that among its other properties this variety of wood was very compact and heavy and was "not liable to rot or become worm-eaten."

One can only surmise that Bush is referring to shipworm, which is not a worm at all but rather a mollusk. But if it was available for such depredation, why would it have thus been especially exempt from universal extinction, or would a pair of these "unclean" beings have paid for their passage by eating away unrewardingly at the ark's (cypress?) timbers. It is not for me to speculate, though it seems likely, based on some readings of Charles Darwin, that it is earthworms who remain the likely source of salvation for our sunless age. Instead, I will conclude, before I have even begun, by recalling the words of an unconvincing savior, who, during his inaugural address, shadowed as it was by the seemingly credible threat of an attack by Somali extremists, said that it is time to "put aside childish things" (1 Corinthians 13:11).

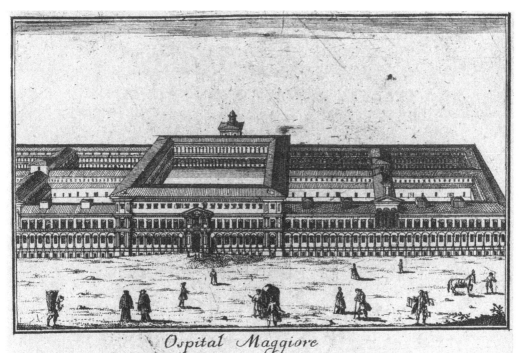

Marc'Antonio Dal Re, engraving of the Ospedale Maggiori from his *Vedute di Milano*, ca. 1745.

THE COURTYARD OF
THE OSPEDALE MAGGIORE

Imagine a conversation sometime in the 1400s between the Duke of Milan, Francesco Sforza; his architect, Antonio Filarete; and his engineer, Guiniforte Solari. They are planning a community hospital for the city of Milan. They agree on an organization for their Ospedale Maggiore. Six hundred years later, their building is still in use.

Ospetale Maggiore is a city block with courtyards. Its "secret" gardens are robust enough to have survived centuries of change, simple enough to continue giving pleasure. By framing the sky, courtyards encourage natural convection. With sunshine and shadow, hot and cool air move in relation to one another. Pleasure and comfort mingle. We trawl precedents to help us invent, to search for characteristics in buildings that have withstood the test of centuries, that are based on common sense—beyond fashion—with proven spatial qualities.

Ospetale Maggiore, with its timeless strengths, continues to influence our work: In Bocconi University in Milan, we imagined a system of raised courtyards, suspended over the "floor" of the city of Milan, to provide light and shade. Each office, protected from wind and direct sun, has windows to a courtyard. Air moves freely. In the School of Economics in Toulouse, France, a vertical twenty-first-century courtyard forms its heart with its own microclimate. Our Institute Mines Telecom complex in Paris, Saclay, is organized around five courtyards, each with its own distinct character; four relate to the seasons and one has mature trees in real ground. Its overall form protects from the wind, and all offices have windows onto the courtyards or the surrounding campus. For architects to engage with climate change, we first need to understand local climates.

The earth is a fragile entity. As architects, our responsibilty is to take care of the natural resources at our disposal. How can we harness, spatially and socially, the *free* resources of light and air? What architectural forms help us do so? We are fascinated by the simple power of the courtyard.

Courtesy of the Lazar Khidekel Society.

LAZAR KHIDEKEL'S LANDSCAPES AT THE END OF THE WORLD

The end of history, like the death of god, was a prematurely announced *coup d'état.* Capitalism's hyper-efficient flexibility tirelessly turns the abstract and ardent subversions of its detractors into commodities. No concept, matter, territory, or state can escape its alchemistic grip. After all, the environment—even in a dilapidated state, approaching demise—can still be manufactured, marketed, and consumed. The Anthropocene, that ultimate conquest of humankind over its environment, makes the technological utopia of the Cubo-Futurist "Victory over the Sun" more than just a fantastic dream. The clashing tensions of the ideological dialectics of the early twentieth century resulted in the imaging of a parallel, almost unheard of universe.

In the landscapes of the Suprematist architect Lazar Khidekel—a series of sketches for futuristic cities he made in 1928–32—totems without qualities levitate like weightless clouds. White, as in José Saramago's *Blindness,* the slender segments of Khidekel's cities fly over mountains and rivers, only slightly touching the ground before defiantly taking off to soar at impossible heights. Desperate times, as many have observed, make it easier to imagine the end of the world than to visualize the end of capitalism. Is the End of the World by Science a crumbling gospel or a latent threat? In our Fevralist fervor, we contemplate these unbearably light mastodons, dreaming of the possibility of absurdist acts of architecture with landscapes radically untouched.

BEING WITH WATER AT THE VILLA LANTE AND ALHAMBRA

As in much of the world, in California our water infrastructure insulates us, not only from extreme conditions such as flooding or drought, but also from seeing and being with water in the landscape. Towers, tanks, dams, pipes, aqueducts, and channels *contain* water, working to keep climate separate from daily life. While the North American Monsoon causes periods of rain and periods without, it is easy for us to fail to see these changes and to ask how our experiences and behaviors involving water might change in concert.

I have drawn inspiration for ways to be with water from two seventeenth-century sculptures. The Cardinal Table, in the Villa Lante gardens in Viterbo, Italy, is a moderate closing gesture to a chain of grand fountains and grottoes. The solid stone table, which holds a trough of water in an oblong centerpiece, integrates moving water with the everyday breaking of bread. The Alhambra's Water Stair, in Granada, Spain, is defined by handrails whose carved channels direct cascading water. When ascending or descending the staircase, one's fingertips can simultaneously feel the water and its years of impact on the stone. For the last four hundred years, both of these sculptural elements have synthesized everyday human infrastructure with water infrastructure. Making water a visible and tactile presence intertwined with daily life, they offer simple yet resonant ways to experience seasons, weather, water, climate, and landscape.

Cardinal Table photograph by Walter Hood.

Le Corbusier in front of his Cabanon in Roquebrune-Cap-Martin. Photograph by Willy Boesiger, © Fondation Le Corbusier / A.R.S. (2016).

LE CORBUSIER'S CABANON

When speaking about climate change or sustainability, usually we hear about energy, materials, and technology. Yet the small (less than 12 square meters) but meticulously designed cabana Le Corbusier built for his vacation house at the Cape Martin presents a more fundamental principle for sustainability and an antidote for a world driven by endless greed and materialism: modesty, or, one step further, asceticism.

Writing this on the first day of December 2015, the most polluted day recorded in Beijing this year, the sight of buildings even just across the street fades away, and I cannot stop thinking about how capitalist desire has expanded the city endlessly. It has exhausted the natural resources for instant gratification, creating almost irreversible environmental consequences.

Modesty isn't about not building but rather about designing and building carefully with absolute necessity. It also means designing great public spaces and building great urban centers with high density and pleasant, affordable living to bring back the people from the periphery. To be modest is also to adjust our notion of comfort, the difference of two degrees in temperature control means a lot to our endangered natural world.

ENGINEERING WITHOUT ENGINES: VENTILATION CHIMNEYS IN SANA'A

In Sana'a, Yemen, a field of thin chimneys rises above the city, capped with large, flat wind funnels—all facing the prevailing winds—to naturally ventilate the six-story buildings below, without any moving parts. Wind, common across the arid Yemeny landscape, is harnessed through a simple act of design to cool hot interiors, modulating the thermal environment with superior efficiency. This is a far cry from the air conditioning and HVAC systems that regulate so many buildings, reducing them to big, blank boxes tube-fed by a whole arsenal of machines.

Bernard Rudofsky once called such vernacular design "architecture without architects." It's a label that might, too, be applied to many feats of engineering. Instead of simply outsourcing environmental performance to engineers and product manufacturers, can't architects again take up the mantle of environmental design to allow the permanent physical design of the building to play an active role in conditioning human habitation?

A RIVER RUNS THROUGH IT: DOMESTIC ENGINEERING IN THE MOUKHTARA PALACE

In the late 1790s Sheikh Bashir Joumblatt, a leader of Lebanon's Druze community, rose to prominence as ruler of the Mount Lebanon range. In a time of sectarian tensions between regional religious groups, Sheikh Bashir embarked on several civic projects that benefited the population regardless of religious affiliation: he built various religious structures, instituted charity for the poor, and undertook large-scale infrastructure building. One such infrastructural endeavor was a canal to bring freshwater into the village of Moukhtara from the Barouk River in 1806. The Moukhtara Palace, the home of the Joumblatt family, was also served by this new canal. In fact, the main water canal runs right through the palace—a marvel of domestic engineering.

Building on centuries of regional water management techniques and expertise (from the Roman Baths to the Ottoman Hamams), the new stream runs underneath and inside the palace complex, branching out to flow into multiple rooms, chambers, and cavities, and performing a variety of uses. The water then emerges as a cascading canal and fountain in a splendid inner garden. In between, a variety of visible and invisible infrastructural systems mobilize the stream's flow for the benefit of the house's internal climate, its agricultural efficiency, and even its political capabilities. Heated by a furnace, the water travels in floor cavities to warm the living quarter of the house in the cold winter; meanwhile the vapor is channeled into the walls of baths and released in steam rooms. In another branch, cascading water powers a mill used to grind wheat, olives, and other food stocks. Besides serving as utility inside the house, channels irrigate the agricultural grounds of the palace where olive and orange trees grow. The fountain within the heart of the complex not only creates a cool microenvironment in the hot summers but also produces a soundscape that eliminates any potential eavesdropping, making political conversation more secure around those spaces. In a final branch, water continues cascading into the village down the hill, serving freshwater to the peaceful town of Moukhtara and the villages beyond.

Abel's case is part of the research project *Different Kinds of Water Pouring into a Swimming Pool*, developed in 2012–13 by Andrés Jaque/Office for Political Innovation with the support of CalArts, California Institute of the Arts (Los Angeles).

FROM COLONIA TO SILVER LAKE: THE RECOUPLING OF ECO-SOCIETIES AND CLIMATE CHANGES

In 1968, Abel left the town of Colonia, Uruguay, where he had been raised. In the course of ten years, farms there—such as the one Abel's family had lived near to up to that point—had gone through a process of consolidation into upscaled agricultural exploitations. The change in Colonia's environment prompted the migration of a significant part of the young *Colonienses*. In his current home in Silver Lake, Los Angeles, Abel has occupied the space between his backyard and the freeway with a stepped vegetable patch. Chatting daily through Skype and exchanging the seeds of their best crops though letters, Abel's *Colonien*

relatives and old friends now living in Venezuela, Italy, or Alaska have reconstructed a transmedia version of their past eco-society, no longer concentrated in Colonia but constituted through networks. Whereas the previous version was perceived by them as natural, place-based, and fixed, the new setting is perceived as constructed, unstable, and technologically mediated—a kind of recomposition in the aftermath of trauma, both personal and ecological.

As the expectation that we might reverse climate change weakens, and as the effects of climate change—in the form of migrations and the decoupling

between human relational ecosystems and their past demarcations—are no longer avoidable, the persistence-through-recomposition of this tiny *Colonien* milieu suddenly stands out as a plausible beginning for a conversation on architecture and adaptation to change.

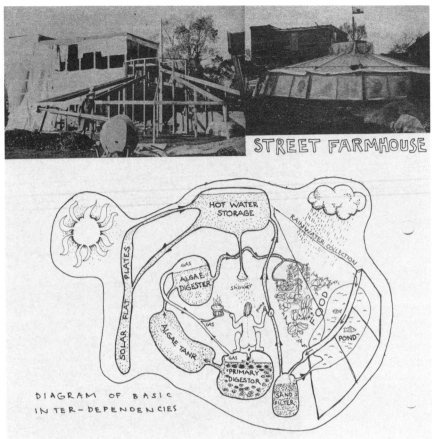

GRAHAM CAINE'S
DIRTY PHYSIOLOGIES

Graham Caine's Ecological House or Street Farmhouse built in London in 1972 is a striking illustration of how architecture intersects with climate change. In his diagram, Caine portrays himself and his family as guinea pigs in order to test the function of several components of the house. He experimented with his waste, his cooking habits, and his use of water, monitoring closely every activity of daily practice until the day the house was demolished in 1974. The architect, therefore, was an indispensable biological part of the house he built, and he portrayed himself as a combustion engine for generating electricity, connected to the house in a diagram where excretion becomes a vital constituent of the system's sustenance. Describing his house as a life-support system, Caine satirically argued that the architect, being involved with the house's biological cycles, might now relate to his own shit.

This illustration of the dirty physiology of the body, which is woven into the ecology of habitation, is far away from contemporary illustrations of climate change depicted in statistical surveys. Commonly, environmental concerns promote a conservationist ethic and a list of cautionary daily practices of scarcity. We tend to think of human waste as a phantom material condition, related to the management of urban resources; yet, it infiltrates the air and water we breathe. The exclusion of the body's physiology from contemporary environmental concerns does not allow us to construct a more nuanced perception of climate change and to motivate behavioral change. As Caine shows us nevertheless, to effect change, one needs to get involved and dirty in architecture production.

THE INFRASTRUCTURAL COMMONS OF W57

Across the street and just downstairs from Bjarke Ingels's W57 project sits a hidden megastructural complex that makes W57's hedonistic sustainability possible. The commission was prompted by Ingels's provocation to developer Douglas Durst: "Why do your buildings look like buildings?" Unlike Durst's high-tech but conventional towers, W57's hyperbolic gesture serves dual purpose, as urban icon and shading device, comforting residents with fresh air and sunlight while sheltering a miniature Central Park in its "courtscraper" form. But W57 doesn't work alone. The building shares sustainability features (such as a blackwater recycling system) with the Helena, a LEED Gold-rated residential tower, also owned by the Durst Organization, which occupies the other half of W57's block. And W57 couldn't have secured rezoning approval without the promise of affordable housing units in *another* adjacent Durst building, called the Frank. W57 cleverly shares the costs of sustainability with its neighbors, capturing scales of efficiency typically not enabled by Manhattan's free market grid. The Durst superblock miniaturizes Manhattan's infrastructures—not unlike the city's district heating system powered by the McKim, Mead, and White–designed steam plant just across the street. W57's iconic form peddles a compelling image of guilt-free luxury, but what if it also gave shape to districts of infrastructural investment? How might we imagine the commons at the brink of climate change?

Original image courtesy of the Durst Corporation, modified by the author. This research on W57 and its neighbors was first published in Janette Kim and Eric Carver, *The Underdome Guide to Energy Reform.*

Courtesy of Chip Lord.

HELLO AMERICA—THIS IS THE FUTURE!

Twizy is not architecture, but it is an object of design and a zero emission vehicle (ZEV). Twizy was designed by Françoise Leboinne and Luciano Bove for Renault and first sold in France in 2012, but it's not available in the US. I found this one parked in my neighborhood in San Francisco, a sign or symbol of the changing times in which we live. Four of the six cars in the photo are Priuses, and the bicycle is, of course, human-powered.

The Paris Agreement negotiated at COP 21 has been called a "turning point" in the fight against global warming. I offer this photo as proof, I hope, of the turning point in the fight against the dominance of the automobile. In San Francisco we have Uber, Lyft, City Car Share, Zipcar, Getaround.com, Bay Area Bike Share, and various Silicon Valley private buses, as well as a workable public transportation infrastructure. And it will take all of these options to remove the privately owned automobile from the central position it holds in urban design and planning. It inspires me to see the beginnings of this movement away from internal combustion, but we can't all drive Teslas.

Courtesy of TYIN tegnestue Architects.

Courtesy of TYIN tegnestue Architects, from *Behind the Lines.*

AID ARCHITECTURE AND CLIMATE CHANGE

In 2014, TYIN tegnestue Architects authored an honest account of the failures of their aid project, *Klong Toey Community Lantern*. In the book *Behind the Lines*, they revealed that their gifted structure had been looted and degraded into a haven for miscreants (bottom), unrecognizable from the utopic photographs of children playing in the created space celebrated in glossy architectural magazines (top). The project stood no chance, as it was built with a lack of engagement and understanding of the community who were to receive and maintain it.

Aid architecture is about people, as much as it is about the design of physical structures; the formula of deploying "foreign" buildings without community involvement engenders indifference and dependence rather than recovery. Climate change guarantees a (social) landscape of post-disaster sites, as catastrophic disasters obliterate not only the physical but the social order of a place. Hence "rebuilding" in this context goes beyond restoring the built environment that existed beforehand. It requires a sensitive rebuilding of the community itself.

While it is an imperative for architects to be more critically engaged in aid architecture as it confronts the effects of climate change, the issue is that it requires our long-term participation, complicated by heavy financial and emotional investments. The challenges we face, as TYIN have, compel us to find viable models of architectural practice that do not set us up to fail, but allow us to sustain.

Courtesy of Felipe Mesa.

ECO-CABINS IN RÍO CEDRO

The journey by car from Medellín, Colombia, to Río Cedro requires a ten-hour drive, crossing the city of Montería before taking a dirt road toward the Caribbean Coast. From there, it's a thirty-minute walk along the beach to the hotel, dragging your suitcases behind you or carrying them with the help of donkeys. The eco-hotel consists of independent cabins located on a small, flat site six meters above sea level. Each cabin has an outdoor bathroom and they all share a roofed dining area. The cabins benefit from the shade of tropical trees and the air currents brought by the sea. Social areas are downstairs with hammocks, tables, and chairs that can be moved freely within the rectangular space. A big room is located on the second floor, at the level of the tree foliage. Air currents, natural light, native fauna, and tropical rain cross the bottom level; the upper level is dark, contained, safer, and less permeable—it's a spatial arrangement found in the traditional buildings of the region.

In order to avoid the use of mangrove wood, an endangered species that has been widely extracted in the region, these cabins employ the abundant *matarratón*. Handrails are built in *palmalata*, following regional traditions, and tied with rubber fibers extracted from worn away tires. The roof is woven in *palma amarga* leaves; they're impermeable to water and have bioclimatic, acoustic, as well as thermal qualities allowing the interior to remain cool. A tank set at the highest point of the site collects rainwater and moves it by gravity to a second concrete tank from which water is taken with the use of calabash containers.

PALAZZO FARNESE AS INTER-SEASONAL ARCHITECTURE

Geography, climate, technique, and culture lie at the origins of typological and morphological evolutions in architecture and urban development. Igloos, compact Nordic buildings, or African wind towers are examples of responses to specific environmental conditions, of a "wise" architecture that contributes to an overall sense of ecological balance.

Palazzo Farnese in Caprarola is one such example that inspires our firm's research through the intrinsic intelligence of its form. Begun by Sangallo and completed by Vignola, it was initially conceived as a fortress, but bastions at each corner gave way to large terraces that look out onto the surrounding countryside. The palace both stands out against the landscape and blends harmoniously into it. A road leads directly from the building to the village center below, visually tying the two together while asserting the palace's preeminent position.

Within, Vignola built a circular courtyard, a true stroke of genius for its heliotropic strategy. The summer usages are located to the north and west, away from the direct rays of the sun, while winter usages lie instead to the south, to benefit from the solar exposure. The thick walls store and emit heat or cool, while the courtyard galleria increases ventilation. The roof and the courtyard collect rainwater for use inside the building and irrigating the grounds.

The building sits atop a true logistical platform: all the building's services—kitchens, storehouses—are linked through a veritable network of "technical" connections. Among the many examples of the building's ingeniousness, the main stairway, known as the *Scala del Cartoccio*, was also used to send messages. The main balustrade in marble was in fact a hollow tube containing sand, down which one could slide a bundle of paper (a *cartoccio*) to the lower floors.

Climate change requires us to be able to understand such examples in order to imagine others, to incorporate their know-how to invent a modern, inter-seasonal architecture that contains multiple climate strategies.

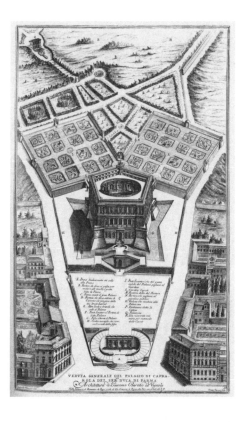

CUMULUS AND THE PLEASURES OF CLIMATE

Architecture that engages with climate often deals with resilience (protection from increasing storms, rising waters, etc.) or with minimization of impact (sustainable materials, energy efficiency). Cumulus, a 2008 proposal by SMAQ for apartments in Oslo, does these things but it does something more: it *celebrates* the climate.

For SMAQ—the architecture firm of Sabine Müller and Andreas Quednau—urban life is inextricable from environmental dynamics. Cumulus is an urban strategy for an existing neighborhood that centers on a square where water cycles are both pragmatic and spectacular. In Oslo, one of Europe's rainiest cities, rainwater collected on roofs and façades is stored strategically to reflect light into apartments and is released in a fall waterfall festival, flooding the sunken plaza for winter ice skating. (Water is also used for watering allotment gardens and for laundries.)

SMAQ not only synthesizes the dynamics of both climate and urban culture but generates a new architectural typology in which this synthesis is legible. They remind us that within an architecture of necessity and pragmatism, there is still space for pleasure.

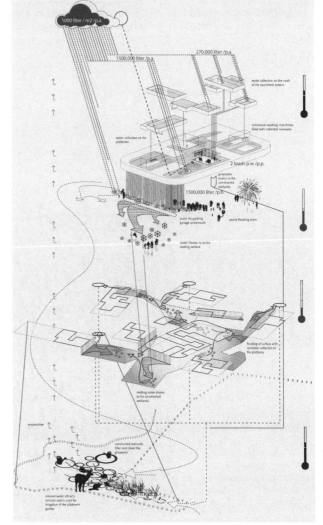

Courtesy of SMAQ.

Courtesy of Diller Scofidio + Renfro.

ENVIRONMENT AS MATERIAL AT *SPIRAL JETTY* AND BLUR BUILDING

On the issue of the relationships between design and climate change, I think of two related but materially distinct projects that instigate entirely new and dynamic relationships with the environment, weather, climate, and atmosphere: Robert Smithson's *Spiral Jetty* of 1970, in the murky pink waters of North America's barren Great Salt Lake; and Diller Scofidio + Renfro's enigmatic Blur Building on the crystalline Lake Neuchâtel for the 2002 Swiss Expo.

Heavy, low-tech, a product of its geologic place and its construction machine making, *Spiral Jetty* sets up a new range of water conditions within its embraces, expressed through changing intensities of pink, gray, and brown. But the jetty can be entirely subsumed by the lake over longer periods of time, as the larger environment responds to broader trends of rainfall and drought. Its disappearance produces traces of its existences and longing for its return; its re-appearance presents a crunchy, crusty surface that, when walked across, produces a sonic and physical experience evoking the larger environmental cycles it engages.

Light, high-tech, at times pure sensation and phenomenon, Blur hovers in the air as a mix of steel and electronic sensors, and the stuff of its immediate environment: water, fog, mist, air, light. It is at once atmospheric and mystical— a true fusion of the climatic conditions of its place and the design and engineering ingenuity that brought it into being. It constantly changes in response to ambient conditions, creating an experience of being engulfed in fog and light that is as all-consuming and yet as fleeting as a summer rain shower.

Both projects register very particular conditions—one tends toward the primordial, the other toward the ephemeral. Through design and cultural intervention, both open our eyes in beautiful and compelling ways to the dynamics of the environment, to the weather, and to the basic stuff of life on earth.

Courtesy of Makkassan / Bangkok / Slum Library for sniffing books / M4-Ann Arbor.

CONCRETE[I]LAND

As of 2012, around 33 percent of the urban population in the developing world lives in slums. Those territories could be described as zones of *non-droit*, informal economies, and criminal organizations, or as places of bottom-up economies and politics, where, through local neighborhood negotiations, residents question the distribution of power in the city. Concrete[i]land plays off of this antagonism.

Concrete[i]land experiments with forbidden substances. By reintroducing human waste in the loop of production, the project harnesses ecosophical entropy, asking its viewers to reconsider the entire footprint of substances both material and immaterial, objective and subjective, sweet and repulsive. It proposes we transform what we cultivate, cook, swallow, digest, metabolize, shit, decompose, recycle, to fantasize a psycho-vitalist energy. The technologies of today are to replace the pigs of the European city in the Middle Ages, metabolizing feces in a physiological and mechanical transformation to face the filthy, grimy, grubby, mucky, drossy condition of our mind ("The River Styx Runs Upstream" of Dan Simmons).

To "change the climate" means to escape from the established post-hygienist discourse, from the agitprop of greenish simulacrum or techno-fetishist ingenuity. To change the climate of climate change is to disqualify the propaganda of neo-petit bourgeois franchise, to refuse the chic, the smart, the fair, the fake vintage *habitus* (to quote Bourdieu) and instead acknowledge its opposite: Nature's life cycle produces nitrogen; it smells, it stinks, and that's the condition for its recurring rebirth.

THE FLOATING DOCK AT THE THAI BUDDHIST TEMPLE THAT NEIGHBORS OUR STUDIO

In 2011, nearly 16 billion cubic meters of water descended upon the Bangkok metropolitan area, causing $45 billion worth of damage. This was no natural disaster—human error was to blame, both in terms of the mismanagement of dam water and the gross misjudgment of the effects of climate change. The 100 acres of Rangsit University, where we teach and have several projects, were inundated with an average of 9 feet of water. It came slowly at first but rose steadily, imperceptibly, and then stayed for nearly three months, paralyzing the university. Devoid of life, the waters were still, ominous, stagnant, and black as oil.

Inspired by this lowly dock that sits on the river near our studio, we designed and engineered a building for the university that could be more proactive when the floods come (and they will come again). We realized that flood-responsive architecture in Bangkok does not need to act like a boat, able to rise indefinitely as a unit, but merely needs to roll up its pant legs as the waters rise. We designed the ground floor slab as a concrete tub that usually rests upon its grade beams. But when the waters come, simple displacement activates the tub; the walls and the contents rise up the columns like a dock. The roof remains fixed and the fabric ceiling is collapsible like a paper lantern.

This design is not universal; it is specific for this type of flood in this context. But as our climate becomes more extreme, our architecture must learn to adapt in uncomplicated and ever more particular ways.

Courtesy of Studiomake.

RALPH ERSKINE'S RESOLUTE BAY, CANADA

In the early 1970s, the Swedish architect Ralph Erskine was invited to design a new town in Canada's far north, where the re-location of an Inuit community and the construction of a major weath-er station and military base in the 1950s had led to badly segregat-ed and impoverished conditions that the national government was hoping to improve with a more formal town plan. Erskine, con-sidered one of the few architects with Arctic-specific experience at the time—in part because of his earlier 1958 project for an "eco-logical arctic town," shown here—was commissioned to develop a cold-climate-responsive design that would integrate a future population of about 1,200 Inuit and Quallanat (the Inuktitut for "foreigner").

The design proposed a horseshoe-shaped, inhabited perimeter wall that was open on one end, which would include apartments for Quallanat and commercial and civic programs, with an array of detached houses in its center for indigenous res-idents. Two major formal moves were intended to respond to the extremity of the climate. First, the living wall was to protect the houses from wind and snowdrift, which often blocked doorways and damaged exterior cladding. Sec-ond, each structure was placed on piles to diminish the buildings' potential to warm the permafrost below. Though only partially real-ized (a significant portion of the living wall was completed before construction was abandoned in 1978), the project embodies an unresolved dissonance between forward-thinking architectural thought on climate and a markedly less progressive social worldview.

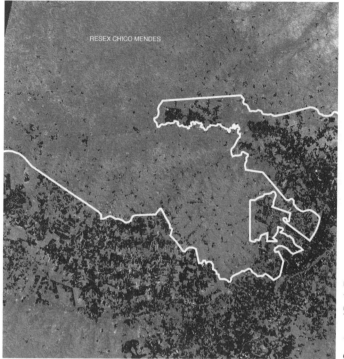

RESEX CHICO MENDES

Courtesy of Paulo Tavares.

RESEX: THE BATTLE FOR THE CLIMATE

In 1988, while the North American landmass was engulfed by devastating droughts, the NASA scientist James Hansen gave testimony before the US Congress to explain that, in his decisive words, "the greenhouse effect has been detected, and it is changing our climate now." Hansen's speech was a milestone, as climate change started to become a public concern beyond scientific laboratories and turn into a question of global politics. In the same year, the United Nation International Panel on Climate Change was established.

Across the Equator in Amazonia, 1988 was also a devastating, violent year, emblematized by the killing of forest defender Chico Mendes in the state of Acre in Brazil. As the development/colonization projects installed by the military dictatorship advanced, forest burnings spread uncontrollably. Deforestation, combined with direct forms of violence such as targeted assassinations, became a weapon for the mass displacement of communities of rubber tappers.

Initiated in the mid-1970s, the rubber tappers movement that Mendes helped to organize surged as one of the main advocates for protecting Amazonia. Its landmark battle, like Hansen's advocacy, was a fight against climate change and changers, even if not voiced in the same language. But it was also

a conflict against the repressive state apparatus by which the military ruled and against the enclosure of the forest; it was a conflict for human and land rights.

In the politics of the rubber tappers, forest conservation was articulated with land reforms based on the creation of autonomous communal "reserves." This proposal was formalized in the mid-1980s in the *RESEX*—Extractive Reserves, a new type of territorial jurisdiction whose principles follow the model of indigenous sovereign lands. Different from the traditional views of conservation, in which strategies of environmental protection are set to limit human intervention, the concept of RESEX presupposes that the protection of the forest stems from its collective management. Its legal-spatial design functions to contain land expropriation and environmental depletion, thereby preserving the carbon-rich forests of Amazonia at the same time as it guarantees the socioeconomic autonomy of disenfranchised populations by

protecting the communal usufruct of land. The conservation of the forest was intrinsically related to the right to the commons; its survival was attached to the life of the people who cultivate it.

This satellite image analysis shows the region surrounding the Extractive Reserve Chico Mendes, a territory covering more than 930,000 hectares of forests that was established in the state of Acre two years after Mendes was murdered. Patterns of deforestation are marked in black. The borders of the communal reserve mark the threshold of deforestation, drawing a barrier between the forest and its elimination. These are the "red-lines" of global climate change, as indigenous peoples repeatedly stated in the Climate Summit in Paris last year—"we have drawn that line with our bodies against the privatization of nature, to dirty fossil fuels and to climate change."

Enclo

sures

The year 1606 marked the passage of the first of the British enclosure acts that would, over the next two hundred years, make private the vast lands that had previously been marked as "waste" or commons. Permitting lords to hem in what had been communal fields, enclosure transformed the British landscape and economy—latticing the countryside with fences, privatized property, and dispossessed peasants, and consequently accelerating urbanization. It also definitively changed the practice of farming as landowners embraced advances in agricultural science and new doctrines of economic efficiency—enabling new technologies and techniques such as crop rotation, fertilization, and controlled grazing, transforming land into a resource whose productivity could be optimized, waste products utilized, and environment capitalized. Today, two centuries after the final enclosure acts, efficiency remains the prevailing spatial currency used to think the relationship between humans and environment, and enclosure continues to be the model through which technologies of efficiency are spatially implemented. But what do enclosures enclose? Considering the multitude of cultural and social productions that architecture contains—from technodeterminism to labor— opens up a field of possibilities for architectural practice that extends far beyond demarcating land.

In calling for a radical reappraisal of "architecture," Amale Andraos casts the discipline as a synthetic practice capable of producing more porous forms and malleable formal thinking. Meanwhile, Daniel Barber revisits a midcentury moment in the history of Avery Hall, telling the story of the Form and Climate Research Group and its index of "adaptive design methods." Pep Avilés looks to the modernists' mediation of the architectural exterior, reviewing Siegfried Ebeling's 1925 *Der Raum als Membran* and the biological properties of the building membrane. Diagnosing the hydrophobic in architecture, Carolina González Vives unearths the secret routes of water, advocating against "dewatering" the urban ground and for a more transparent hydrologic infrastructure. In London, May Ee Wong assesses the techno-optimism of *Our Urban Future* with a visit to an exhibition on sustainable design at the Crystal, Siemens' Center of Competence Cities. Drawing a genealogy of risk that extends from the LEED Platinum–certified Bank of America Tower, Reinhold Martin returns us to the impossibility of enclosure, showing that the very systems we believe to be closed—from air-conditioning to solar paneling—are, in fact, actors in planet-spanning networks. In a similar spirit, Gökçe Günel alights in the Emirati desert to examine the unsustainability of social isolation in Norman Foster's Masdar. Finally, David Gissen offers an experimental and algorithmic tale of pollution and perception.

What Does Climate Change?
(For Architecture)

AMALE ANDRAOS

1

Klein's rallying cry has taken the form of a film, a book, and a global series of events. See Naomi Klein, *This Changes Everything: Capitalism vs. the Climate* (New York: Simon and Schuster, 2014).

2

One such effort can be seen in the conference *Climate Change and the Scales of Environment*, which took place at Columbia's Graduate School of Architecture, Planning and Preservation in the fall of 2015. The conference attempted to frame the broader effects of climate change across various academic disciplines, and several of the speakers have since developed their texts for this book.

Climate change changes everything. As Naomi Klein and others have observed, the realities of climate change—from our understanding of the human impact on the planet under current economic and political conditions to our settlement patterns and our energy and agricultural models, from our historical rereading of industrialization and modernization to contemporary visions of our future as one species among many—have altered the conceptual coordinates of inhabitation on this planet and complicated their social frameworks.[1] Now more than ever, there is a clear need to enlist emerging, diverse, and multifaceted forms of practice and action in a renewed engagement with the challenges of the present. To do this requires framing the broader effects of climate change on various disciplines, bringing together climate and social scientists, engineers, technologists, artists, writers, scholars of the humanities, lawyers, historians, and, importantly, architects.[2] For architecture in particular, climate change is recasting the boundaries and interconnections that define the field—affirming that architecture is, in fact, a synthetic discipline.

The overwhelming and impending transformations of our built and natural environments—the vulnerability of flood-prone coastal cities, water shortages, changes to worldwide food production, and the resulting conflicts and large-scale forced migrations—is matched by a collective realization of the herculean response needed for humanity to ensure its survival. The staggering quantitative and qualitative projections for our planetary future inevitably and instantly recast everything about our way of life: the very foundations of our global economy; the urgency of political action in support of technological innovation, regulation, and implementation; and the need for radical social and entrepreneurial transformations. This recasting demands a mobilization across all fields at unprecedented scale and speed, beyond even the most ambitious post–World War II reconstruction. The building of new housing and infrastructure, the launching of radical transnational and developmental initiatives, and the constitution of new diplomatic and nongovernmental institutions are called for.

So what does climate change change for architecture? For well over a decade already, architecture's engagement with climate change has taken on various forms. Most evidently, the notion of *architecture as technology* has led to innovation in energy systems, material performance, and energy regulations and certifications, consolidating the building (in terms of both construction and operation) as the optimal frame through which to regulate architecture's impact on carbon emissions. Combined with the increasingly important field of data science, new developments in sensing technologies, and the possibility of reacting to use patterns through feedback loops have become the leading frontier of building performance, cementing the understanding of architecture as a technological system reliant upon engineering problem-solving. At its most experimental, this notion of architecture as technology has built on the history of biomorphic design to move beyond formal and structural analogies to natural systems and enlist instead recent computational advancements to "grow" organic and bioengineered materials with carbon footprints close to zero.

Scaling up this systems approach is the idea of *architecture as infrastructure,* which has expanded on Team 10's concept of *building as urbanism* to consider how large urban and ecological systems are intertwined (and can be made more so). This infrastructural focus unites the smaller material scale of building parts with a larger environment. The turn of the twenty-first century saw an expanded definition of "architecture," and of its attendant territory of intervention. The traditional boundaries of the building dissolved, making clear that buildings no longer simply constituted autonomous objects but that their scale stretched to come to that of territorial and ecological systems. This strategy has posited a form of continuity between the urban and the natural, leading to new typologies such as the "landform building" and new forms of strategic infrastructural and ecological interventions.[3]

This conceptual dematerialization of building boundaries has also revealed a new understanding of the material qualities of buildings, shedding new light on the possibilities of making and revealing *architecture as embodied energy.* As Kiel Moe suggests in his *Convergence: An Architectural Agenda for Energy,* architecture needs a more ambitious agenda for energy.[4] As buildings weave together complex systems, parts, and materials, each element extends beyond its presently self-contained and bounded state to touch on the geographies and processes of its extraction, production, transport, and assembly, connecting buildings to the vast territories and scales, spatial as well as temporal, of their making. With every step, energy is consumed, produced, or exchanged, accumulating within the "finished product" as traces of a networked and always-in-transformation life cycle. Seen as assemblages of energy in this way, buildings are imagined as registers of intertwining life cycles, not only consuming energy but also possibly producing it, as the excesses that differentiate building from architecture are transformed into positive externalities, contributing to the life beyond their walls. As architecture registers and manifests the material life it is made of, exposing the complexity of its systems, buildings become quite powerful ecologically.

The need to render such systems legible also points to a renewed interest in *architecture as visualization,* which has intensified drawing and ushered in new mapping practices critically engaged with the nature of data. Lines are no longer drawn as walls but as vectors, making the interconnected and scalar relationships of networks of exchange manifest across extensive landscapes and territories.[5] Drawing climate change's intricate web of causes and effects across geographical as well as historical scales, these new forms of visualizations create layered understandings of planetary politics (resource extractions, forced migrations of humans and animals, labor movements, conflict, and high-speed urbanization) in tandem with the spatial and temporal transformation of our built and unbuilt environments.[6] Interestingly, this reconceptualization of architecture as visualization engages the discipline's capabilities as a parametric practice—not to produce the endless and self-referential form-making that early parametric design invited but rather to open up new possibilities for *architecture as narrative,* a practice that is at once analytical, informational, and projective as it is critical, aesthetic, spatial, and

3

Stan Allen and Mark McQuade, *Landform Building: Architecture's New Terrain* (Princeton, NJ: Princeton University School of Architecture and Baden, Switzerland: Lars Muller Publishers, 2011).

4

Kiel Moe, *Convergence: An Architectural Agenda for Energy* (London: Routledge, 2013), in particular the introduction "Matter Is but Captured Energy," 11–32, and Chapter 1 "Energy Hierarchies and Architecture," 33–106.

5

See Neil Brenner's work on "Operational Landscapes" in the exhibition *Operational Landscapes: Towards an Alternative Cartography of World Urbanization,* curated by Brenner and the Urban Theory Lab, ALKF Gallery, University of Melbourne, 2015.

6

See Laura Kurgan's work on conflict urbanism with the Center for Spatial Research.

experiential. From this, new overlapping representations of the world are produced, enabling new forms of collaboration, of politics, practice, and action.

Beyond the legibility of its systems and its making, it is *architecture as form* that climate has yet to change radically. As it continued to resist the question of its contingency, the discourse on form became increasingly isolated, with dire consequences for its ability to contribute to today's shifting concerns. Form's reduction to the abstract and the visual alone found its limit as it became real, with the recent hardships in materializing its smooth virtual surfaces unleashing an explosive response—from the protest surrounding the labor conditions it produces, to the assail of the material and resource excesses of its assemblies, to critique of the complexities and absurd scale of its construction. Only by reconnecting the formal to the informal, the tangible to the intangible, and the visible to the invisible networks of building "form" can the new forms of our contemporary condition be discovered in a renewed collapse of "the real" and its representation.

Finally, and perhaps most centrally for the essays contained in this book, climate change has opened up new lines of inquiry for *architecture as discourse* and as a form of political engagement. If architecture has long been seen as strictly anthropocentric—in its history as well as its projections into the future—design thinking is now considering other species, plants, and animals as equally entitled to shelter and livable environments, undoing the age-old separation of nature from culture and pointing instead to the imbrication of all things in the production of contemporary life. The architectural responses to this awareness consist of shrinking and compressing the human footprint as well as working to improve the lives of other species, moving beyond the common notion that "life on the planet has overall gotten better" and instead acknowledging that it has, in fact, only gotten better for humans—not for the planet and its wildly diverse and equally important "other" forms of life.[7] To refocus architecture on a wildly diverse set of actors can only lead to exciting new possibilities for the field—as discourse and as practice. Moreover, as the concept of the Anthropocene continues to redefine disciplinary boundaries across the sciences and the humanities, it invites us to consider new spatial and temporal scales as frames of inquiry, new material agencies, and new intersections between human and natural histories. The study of the Anthropocene urges an understanding of architecture as a geological agent able to mobilize Earth's resources and alter its atmosphere, and thus argues for renewed critical thought that brings together planetary politics with the design of the built environment.[8]

In many ways, climate change has already transformed architecture, charging and intensifying its expanded field to focus and qualify certain directions, while opening up further territory for critical engagement and for new modes of practice. Architecture was never a single object; today it is more than ever a form of knowledge that can enable the convergence of physical space and historical time. This involves an expanded notion of architecture's "subjects" as well as a reflection on how it is simultaneously constituted as an expanded object, a network, and a field. And yet, despite this promise of convergence, climate change has yet to undo the familiar constructed oppositions between discourse and practice, art and life, aesthetics and performance, communication and technology. At one extreme are discourses and practices focused on technocratic solutions and the firm belief in architecture as a solution advanced by new technologies. On the other are discourses and practices that continue to prop up increasingly fragile walls around the idea of the "art

7
This is a common line often most heard at World Economic Forum–type gatherings. For example, this view was upheld by Eric Anderson, chairman of Planetary Holdings, this year (2016) as part of the panel "Life in 2030: Humankind the Machine," which also included Jennifer Doudna, professor of chemistry and of molecular and cell biology, University of California, Berkeley, United States; Nita A. Farahany, professor, law and philosophy, Duke University, United States; Toomas Hendrik Ilves, president of Estonia, Global Agenda Council on Cyber Security; Andrew Moore, dean, School of Computer Science, Carnegie Mellon University; and moderated by Zanny Minton Beddoes, editor-in-chief, the *Economist*, United Kingdom.
See also Doyle McManus, "Bill Gates: The World Is Better than Ever," the *Los Angeles Times*, February 8, 2014.

8
See Etienne Turpin, *Architecture in the Anthropocene: Encounters Among Design, Deep Time, Science and Philosophy* (Ann Arbor: Open Humanities Press, 2013).

of architecture" as the creation of autonomous objects, moved by formal processes and nostalgia for an imagined shared "discipline." This continued polarization leaves architecture's possibility for a renewed and more meaningful engagement with the material realities of the present unsatisfying and devoid of the boundary-defying thinking occurring in other fields.

Yet there is hope for architecture still. In his seminal essay, "The Climate of History: Four Theses"—and, subsequently, at a conference on architecture and climate change hosted by Columbia University GSAPP in December 2014 (a discussion continued in an interview included in this book)—historian Dipesh Chakrabarty offers several insights into how diverse fields, especially those in the humanities, might meaningfully enter into the climate change conversation. What Chakrabarty defines as one of the principal challenges presented by anthropogenic global warming to history—a discipline bound up in a particular narrative of development as freedom, enabled by fossil fuels—is the "collision" of three histories that have traditionally been treated as separate processes: the history of earth and its systems; the history of life on the planet, including that of human evolution; and the history of the industrial way of life we often equate with the era of capitalism. Faced with the vastly different spatial as well as temporal scales of these disciplines and their transversal applications across the humanities and sciences, Chakrabarty finds his framework as a historian insufficient. A more enabling frame would allow one to think across differing scales of time and space—from the history of our immediate lived experiences to the deep history of geological transformations. It would enable holding together our conflicts and differences, with the consciousness of being collectively one: an endangered species among others. Finally, it would imply not the *certainty* of risk management but life faced with *uncertainty* and our inability to model it.

As he redefines the terms of a historian's engagement, Chakrabarty invites other disciplines and fields to reconsider their own boundaries, modes of knowledge, forms of practice, and terms of action. "I knew nothing of the history of earth and the planet system," Chakrabarty remarked at a conference in Berlin, before entering into a detailed explanation of the key moments of that history he found transformative for his own thinking about climate change and the radical paradigmatic shift it presented for the understanding, production, and practice of human history.[9] The siloing of disciplinary expertise has long become a part of universities and governments, from the economists waiting for the scientists to give them the "stable data and projections" they need to compute probability (and thus continue their usual embrace of "risk management" as a model for life on the planet) to the computer scientists at work in the development of automated cars and awaiting "others"—lawyers and ethically concerned members of "society"—to tell them how computerized cars should be programmed to choose between saving various forms of life.[10] This deferral to other disciplines can be heard from the architects who maintain that climate change has nothing to do with "our discipline" as they urge us to wait for industry regulations to dictate what to do—a common historical refrain, already upheld with the adoption of ADA guidelines some twenty-five years ago, with important and at times greatly problematic consequences. Collaboration across expertises, which architects have rightly embraced in recent years as a creative and productive mode practice, is also an invitation to dive deeper into other disciplines, zooming far beyond their contemporary boundaries to retrace old connections and create new ones. Climate change has radically upended what we thought were stable paradigms though which to see the world. In so doing, it presents an incredible chance to reimagine what architecture is, as discourse and as practice.

9

Dipesh Chakrabarty, Keynote, "Anthropocene Project: An Opening," Haus der Kulturen der Welt, Berlin, January 13, 2013.

10

See the presentation "Life in 2013: Humankind and Machine" at the World Economic Forum 2016, in which Andrew Moore, dean of the School of Computer Science at Carnegie Mellon University shared the difficulties computer scientists are facing as they program for automated cars to choose between saving lives: for example, choosing between a male driver and a mother and child crossing the street or between a driver and a dog. Moore's position was that scientists were waiting for others to decide those ethical questions.

11

See note 9.

The second invitation extended by Chakrabarty's thought, one with deep consequences for architecture, is his call to "cope with the problem of scale."[11] As one considers not only architecture but all of the disciplines of the built environment, one is confronted with academic disciplinary boundaries and professional expertise that impede the possibility of connecting, thinking, and acting across scales: from the material and structural scale for engineers, to the building scale for architects, the neighborhood and master plan scale for urban designers, and that of cities and regions for urban planners.[12] These

12

Henk Ovink, the former senior adviser to the secretary of Housing and Urban Develop-ment, for instance, advocates against strict professional boundaries, promoting more active regional planning and urban planning for water engineering and infrastructure.

boundaries follow both spatial and temporal scales, visible in the constructed boundary between "architecture" as a discipline and "historic preservation" as another—as if past, present, and future could ever be discontinuous. This embrace of scale was the driving principle for one of the most seminal architecture books of the twentieth century, Rem Koolhaas's *S,M,L,XL*, in which scalar discontinuity was at once structural as a main framing device for building and undermined as continuous ideas, forms, and effects cut across all of the scales. Today, following the notion of "scales of environment" it is the tracing of people, animals, objects, materials, and energies that could become structuring, and restructuring of architecture and the disciplines of the built environment.[13] How do we design "scales of environment" at once connecting the scale of a brick with that of a building, a city or an entire territory always in simultaneity? How do we rethink the past and design for the immediate future as well as for the longer future of geological time?

13

This term was first coined by Kate Orff as part of her design studio The Urban Estuary: Scales of Environ-ment; it was then used in the title for a conference held at GSAPP.

Finally, and maybe the most compelling and challenging of Chakrabarty's invitations is to consider what it means to live with uncertainty. For architecture and the disciplines of the built environment, this question has been posed as "how do we design or plan for uncertainty?" But the issue should be larger than the challenging of design approaches and planning frameworks, even if both need to be recast. The history of architecture is one of certainty: from the classical representations of power and authority to the modernist embodiment of progress, the postmodern claims about disciplinary origins and boundaries or the more recent abandonment to formal pleasures and scalar excesses. Throughout this evolution, even the most critical of practices failed to undermine architecture's Vitruvian "firmitas," despite having the ground gradually pulled from under its foundations.[14] And yet, as rising waters redraw edge conditions, as migrants erase territorial boundaries, as time is stretched to that of geological transformation, and as seemingly endless flows of information recast our concept of context, there is an urgency to move beyond the stability and certainty offered by oppositions, to consider instead weaving together *uncertain* grounds and positions from which to project new forms of knowledge, of engagement, and indeed, of architecture. As architecture becomes an expanded, perforated, and porous object, whose edges in space and time are always in flux, we can plan for redundancies and design for resiliency, or at times, engage in mad alchemy as we rewrite architecture as the art and science of the unknown.[15]

14

Vitruvius Pollio, *Vitruvius on Architecture*, trans. Thomas Gordon Smith (New York: Monacelli, 2003).

15

See Hilary Sample's lecture at Columbia University Graduate School of Architecture, Planning and Preservation, New York, March 21, 2016.

Amale Andraos is dean of Columbia University's Graduate School of Architecture, Planning and Preservation (GSAPP) and co-founder of WORKac, a New York–based architectural and urban practice with international reach. In addition to Columbia, Andraos has taught at universities including Princeton University School of Architecture, Harvard Graduate School of Design, University of Pennsylvania Design School, and American University in Beirut. Her publications include *The Arab City: Architecture and Representation* (2016), *49 Cities*, *Above the Pavement, the Farm!*, and numerous essays. WORKac is focused on reimaging architecture at the intersection of the urban, the rural, and the natural. It has achieved international recognition and was named the 2015 AIA New York State Firm of the Year.

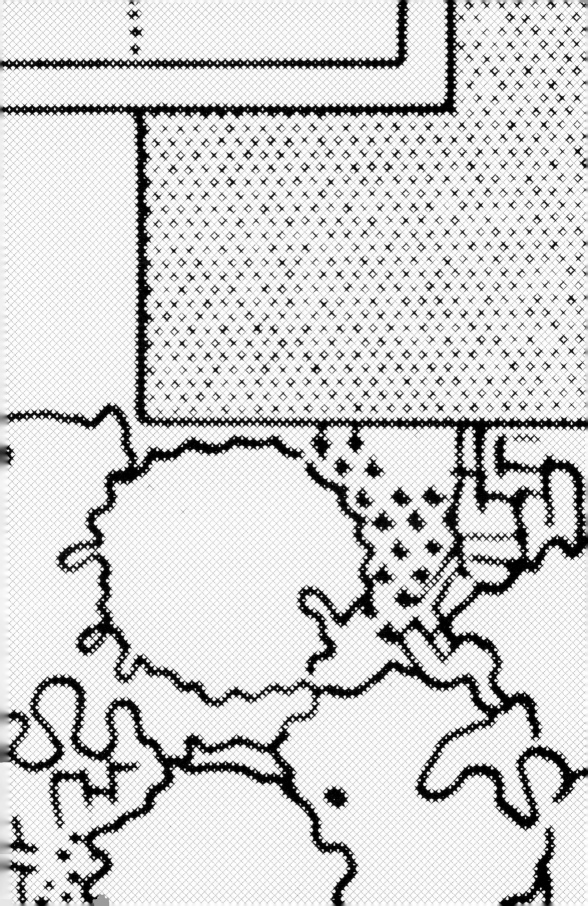

The Form and Climate Research Group, or Scales of Architectural History

DANIEL A. BARBER

Architects have long had an interest in the relationship between their design interventions and surrounding climatic conditions. Concerns over site, orientation to the sun, and the coordination of materials to heat and humidity are all embedded in vernacular design traditions and have been essential to the provision of human shelter for centuries. With the emergence of modern architectural techniques beginning in the mid-nineteenth century, the relationship of a building to its climate, and the means to understand and engage this relationship, underwent significant transformation. Climate has been an important concept in the historical trajectory of modern architectural projects and ideas.

After World War II, as various scientific concepts and frameworks were integrated into design methods—from the meteorological to the biological, from the managerial to the behavioral—some architects began to engage in more detail with how a building could relate to its climatic surround. These investigations, though as yet under-recognized in historical narratives of modernism, were widespread. Some took place at what was then called the Columbia Graduate School of Architecture, where in the early 1950s a group of students organized the Form and Climate Research Group. Working with a number of faculty members, they sought to develop techniques for refining the design process according to climatic adaptability.

In order to make sense of this group and of the more general interest in climate in the 1950s, and to inflect the theme that brought scholars, scientists, and architects together at Columbia University in December 2015 to discuss "the scales of the environment," this essay considers the relationship of climate to the scales of architectural history. It does so not in terms of time frame, the long view vs. the short view—though a rigorous *longue-durée* analysis of architectural-climatic techniques would certainly be interesting— but rather in terms of discourse. A large scale, in this sense, is the disciplinary scale: how do we account for this interest in climate and its significance to the development of modern architecture, and how does it allow for a reconsideration of the contours of that history? Can these contours be re-drafted according to the increasing uncertainty of our environmental future? In this essay, I will oscillate between this disciplinary scale and the smaller scale of the Columbia research group as a means to explore how climate emerged as a consideration for architects.

The stakes here are historiographic. The innovations in design, technology, and materials that we often gather together as "modern architecture" were profoundly inflected by climate considerations, and a revisionist history is clearly forthcoming. Such considerations clarify that, while contemporary interest in the materiality of the climate crisis, and especially the energy

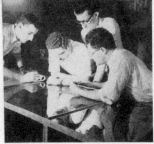

photographs by h. salm

A research group to study the influence of climate on the macroform *(general planning) and* microform *(architectural details) was organized at the Graduate School of Architecture, Columbia University, in February, 1951, by Valerius Michelson and Raymond Guay. Rapport developed among students from tropical or semi-tropical countries as well as Americans interested in such countries. This year, graduate students in the Technical Research course under Professor Kenneth Smith were encouraged by him to work with the group. At left: Costa Decavalla, Abraham Shapiro, Richard Fleischman, Valerias Michelson.*

The "Form and Climate" research group

At the air-flow chamber

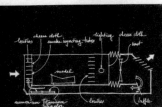

Longitudinal section of air flow chamber at Columbia University. 4'x4'x8' (exclusive of motorized equipment), it has glazed sides and top for observation of the test model inside.

The *Form and Climate* research group was confined to laboratory-type experimentation, that is, work on models. A student of the subject will know that the feasibility of using models in this kind of research was proved by several technical institutions in this country, notably the Texas Engineering Experiment Station.

The group experimented with *negative* methods—the protection of the interior from nature's undesirable manifestations, and *positive* methods—harnessing natural phenomena. They proceeded along two main lines: 1. To discover how the sun would behave toward architectural forms they tested—in any given latitude and altitude during all hours of the day and all seasons of the year. There are many tables and graphic devices that give fairly accurate pictures of the sun's angle and the number of hours it will shine on any given point of the globe any season and hour, but an experimental device developed some years ago by Henry Wright, and available at Columbia University (illustrated on page 53), made it possible for them to see a miniature sun throw light and leave shadow on the surfaces and recesses of their architctural models. 2. Their second line of investigation concerned natural air flow, whether from existing winds, if any, or, if not, from artificially induced convection. This was studied in a specially built air flow chamber.

The medium which indicates the flow of air in the air flow chamber is the same

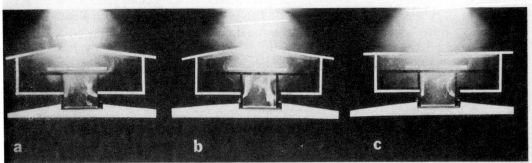

a) b) c)

a) Hip roof with clerestory windows. Sunken living room properly ventilated, but there is no air movement in the living zone at periphery of the split level (including bedrooms, kitchen, etc.)

b) Hip roof with ridge ventilating strip. Air movement speeded up, but the split level rooms are still not properly ventilated by the upward flow of air from the cool underside.

c) Flat roof with ridge ventilation. Good distribution of moving air all through the building, cooling whole on journey to relatively small escape openings made by cutting flues in roof.

The Form and Climate Research Group working on their wind tunnel, as published in *Interiors*, 1953.

1

"The Form and
Climate Research
Group" in *Interiors*, vol. 112, no. 7
(August 1953): 52.

2

Stamo Papadaki,
a Greek architect
in Brazil from
the early 1920s,
claimed some
primacy in the
invention of the
modern use of
the independent
shading device. His
proposed *Christopher Columbus
Memorial Lighthouse* competition
entry of 1928 has
a south façade of
building-length
horizontal fins, and
is cited by Jeffrey
Aronin and Colin
Porteous as the
first *brise-soleil*.
Papadaki also built
a small house and
studio in Athens
in 1930 that used
a *baldequin* and
brise-soleil system
similar to Le Corbusier's 1930 Villa
Baizeau. Aronin
and Victor Olgyay
both indicate the
importance of
Papadaki's books
on Le Corbusier (1948) and
Niemeyer (1950)
as central to the
dissemination of
the *brise-soleil*
idea after the war.
As art director at
Progressive Architecture in the US
starting in 1950,
Papadaki helped to
promote Brazilian
modernism for
the international
architecture
public. Citing the
impact, in Brazil,
of the 1946 *Brazil
Builds* exhibition
at MoMA, Lucio
Cavalcanti claims
that early Brazilian work was
strengthened by a
US-inspired rejection of Europe and
embrace of its own
traditions, and that
the resultant Brazilian modernism
received an international audience

largely due to the
efforts of American
editors (56); thus
Papadaki may
have had oversize
influence on the
dissemination
of Brazilian
modernism as an
important component of postwar
architectural production. See Jeffrey
Aronin, *Climate
and Architecture*
(New York: Reinhold, 1953); Colin
Porteous, *The New
Eco-Architecture:
Alternatives from
the Modern Movement* (London:
Taylor and Francis,
2001); Lauro
Cavalcanti, *When
Brazil Was Modern*
(New York: Princeton Architectural
Press, 2003).

performative capacities of buildings, continues to be the concern of architects, engineers, and architectural curricula, there is also a profound *immaterial* and discursive aspect to these challenges. As we collectively imagine and instantiate methods for engaging an uncertain future, the way that architectural discourse considers climate issues and the environment more generally will become equally important as the technological means to manage it.

ADAPTABILITY

Across these two scales—the disciplinary and the case study—I want to offer two framing concepts for how the history of architecture can begin to be reconsidered. The first is *adaptability*: what architecture offered, in the early and mid-twentieth century, was a socio-technological apparatus—of materials, technologies, and concepts—appropriate to a wide range of situations, conceived of in terms of the capacity to adapt a general principle for a specific site or condition. This is a well-known premise of modern architecture, perhaps best emblematized by Le Corbusier's 1914 "domino" diagram. This diagram is seen to suggest, according to a relatively simple set of new parameters—steel columns and reinforced concrete floors—a wide range of applications and adaptations. With it, Le Corbusier attempted, quite directly, to articulate a new idea at the scale of the discipline, one that would transform architectural approaches to the built environment. The general premise of adaptability is expressed in many other framing concepts that, in the inter-war period, encouraged new ways of thinking about the relationship between design, materials, and technology. These concepts have since been crucial to educating architects about the history of architecture's modernization.

The Form and Climate group at Columbia focused on a specific set of adaptive design methods relative to climate, design tools that were referred to as "negative methods" in the 1950s: the use of the roof or of shading devices to strategically *prevent* solar radiation from entering a building.[1] The basic premise of a negative method, also often attributed to Le Corbusier, is that a well-designed shading fin or extended eave can prevent the sun from entering the interior in the summer and allow it in during the winter. There was heightened interest in such shading techniques from the 1930s to the '50s; Le Corbusier's development of the *brise-soleil*, or sun-breaker, in 1928 was widely published, and quickly spread—or indeed, had spread already.[2] The principle of the *brise-soleil* can be placed next to the domino diagram as another significant intervention on this disciplinary scale—in 1969 Reyner Banham referred to the *brise-soleil* as one of Le Corbusier's "most masterly inventions, and one of the last

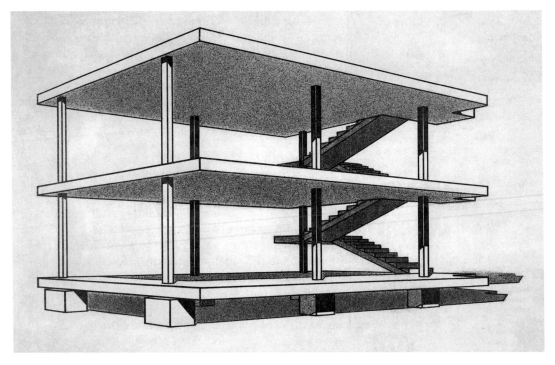

Top: Domino diagram, Le Corbusier, 1914. (c) F.L.C./ADAGP, Paris / Artists Rights Society (ARS), New York, 2016.

Bottom: Examples of *brise-soleil* types, 1936.

structural innovations in the field of environmental management."[3]

A primary arena for the integration of sophisticated shading techniques into the emergent principles of architectural modernism was, generally, the Global South, and specifically, in the building and modernization programs of Brazil. In fact, the means by which modern architectural strategies *globalized* in the first half of the twentieth century—before mechanical systems of heating, ventilation, and air conditioning (HVAC) were widely available—was largely through the shading device, an adaptive method that could adjust a building to its climatic location, with clear benefits for the experience of the interior.

That the locations for these innovations were often regions then being subjected to new forms of economic management and industrial development is not incidental. The premise of modernization was simultaneously cultural, industrial, and political, and these climatic buildings were often developed in concert with wide-ranging government initiatives. The well-known Ministry of Education and Health, for example, designed by Lucio Costa, Oscar Niemeyer, and a team of Brazilian architects in Rio de Janeiro in 1936, brought together a new set of national

3

Reyner Banham, *Architecture of the Well-Tempered Environment* (Chicago: University of Chicago Press, 1969), 158.

Instituto de Resseguros do Brasil (Brazilian Insurance Institute), M. M. Roberto, Rio de Janeiro, 1942.
Drawing courtesy of the Research and Documentation Center, Universidade Federal do Rio de Janeiro.

services in a dramatically modern, and delicately shaded, built setting. Many
of the modern, *brise-soleil* clad structures built in Brazil in the 1940s and '50s
were for insurance companies, the industry most characteristic of the risks
and responsibilities being taken on by the new forms of capital and governance
that would develop into the neoliberal institutions we know today.[4] For many
examples of climatic-design innovation, such as M. M. Roberto's Instituto de
Resseguros do Brasil (Brazilian Insurance Institute), built in Rio de Janeiro
in 1942, not only shading but also the careful manipulation of the horizontal
window, with direct reference to the scientific basis for these adjustments,
were deployed to render more comfortable the emergent space of a global
interior, replete with flows of capital and risk.

4
Luiz Felipe
Machado Coelho
de Souza, *Irmaos
Roberto Arquitetos*
(Rio de Janeiro:
Rio Books, 2014).

Though largely lost to the historical record, this global view of climate and architecture was, from the 1930s to the 1950s, well known. Journals and exhibitions celebrated the architecture of Brazil, and related developments in West Africa, Morocco, Indonesia, India, Australia, and elsewhere. These areas were seen as new frontiers for design and also sites for experimentation in the felicitous engagement between architectural techniques, government programs, and social needs, all seemingly resolved through climatic adaptability.

This entanglement of geopolitics and geophysics entered into the Form and Climate group's research through their analysis of Richard Neutra's wartime work in Puerto Rico. In 1943, Neutra was commissioned to build schools and hospitals around the island. He developed a number of prototypes and methods, mostly focused on induced ventilation, to best accommodate the regional climate. The work was published in São Paulo

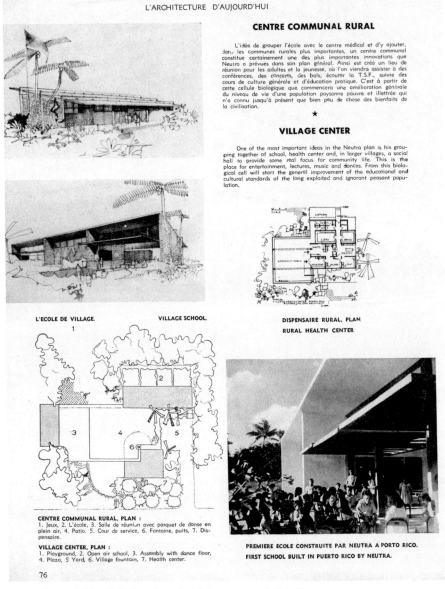

Projects for Puerto Rico, Richard Neutra, in *Architecture d'Aujourd'hui*, 1944.

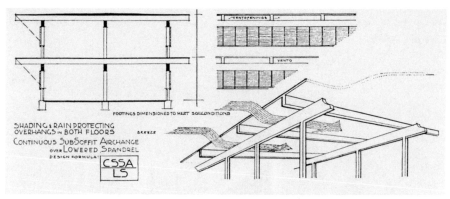

Neutra's "Continuous Sub-Soffit Airchange over Lowered Spandrel" system used in hospitals and dormitories.

5

Richard Neutra,
*Architecture of
Social Concern in
Regions of Mild
Climate* (São
Paulo: Gerth Totd-
mann, 1948).

6

See, for example,
Richard Neutra,
"Comments on
Planetary Recon-
struction," in *Arts
and Architecture*,
vol. 61, no. 12
(December 1944):
20–22; "Projects
of Puerto Rico:
Hospitals, Health
Centers, and
Schools" in
*Architecture d'Au-
jourd'hui*, vol. 16,
no. 5 (May 1946):
71–77 and Richard
Neutra, "Designs
for Puerto Rico
(A Test Case),"
unpublished,
1945, Richard
and Dion Neutra
Papers (Collection
Number 1,179),
Department of
Special Collec-
tions, Charles E.
Young Research
Library, UCLA.

in 1948 under the title *Architecture of Social Concern for Regions of Mild Climate*. Here, as the title suggests, the issue was not to manage climatic extremes but rather to use design techniques to ameliorate social conditions with the least financial and infrastructural outlay.[5]

At the time, Neutra was the US representative for CIAM and was also the self-appointed CIAM representative at the San Francisco meeting that inaugurated the United Nations in 1945. He was deeply enmeshed in emerging, formative questions regarding global systems—governmental, social, and ecological. He was also concerned with the destructive effects of the war and with how architects and planners would be called upon to rebuild and build a society more attuned to the socially beneficial prospects of modernist architectural ideals, as he interpreted them.

In this context, Neutra discussed his work in Puerto Rico as a "Planetary Test."[6] The term is fraught with the experimental attitude that many architects (and engineers, planners, government agents, and corporate researchers) would take to what were soon to be termed "developing economies"— the inhabitants of which were seen to be subject to an experimental mode of design and planning. Such experimentation was necessary, according to Neutra and others, in order to raise the quality of life in these regions and to integrating their populations into the global economy. Neutra's schools and hospitals intended to outline new parameters for life improvement through a series of architectural techniques. Climate was a preeminent design device in proposals to ameliorate learning, healing, and administrative environments. The region was being "tested" to see if it could better approximate a Euro-American managerial model of social organization.

In this sense, Neutra's "Planetary Test" offers a potent alternative phrasing to the "International Style" as a term to consider the adaptable and seemingly universal premise of modern architecture on disciplinary terms. His shift from *international* to *planetary* opens up the analysis of the globe to the geophysical and environmental conditions with which architects were increasingly engaged. The shift from *style* to *test* rereads familiar terms of modernist functionalism toward a new sort of operationalism. Architectural strategies of climatic adaptability were always enmeshed in the process of modernization, alongside cultural innovations and infrastructural interventions, and according to corporate or governmental aspirations.

From this broad perspective it is likely that the British Petroleum headquarters in Lagos, designed by Fry, Drew, Lasdun, and partners in 1960, will emerge as an important case study in the architectural histories of the

7

Le Corbusier,
*Precisions on the
Present State of
Architecture and
City Planning*
(1930; repr. Cam-
bridge, MA: MIT
Press, 1986), 66.

8

See Michelle
Murphy, *Sick
Building Syndrome
and the Problem
of Uncertainty*
(Durham, NC:
Duke University
Press, 2006), 12ff.

Headquarters for British Petroleum, Fry, Drew, Lasdun, and Partners, Lagos, Nigeria, 1960. Courtesy of RIBA Collections.

future—for the clarity of an international form of corporate organization; for the precise use of a complex shading system; and as evidence of how former colonial powers were refashioned as corporate entities and redirected toward resource extraction. More generally, the diagrammatic, generative aspect of the *brise-soleil* opens up our understanding of architectural history to better engage the geopolitics of the Cold War, the increase in scientific knowledge and the uneven consequences of its application, and the complex means by which *environments* became available to *economies*, allowing for seemingly peripheral projects such as those mentioned above to be recognized for their expression of some of the field's core concerns.

NORMATIVITY

9

Fitch was not yet involved in preservation, an interest he would develop from the early 1950s as he began to teach at Columbia full time.

10

At around this same time, Fitch began to teach at the night school at the Columbia architecture school, eventually taking a full-time position in 1954. James Marston Fitch interview with Suzanne O'Keefe, 1978. James Marston Fitch papers, Department of Drawings and Archives, Avery Architectural and Fine Arts Library, Columbia University.

11

James Marston Fitch, "Micro-climatology," in *Architectural Forum*, vol. 36, no. 2 (February 1947): 18; see also G. Manley, "Microclimatology: Local Variations of Climate Likely to Affect the Design and Siting of Buildings," in *RIBA Journal* (May 1949): 317–323; W. E. Graham, "The Influence of Micro-climate on Planning" in *Planning Outlook* (March 1949): 40–52; Helmut Landsberg, "Micro-climatology," in *Architectural Forum*, vol. 36, no. 2 (March 1947): 114–120.

The second figure of thought that climate places in relief is *normativity*—an opportunity to reassess Le Corbusier's 1930 dictum that "every building, around the globe, will be 18 degrees [Celsius—about 65 degrees Fahrenheit]."[7] One could fly from Rio to Lagos, for example, and experience the uniform interior space of modernity. This basic conception of manipulations of interior climate as essential to the development of a certain type of civilization, and a certain means for the management of security, territory, and population, suggest that climate was both a challenge—a complex set of factors that architects had to encounter— and an ambition, an end for which a careful approach to design was the means.

This aspiration for a universal interior was an important driver for many early climate design methods. It relied on a proposal, which by midcentury was well examined through industry experimentation, that all humans have an optimum climatic state, and that architecture can find ways to produce it.[8] This research was brought into the purview of the Form and Climate research group through the writings and teaching of the historian James Marston Fitch. Fitch, yet to explore the interest in preservation for which he would become known, was tightly focused on how a building could be designed to best accommodate itself to the precise climate of the site.[9] He had spent the war as a meteorologist, becoming fascinated by the visual tools used to represent climate patterns. In 1947, just as his magnum opus, *American Building: The Forces that Shape It*, was published, he cowrote an article on microclimates and house design with one of his wartime colleagues.[10]

The intense specificity of the climate problem was, to Fitch and his colleagues, quite daunting: "Although everyone is aware of the general climate of his locality," Fitch wrote, "no one knows much about the climate of his own backyard."[11] Details such as elevation, proximity to water, soil conditions, grass species, paving materials, and arrangement of hedges and trees, among many others, he explained, had to be taken into account. Pollution, the grass

HOW TOPOGRAPHY AFFECTS MICROCLIMATE

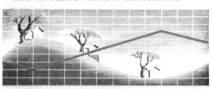

Your heating bill will be less if you move your house uphill

But cold winds at crest may offset higher temperatures there

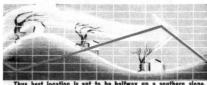

Thus best location is apt to be halfway up a southern slope.

HOW WATER BODIES MODIFY MICROCLIMATE

In daytime, an offshore breeze may cool air by as much as 10° F.

At night, breeze is reversed but still has a cooling effect

Yet force of breeze is limited. Tall buildings may block it.

Illustrations of microclimatic factors from Helmut Landsberg, "Microclimatology," in *Architectural Forum*, March 1947.

species of lawns, and numerous other factors were also discussed. Careful treatment of trees and other landscape elements were seen to be an especially effective means to influence microclimate and also a means to integrate architectural and landscape architectural practices.

Microclimatic analysis, as Fitch modeled it, was dizzyingly complicated. One challenge was that wartime analyses, and then those of industry, tended to focus on upper-air observations and models of general atmospheric dynamics. Privileging the global over the local, such strategies focused on theoretical models, with a secondary concern for the capacity to use this knowledge for local or regional prediction on the ground. As the macroclimate became of increasing interest, observation stations moved to airports and other sites at a distance, relatively speaking, from population centers, and often elevated. Thus the phenomenal increase in weather data after the war was, generally speaking, of limited use to architects. As a result, they needed to employ specialists or make more relevant observations.[12] Collaboration was essential. "Cooperation between architects and climatologists," Fitch concluded, "will yield designs better adapted to their environment."[13]

12

Roger Turner, "Weathering Heights: The Emergence of Aeronautical Climatology as an Infrastructural Science," PhD diss, University of Pennsylvania, 2010: 11–14.

13

Fitch, "Microclimatology," 21.

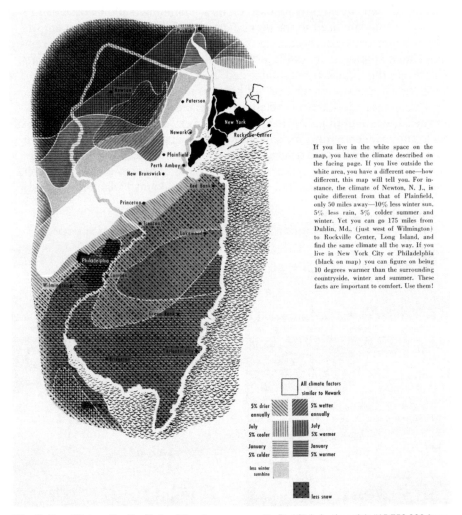

If you live in the white space on the map, you have the climate described on the facing page. If you live outside the white area, you have a different one—how different, this map will tell you. For instance, the climate of Newton, N. J., is quite different from that of Plainfield, only 50 miles away—10% less winter sun, 5% less rain, 5% colder summer and winter. Yet you can go 175 miles from Dublin, Md., (just west of Wilmington) to Rockville Center, Long Island, and find the same climate all the way. If you live in New York City or Philadelphia (black on map) you can figure on being 10 degrees warmer than the surrounding countryside, winter and summer. These facts are important to comfort. Use them!

All climate factors similar to Newark

5% drier annually | 5% wetter annually

July 5% cooler | July 5% warmer

January 5% colder | January 5% warmer

less winter sunshine

less snow

Climatic Map of Metropolitan New York and New Jersey, prepared by Paul Siple for the article "15,750,000 Americans Live in this Climate" that accompanied the presentation of the 1949 Pace Setter House in *House Beautiful*, November 1949.

14

Memorandum in
Hoyt C. Hottel
Papers, Box 11,
Folder 7, Archives
and Special
Collections, Mas-
sachusetts Institute
of Technology.

15

"Proceedings
of the Research
Correlation
Conference on
Weather and the
Building Industry"
(Washington, DC:
National Academy
of Sciences, 1950).

16

See, among others
Aronin, *Climate
and Architecture*;
Aladar and Victor
Olgyay, *Solar
Control and Shad-
ing Devices* (New
York: Reinhold,
1957); Victor Olg-
yay, *Design with
Climate: A Biocli-
matic Approach
to Architectural
Regionalism*
(Princeton, NJ:
Princeton Univer-
sity Press, 1963);
Groff Conklin,
*The Weather-
Conditioned House*
(New York: Rein-
hold, 1958); and
somewhat later,
Baruch Givoni,
*Man, Climate, and
Architecture* (New
York: Elsevier,
1969).

17

Walter A. Taylor
and Theodore
Irving Coe,
"Regional Climatic
Analysis and
Design Data,"
*Bulletin of the
American Institute
of Architects*
(September 1949):
11–17, 15.

Expansion of housing into the suburbs, on the one hand, and interest in the applied possibilities of scientific research, on the other, informed a number of discussions about architecture and climate in the American context in the 1940s and '50s. Fitch, at Columbia, was only one of a number of faculty at promi-nent institutions who were involved. In 1947, the Housing and Home Finance Agency (HHFA)—a successor to the wartime National Housing Agency—began a "Research Program for Applied Climatological Data in Dwelling Design, Site Selection, and Planning" with architects at the Massachusetts Institute of Technology.[14] In 1949, the Building Research Advisory Board (BRAB), a part of the Division of Industrial and Engineering Research of the newly formed National Research Council, organized a Committee on Climatic Research and began to solicit research proposals and other means of studying the relationship between "Weather and the Building Industry," as their January 1950 "Research Correlation Conference" was titled.[15] Other reports and analyses would follow, in architectural and engineering journals, at conferences, and in the popular design press.[16] The goal of this research, broadly considered, was to develop methods and tools that architects could take advantage of in order to best under-stand site-specific climates and then design accordingly. This general trend was given its most public iteration in the "Climate Control Project," a collaboration between the magazine *House Beautiful*—which was to "represent the con-sumer"—and the American Institute of Architects, which collected, organized, and disseminated relevant technical information to design professionals.[17] Fitch was hired by *House Beautiful* to help edit and direct the project.

Though originally conceived to present general regional data, Fitch's insis-tence that climate data was only useful when precise pushed the researchers toward a smaller scale. In the end, the AIA published thirteen pamphlets analyz-ing the climatic aspects of thirteen different metropolitan regions. Even here, as seen in Paul Siple's "Climatic Map of Metropolitan New York and New Jersey,"

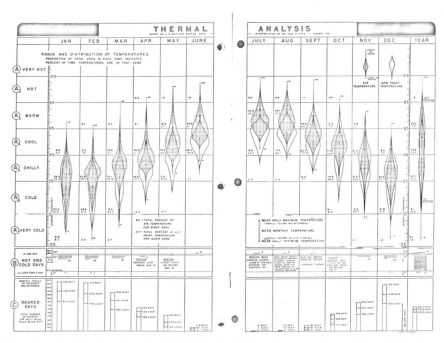

Thermal Analysis of the Boston Area, from the AIA's *Regional Climate Analyses and Design Data*, in the March 1951 *Bulletin*.

only those within the white band are similar enough for uniform application of the data presented; other micro-regions require additional adjustments, as noted in the key. The AIA also provided charts and diagrams—of solar and wind patterns, of summary thermal conditions—along with a list of techniques to mitigate potential effects. *House Beautiful* provided more schematic illustrations and an extensive series of articles that sought to convince the reader—the potential client—to engage questions of climate in their own house.

House Beautiful's Pace Setter Homes program also integrated climate as one of its most important principles. The second Pace Setter House, designed by Emil Schmindlin and built in Orange, New Jersey, in 1949, demonstrated

Cold, slushy weather demands warm floors, resistant to mud and easy to clean. This house has radiant heating coils embedded in concrete slabs. Surfacing is integrally colored cement, kept waxed.

Plants get special climate in this bay —lots of sun, cool, moist air. If you didn't want plants, you would omit inner doors, throw this space into living room and use insulating glass on outer doors.

Even in deep winter, you can afford an unobstructed view with insulating glass and weatherstripping. Door-high curtains for night use are out of sight at right. The triangular gable is uncurtained.

Sealed for winter, louvered gable end of living room and ceiling panel in dining room conceal insulated doors in attic. They are open all summer. A big attic fan pulls hot air up through these openings and exhausts it. Note that dining room gets morning sun even in the winter.

A lesson
in Climate Control

(continued)

The premise of the second Pace Setter House as illustrated in the introductory issue of the Climate Control Project in *House Beautiful*, October 1949.

the ease with which modern idioms of outdoor living, expansive living rooms, and simple materials could be made comfortable across the varying seasonal demands of temperate climates.[18] Schmindlin's house, published in a lavish set of spreads just a month after the Climate Control Project was introduced, was directly aimed at improvements to living—"167 pace making ideas ... including the new field of Climate Control" were shown on these terms. The house used insulated glass on the large south-facing wall, with integrated screens to block the sun in the summer. Ventilation inducement was effected through strategically placed openings, and a large tree was used for seasonal shade. Alongside the photographs of the main living room and its fully glazed façade—in summer, in winter, and night, from inside and out—diagrams laid out the principles by which eaves, trees, and blinds excluded the sun in the summer and let it in during the winter, while maintaining privacy year round.

The ambitious Climate Control Project clarifies a few important aspects of the postwar interest in designing with climate: first of all, much of the discourse focused on methods that could produce a consistent, normative interior despite the vagaries of conditions outside. Second, it was conceptually framed through a premise of determinism. A stand-alone image in the first *House Beautiful* issue dedicated to the project played this out. It showed six books, arrayed across a lined landscape with clouds above, and a man, naked above the waist, looking at them.[19] Most of the illustrated texts fell squarely in the historical narrative of climatic determinism, claiming that cultures developing in temperate climates benefited from their geographic conditions in the

18

"Presenting *House Beautiful*'s Pace-Setter House for 1949" in *House Beautiful*, vol. 91, no. 11 (November 1949): 195–201.

19

"It May Be News to You, But..." in *House Beautiful*, vol. 91, no. 10 (October 1949): 142.

The image for the article, "It May Be News to You, But..." in *House Beautiful*, October 1949.

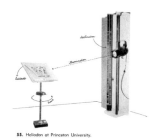

55. Heliodon at Princeton University. 56. Sunmachine at University of Kansas. 57. The Solarscope. 58. Thermoheliodon at Princeton University.

Heliodons in *Solar Control and Shading Devices*, Victor and Aladar Olgyay, 1957. The collection includes the thermoheliodon the Olgyays built at the Princeton Architectural Laboratory with a grant from the National Science Foundation in 1957.

formation of knowledge and civilized social practices. Other cultures, it followed, especially those in the tropics, were seen to develop more slowly and less completely. Ellsworth Huntington's *Mainsprings of Civilization* (1945), a follow-up to his *Civilization and Climate* (1915) and *The Red Man's Continent* (1927), was in the image and is emblematic here. Huntington saw climate as a sort of filter to assist the process of evolution, with Europeans and Americans made more favorable for selection. Other texts argued similar theses on social and political, rather than biological, terms, and a much smaller group examined how these broader climatic patterns resonated in building practices. Fitch's *American Building*, though not an orthodox expression of this tendency, was also shown. In short, by identifying climate as a major factor in the comfort on the American home, concerns over race, xenophobia, and cultural elitism were also placed in the frame. The production of a uniform interior climate was discussed on these terms—as the need to use architecture to facilitate a certain kind of productive and normative way of life and thereby to render the suburbs, and other emergent territorial conditions, as exemplary of Western civilization's promise.

The Climate Control Project also begins to suggest the importance of simulation to the midcentury interest in climate. The ideal climatic interior was a space of aspiration, an image of a possible future in which design methods could maximize comfort, with all of the complex resonance of this term relative to the determinism just described. The Form and Climate Group was experimenting with heliodons to determine ideal site orientation, and with wind tunnels to test roof shape. The first heliodon in the United States, built by the planner Henry Wright in 1936, was a relatively simple device: the sunlamp was calibrated along the vertical calendar to provide the seasonal height, and the building was placed on a platform angled according to latitude.[20] The platform could spin to simulate diurnal patterns relative to the sun's location.[21] Other heliodons were built at the University of Kansas, Princeton, and at a number of the research stations in former British colonies.[22]

The most elaborate of such devices was at the Princeton Architectural Laboratory, another locus for climate research. This device elaborated on previous models by adding misting jets for humidity and a means to approximate high atmosphere pressure changes; each building could also be tested with soil from the building site so as to approximate the thermal conditions of the ground.[23] The thermoheliodon, as it was called, was essential to refinement of the climate design method of Victor and Aladar Ogyay, later published as *Design with Climate: Bioclimatic Approach to Architectural Regionalism* in 1963. The method seemed to answer Fitch's imperative for specificity: it was a quasi-scientific means to understand the climatic conditions of any given site and to provide parameters to design the building accordingly. It also was aimed precisely at the production of a uniform interior space of stability, of a reified sense of the human and its purported ideal state.

This discourse on climate methods hummed with an imperative for stasis, uniformity, and normativity as the ideal conditions for existence. Architecture was seen as a site for the possible re-alignment of a number of analogous relationships: that between the interior of a building and its site; between the inhabitants of a building and the weather outside; between climatic analysis and the forms, materials, and orientations than can lead to a normative climatic condition. These simulation devices appealed to the technological disposition and the aesthetic intentions of the midcentury designer, encouraging them to realize an architecture of human shelter in this wide sense—as the provision of comfort that can, in turn, improve the lives of the humans that inhabit it. It

20

Olgyay and Olgyay, *Solar Control*, 37.

21

Wright began to build the device at Columbia, where he taught until his death in 1936; it was completed by his son, Henry Wright Jr., later that year. Wright's design was based on those published by RIBA in 1935; it was used by Clarence Stein, and also by Raymond Unwin in his planning studios. See "Heliodon Installed by Architecture Students for Aid in Building Design" in the *Columbia Spectator*, Monday, October 12, 1936, 4; and A. F. Dufton and H. E. Beckett, "Orientation of Buildings-Sun Planning by Means of Models," in *RIBA Journal* (May 1931): 509ff.

22

George Atkinson, "Building in the Tropics: Research into Housing in Tropical Countries, especially in the Commonwealth," in *RIBA Journal* (June 1950): 313–320.

23

See Daniel A. Barber, "The Thermoheliodon: Architecture at the End of Calculation," in *ARPA Journal* 1 (2014).

24

Such prospects
were widely
examined in
the immediate
postwar period:
the physiological
assumptions on
which the Olgyays
and the Form and
Climate Research
Group relied were
themselves the
subject of much
debate. For physi-
ologists interested
in the possibilities
of architecture and
climate, the nor-
mative conditions
of the designed
interior went hand
in hand with the
integration of
new subjects and
economies into
an increasingly
managed global
social order.
See, for example,
D. H. K. Lee's
review of the
report on the MIT
research: D. H. K
Lee, "On *Appli-
cation of Climatic
Data to House
Design*" in the
*Geographical
Review*, vol. 45,
no. 2 (April 1955):
307–308.

25

Ian Hacking,
"Canguilhem
Among the
Cyborgs," in *Econ-
omy and Society*,
vol. 27, nos. 2–3
(May 1998):
202–216.

26

ASHRAE stands
for American
Society of Hearing,
Refrigeration, and
Air Conditioning
Engineers. The
Society was
formed out of a
number of regional
and smaller indus-
trial societies in
1959 with, despite
the *American* in its
name, a decisively
international
focus. The thermal
comfort standards
ASHRAE has
produced since
this period have
been international
in nature and
purview.

27

Isabelle Stengers,
*In Catastrophic
Times: Resisting
the Coming Bar-
barism* (Lüneburg,
Germany: Meson
Press, 2015), 39.

is an idealist premise: an argument for using technology to contain the human and isolate the species from the unpredictability of the natural world.[24]

At stake across these investigations was the instantiation of a flexible method not simply for adapting to a given set of data-reliant environmental conditions but also for facilitating species development on an endless forward trajectory.[25] The Form and Climate group did not participate in these debates so much as reinforce specific conclusions in an architectural context, exploring a constructed milieu in which this stable human subject could, it was proposed, most effectively operate.

—

To reconsider the history of modern architecture on the terms of climate highlights the fact that, by the end of the 1950s, most of these methods had been eclipsed. As fossil fuels became more available in the West, so did mechanical HVAC systems that ran on them, dramatically changing the methods for acclimatizing the built environment. The history of architecture and climate is also that of the wild proliferation of mechanically air-conditioned buildings.

Buildings are among the primary accelerators of the "great acceleration" through which the Anthropocene has emerged. It is not so much that the methods described above didn't work, just that they worked differently—the absorption of the factors, methods, and simulation tools of midcentury climate research by the HVAC industry, through ASHRAE regulations, clarified the architectural and physiological terms of the "comfort zone" that we are now all forced to inhabit.[26] We have, collectively, failed the test, it seems, of how to live on this planet, in part because of our reliance on HVAC systems and the reliance of these systems on fossil fuels. These consequences could not have been imagined when these climate-design methods were under investigation in the 1950s. In reflecting on events and projects such as those described here, Isabelle Stengers has recently written, "What is proper to every event is that it brings the future that will inherit from it into communication with a past narrated differently."[27] This essay suggests that the history of the Form and Climate Research Group, amid the wide range of related research and building practices, operates on a disciplinary scale, narrating a different past so as to communicate with an as yet undetermined architectural future. Such histories can facilitate new questions, new experiments, and new capacities for informed and cautious approaches to adaptability as environmental pressures increase.

Daniel Barber is the Currie C. and Thomas A. Barron Visiting Professor in the Environment and Humanities at the Princeton Environmental Institute (2015–2016), the Alexander von Humboldt Foundation Research Fellow at the Rachel Carson Center for Environment and Society, and assistant professor of architecture at the University of Pennsylvania School of Design.

On Membranes, Masks, and Siegfried Ebeling's Environmental *Raumkubus*

PEP AVILÉS

The author grate-
fully acknowledges
Siegfried Ebeling's
family, in partic-
ular the support
of Wolfgang
Ebeling and Rago
Torre Ebeling,
as well as Walter
Scheiffele, for their
contributions to
the present essay.
A previous version
of this essay was
published in the
Spanish edition of
Siegfried Ebeling's
*Space as Mem-
brane [El espacio
como membrana]*
(Barcelona: Mudito
& Co., 2015).

1

Siegfried Ebeling,
*Der Raum als
Membran* (Dessau:
C. Dünnhaupt
Verlag, 1926).

2

For a lengthier
biographical
account of Ebel-
ing's activities, see
the essays by Wal-
ter Scheiffele and
Spyros Papapetros
included in the
English translation
of *Space as Mem-
brane by Siegfried
Ebeling*, ed. Spyros
Papapetros, trans.
Pamela Johnston
(London: Architec-
tural Association,
2010), i–xxiii.

3

Siegfried Ebeling,
"Kosmologe
Raumzellen:
Ideen zur Ethik
des konstruktiven
Denkens," *Junge
Menschen, Monat-
shefte für Politik,
Kunst, Literatur
und Leben aus dem
Geiste der Jungen
Generation der
zwanziger Jahre*,
special issue on
the Bauhaus in
Weimar, vol. 5,
no. 8 (November
1924): 153–54.
Here, Ebeling
already speaks
about a membrane
placed between the
body and external
energies.

ENVIRONMENTAL MEMBRANES

In 1926, Siegfried Ebeling published *Der Raum als Membran* ("Space as Membrane"), a long essay championing
a new architecture that could optimize the relationship
between man's physical and psychological capacities and
the environment.[1] Polymath artist and architect, Ebeling
had studied physics, theology, art history, and dance before
enrolling at the Bauhaus in Weimar, where he took classes
from 1922 to 1925. A student of Paul Klee and Wassily
Kandinsky, he remained in the school's orbit once it moved
to Dessau, thanks to his involvement with the Junkers air-
craft factory.[2] Biologically inspired, Ebeling understood the
environment as a combination of the visible (nature, light)
and the invisible (climate, energy). The environment was a
continuous field of "vibrations" and "radiations" affecting all
forms of life on earth. Architecture, Ebeling argued, could
behave as a biological membrane within this field—a skin
creating a seamless physical continuum between humans and
their environments. Reconceiving the material and psychological performance of the façade as an organic extension of
man, Ebeling saw this membrane as maximizing health benefits while minimizing the pernicious effects that the multiple
climates, magnetic fields, electric radiations, and telluric
energies had on humans.

Ebeling's analytic and prescriptive essay expanded an
article that had appeared two years earlier in the magazine
Junge Menschen titled "Kosmologe Raumzellen" ("Cosmological Spatial Cell"), in which he wrote about the impending
evolution of a biological architecture.[3] This new approach to
space was further conceptualized in *Der Raum als Membran*,
unfolding the synthetic idea of architecture as an environmentally responsive epidermis, performing almost clinically.
With this membrane, the space that architecture defines
becomes an extension of its own materiality, a prosthetic
device adjusting the subject's physical and psychological
status according to the flow of energies on earth. In order
to visualize his concept of architecture, Ebeling proposed
a *Raumkubus*, a cube whose most remarkable quality was
not the space it created but the set of metabolic relations

Cover of *Der Raum als Membran*, Siegfried Ebeling, 1926.

4

This thickened architectural space extended the earth's total psychological field—a geographical interpretation of Jung's collective unconscious reminiscent of Gaia philosophy as the mental unity and correspondence between organic beings. The influence of *Lebensphilosophie* theories was very common among partisans of *Jugendreformbewegung* and *Lebensreformbewegung* groups such as the *Bündische Jugend*, *Wandervogel*, and the *Freideustche Jugend* with which several members of the Bauhaus were associated. So it is not unlikely that Ebeling was familiar with these heterodox groups and their philosophies of life and their sympathies toward nature.

5

Walter Scheiffele, "Membrane and Ecological Architecture," *Space as Membrane* by *Siegfried Ebeling*, 1. Hans M. Wingler included his name among the students of the first period of Weimar in his monumental *The Bauhaus: Weimar, Dessau, Berlin, Chicago*, ed. Joseph Stein, trans. Wolfgang Jabs and Basil Gilbert (Cambridge, MA: MIT Press, 1969), 616.

facilitated by the membrane defining its perimeter. Ebeling's bio-Cartesian synthesis was explicitly manifested on the cover of his treatise, where every single environmental constraint adding to its intellectual construction was present either in graphic or symbolic form. On the cover, one can see an active façade operating as a regulatory medium for atmospheric and geological conditions—climatic variations, thermal radiation, electric waves, telluric energies, etc.—optimizing their impact on the interior space that the subject inhabits.[4] The *Raumkubus* was therefore the materialization of a biological idea: the management and mediation between interior and exterior as a form of both *protection from* and *reconciliation of* the subject with the countless environmental and climatic contingencies to which this subject was exposed in the 1920s.

Originally, Ebeling's utopian bio-technological project was received with "little fanfare and enjoyed only modest" success among his contemporaries.[5] Like many observers of the social and cultural changes taking place during the

6

See, for instance, the multiple reviews on architecture that Siegfried Kracauer published in the *Frankfurter Zeitung* during the 1920s and 1930s, particularly his review of the Stuttgart exhibition on "Die Wohnung" organized by the Deutscher Werkbund and curated by Mies van der Rohe and that took place in parallel to the construction of the Stuttgart Weissenhof Siedlung. Siegfried Kracauer, "Das neue Bauen: Zur Stuttgarter Werkbund-Austellung 'Die Wohnung,'" *Frankfurter Zeitung*, July 31, 1927. For a discussion on functionalism see Stanford Anderson, "The Fiction of Function," *Assemblage* 2 (February 1987); Colin St. John Wilson, *The Other Tradition of Architecture: The Uncompleted Project* (London: Academy, 1995); Larry L. Ligo, *The Concept of Function in Twentieth-Century Architectural Criticism* (Ann Arbor: UMI Research Press, 1984).

7

Gropius had engaged in a shift toward a new unity of art and technique after Johannes Itten's departure that left very little space for biological speculations. As a result, Ebeling's work is neglected in Hans Maria Wingler's monumental work on the Bauhaus, referring to Ebeling only as a student. Wingler, *The Bauhaus*.

8

Consider, for example, the studies of the scholar Fritz Neumeyer unveiling the multiple intellectual genealogies beneath Mies van der Rohe's "skin and bones" metaphors.

9

The aesthetic values of Catholic author Romano Guardini are not to be discounted with regard to the work of Francé. For more on these developments, particularly vis-à-vis the architecture of Mies van der Rohe, see Fritz Neumeyer, *The Artless Word: Mies van der Rohe on the Building Art*, trans. Mark Jarzombek (Cambridge, MA: MIT Press, 1991), 171–179. See also Detlef Mertins, *Mies* (London: Phaidon, 2013).

Weimar Republic—Kracauer comes to mind here—Ebeling reacted critically to narrow interpretations of functionalism such as those brought about by the *Neue Sachlichkeit* movement.[6] As a consequence, Ebeling and others faced disdain, if not contempt, from some of his contemporaries, Walter Gropius the most notorious among them.[7] Therefore, interest in the work of Ebeling since the 1980s has only grown surreptitiously, and in relation to the ongoing hagiography of the modern masters.[8] A heavily annotated copy of *Der Raum als Membran* in Mies's library demonstrates the relationship between his work and Ebeling's theoretical universe. In fact, Ebeling introduces an entire constellation of authors and intellectuals in his essay that permeate van der Rohe's thinking. The depiction of nature as a quasi-utilitarian mechanism by biologist and science promoter Raoul Francé dovetailed with Friedrich Nietzsche's materialist interpretation of nature, contributing to a growing interest on the part of Mies in the synthesis between the natural and the artificial, between the biological and the physical.[9] All these references were already contained in Ebeling's treatise. In addition to a common appreciation of Nietzsche's philosophy, Ebeling demonstrates a profound interest in the works of the psychologist and politician Willy Hellpach and, perhaps more significantly, the physicist and director of General Electric's Lighting Research Department Matthew Luckiesh—whose work would constitute an important point of reference for later Bauhaus-related artists and pedagogues such as György Kepes during the 1950s. Ebeling's multiple references to these authors and intellectuals transforms the book itself into a selective membrane, filtering and distributing scientific and philosophical knowledge among architects and designers.

Siegfried Ebeling as a soldier in World War I. Courtesy of Rago Torre Ebeling.

There are two additional aspects, though, that motivated the resurrection of Siegfried Ebeling's text, given the growing presence of ecological thinking in architecture since it was written. First, Ebeling's work in interwar Weimar fed into the increasing sophistication of the environment as a topic of central concern for the techno-scientific architectural culture that developed during the second half of the twentieth century.[10] We can already identify in Ebeling's writings the theoretical framework that inspired some of the illustrations that the technophile engineer Françoise Dallegret made for Reyner Banham's article "A Home Is Not a House"—although devoid of the latter's seductive, erotic, and narcotic overtones.[11] Second, it is possible to detect in Ebeling's concerns a formal and theoretical prelude to the post-avant-garde of the 1960s and 1970s.[12] The attention he pays to the atmosphere as a medium, his research on the influence of electrical waves on human behavior, his multiple references to magnetic and geological fields, and his profound interest in climate and its radiations (*Strahlung*) resonate today with debates on sustainability in many of its multiple forms and geographies. It seems pertinent then to consider Ebeling's work in the context of a long tradition of criticism concerning the foundational moments of European modernity. And, accordingly, to consider how *Der Raum als Membran* must have also contributed to modern architecture in its "heroic" period. If these theories and the architecture with which they engaged affected the development of a shared epistemology situated at the center of biological debates, what was the cultural climate in which such a recalibration could take place? To avoid furthering the all-too-often monolithic and teleological treatment of such a tradition, the inclusion of *Der Raum als Membran* among the foundational texts within the archaeology of environmental studies must necessarily reference the cultural and political milieu of the post–World War I period in which the text blossomed. Without pretending to marginalize the reading of Ebeling as a pioneering environmental and climatic scientific thinker within architecture, it is important to underscore the historical context in which *Der Raum als Membran* came to light.

ENDANGERED CLIMATES

Peter Sloterdijk marks the arrival of the twentieth century with surgical precision on April 22, 1915, at 6 p.m. At that time, the Central Powers—Germany and the Austro-Hungarian Empire—unleashed 150 tons of chlorine gas on the Franco-Canadian infantry positions at Ypres in Flemish Belgium. An enormous yellow cloud 6 kilometers long and 500 meters high emerged out of more than 1,500 bottles of gas that had been installed in the area days before the battle. Though the number of casualties remains uncertain, the success of the military operation was irrefutable: the German coalition opened a breach in the enemy's positions, allowing them to enter triumphantly into the city of Langenmarck.[13] For Sloterdijk, this moment was the first time in history when the systematic attack on the physical and psychological environment of human beings was a priority for techno-military intelligence.[14] This kind of attack is based on a shared agency, requiring the active participation of the subject in his own injury: humans inhale the poisonous gas following unconscious mechanic respiration. Here the atmosphere functions as a decisive medium in the modern perception of horror: *atmo-terrorism* as an epistemological frame under which invisible aggressions toward human and non-human

10

This preoccupation with ecological habitats displaced the psychological abstract space à la August Schmarsow upon which the foundations of modern architecture were secured.

11

Reyner Banham, "A Home Is Not a House," *Art in America* 2 (April 1965), 70–79.

12

Banham, "A Home Is Not a House," 70–79; "The Triumph of Software," *New Society* (October 31, 1968), 629–630.

13

Reported injuries ranged from six hundred to fifteen hundred, and fatalities from three to five thousand, depending on the source.

14

Peter Sloterdijk, "Der Gaskrieg-oder: Das atmoterroristische Muster," *Schäume: Plurale Sphärologie, Band III* (Frankfurt am Main: Suhrkamp Verlag, 2004), 89–125. See also Dieter Martinetz, *Der Gaskrieg 1914/18: Entwicklung, Herstellung und Eintsatz chemische Kampfstoffe; das Zusammenwirken von Militärischen Führung, Wissenschaft, und Industrie* (Bonn: Bernard und Graefe, 1996).

Shock Troops Advance under Gas (Sturmtruppe geht unter Gas vor) from *The War* (Der Krieg), Otto Dix, 1924.
© Museum of Modern Art, New York, licensed by SCALA / Art Resource, NY.

life can take place. From then on, the air we breathe—poisoned with spurious interests of class, gender, race, and nation—was no longer reliable.

Such atmospheric instability contributed to an increasingly suspicious and somber mechanic civilization. Celebrating war as an aesthetic experience was among the cornerstones of modernism: Italian futurism, Le Corbusier and László Moholy-Nagy's fascination with the conflict, and a formal interest in fighter planes and bombers are just a few examples of the early-twentieth-century trivialization of horror. However, such unidentifiable dangers in the environment added a different existential layer to the harassment modern subjects were already experiencing in the Weimar Republic. This volatile existence was recorded in the plays *Gas I* (1918) and *Gas II* (1920) by expressionist playwright Georg Kaiser, for example, which depicted the space of the factory as the locus of class and techno-chemical struggles. Unlike later fictional plots addressing similar topics in which social conflicts intertwine with fantasies of technological domination—for instance, Bertolt Brecht's *Saint Joan of the Stockyards* or Fritz Lang and Thea von Harbou's *Metropolis*—Kaiser identified the origins of the fatal destiny of labor in the pernicious formula of gas produced in the factory. German society was beginning to be aware that threats to its physical integrity did not always make their presence known in solid, visible form.

Social suffocation was not only limited to air pollution: After the Great War and the signing of the Treaty of Versailles, the early 1920s inaugurated a period of hyperinflation in the German economy, the impact of which is still perceptible in the construction of the European Union today.[15] Prices doubled every three to four days, and the constant introduction of new currencies to ameliorate the situation—such as the *Papierenmark* or *Retenmark*, equivalent to a trillion of the original *Deutsche Mark*—made evident the devastation of German finances. This issue was present in the Bauhaus, as Herbert Bayer's designs for different bank notes from 1923 demonstrate: the abstraction and

15
The etymology of the word *hyperinflation* does not conceal its environmental overtones. From 1921 to 1923, this phenomenon was the result of the debt Germany amassed to support its military campaigns, worsened by the debt imposed on the country by the victorious allies.

simplicity of the design paralleled capital's volatility as well as its prognosis. It is not surprising then to find Ebeling complaining in *Der Raum als Membran* that "money" and "human beings" were perceived as already equivalent magnitudes (*Größen*) within a technically oriented German society. Ebeling sees the environmentally alienated subject as embedded within the same logic that immaterial, abstract capitalism had brought about.[16]

As has been frequently argued, the economic climate asphyxiating Germany's population was among the major causes of the widespread discontent that brought Hitler to power. And subsequently, the Nazi regime went even further in maximizing scientific management and chemical development to increase the drama—and efficiency—of mass annihilation. In 1924, on the same day that Ebeling's first article on cosmological spatial cells was published in Hamburg, the consortium of companies led initially by the Nobel Prize laureate Fritz Haber, moved the production of the gas Zyklon-B—the

16

Ebeling, *Der Raum als Membran*, 25.

Banknotes designed by Herbert Bayer.

infamous hydrogen cyanide insecticide responsible for the final stages of the *Endlösung* [Final Solution]—to Dessau. The aeronautical industry that made the 1926 migration of the Bauhaus from Weimar to Dessau economically feasible existed in the same geography as these chemical plants. The 1920s witnessed the experience of space and time not only as a form of liberation but also as a trap and as a refuge in which the modern understanding of environment and climate had a significant role.[17] The chemical industry at the time was built to replicate the devastating effects of earlier plagues, capitalizing on the no-longer-safe environment of the modern subject. Very soon, the menace of aerial, climatic radiations would pale in comparison to that which uranium's radiation represented. Against this panorama, assimilation or retreat were the only plausible responses.

TECHNOLOGICAL ISOLATION, PROTECTIVE MASKS

In this context, Ebeling's *Raumkubus* was as much an affirmative response to the problem of housing as it was a cautionary reaction to its own cultural climate. Ebeling's man intuits that he needs a more sophisticated and effective protection, and not only against the economic model that Walter Gropius and the *New Sobriety* championed, represented on the cover by the taut rational logic of the adjacent blocs.[18] The man Ebeling portrays is an interwar subject on the cusp of a triple assault: material, biological, and cultural.[19] The aesthetic and philosophical substitutions that modernism brought about were followed by other threats: economic volatility (assets were increasingly uncertain); environmental attacks (the atmosphere as a place full of daunting molecules and radiations); and finally, the political destabilization of national values: German nationalism was nothing more than the hyperinflation of the notion of belonging.

If the architectural membrane that Ebeling described behaved as a shell protecting the postwar subject from countless environmental threats, the *Raumkubus* functions as a fully developed immune system, an apparatus mediating and domesticating the multiplicity of aerial and geological radiations, heretofore physically and psychologically indiscernible.[20] In so doing, the space defined by Ebeling's cube enlarges the performativity of protective gas masks, so common during World War I, to an architectural scale. A thick membrane preserving healthy biological interchange between man and exterior, the *Raumkubus* functions at the expense of cultural isolation. In it, one can find the advent of the modern fortress, individual spaces and singular constructions that nonetheless remain intensively and extensively connected with the exterior. Ebeling went even further in his architectural solutions soon after: the *Kugelhaus* (1926)—a spherical domestic

17
Historian Timothy Snyder has recently underscored the role of environmental arguments triggering external political domination and race stigmatization in 1930s Germany. Timothy Snyder, *Black Earth: The Holocaust as History and Warning* (New York: Crown, Tim Duggan Books, 2015).

18
Although *Neue Sachlichkeit* is usually translated as "New Objectivity," I use the translation of *Sachlich* as "sobriety" to enlarge and complicate the economic and historical significance of the term.

19
Ebeling had to drop his studies at the Bauhaus in 1925 due to economic struggles.

20
Ironically, Ebeling always had in mind a metal membrane, the result of his collaboration and research at the Hugo Junkers factory during 1926. Walter Scheiffele identifies the man inside the *Raumkubus* with Junkers, a character who deeply fascinated Ebeling. Walter Scheiffele, "Membrane and Ecological Architecture," *Space as Membrane* by Siegfried Ebeling, vi–vii.

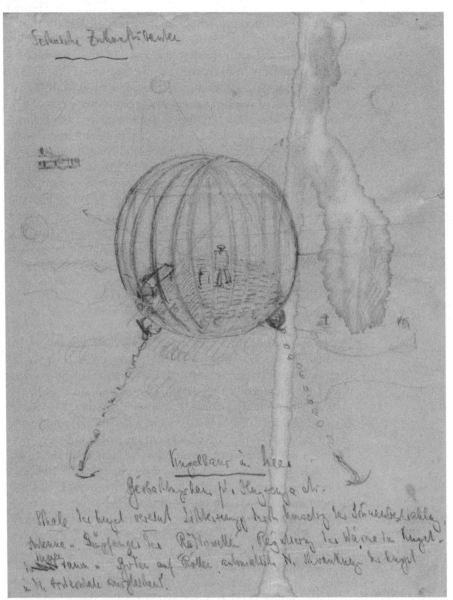

Sketch for the *Kugelhaus*, Siegfried Ebeling, 1926. Courtesy of Rago Torre Ebeling.

space floating above the sea, chained to an undetermined site, permanently or temporarily—was Ebeling's draft response to the question of rootedness and mobility. With this globular construction, Ebeling placed his subject in an existentially controlled isolation intended as generative of a new universal culture. As the last segment of his treatise stated, the goal is to cultivate a subject that "strives for, and radiates, a new evaluation of himself and his connections" to nature and the cosmos, from which a new style could emanate.[21]

Ebeling's proposals appear then as an early biopolitical response to the issue of globalization. These architectural forms—the shell, the pod, the capsule, the cell—are the same ones the architectural imaginary will incorporate seamlessly during the following decades, resulting in the bunkerization of the individual against supposedly external aggressions. Indeed, both his *Raumkubus* and *Kugelhaus* designs avoid physically stable foundations by

21
"Es kann daher vorläufig die größte Aufgabe des nächsten Jahrhunderts nur darin gesehen werden ... erst einmal den neuen Menschentypus anzubauen, indem jeder Erkennende für sich nach einer neuen Einschätzung seiner selbst und seiner Zusammenhänge mit dem Weltgeheimnis strebt und diese ausstrahlt." Ebeling, *Der Raum als Membran*, 32.

relying on technology. They constitute Ebeling's solid and material reaction to geographic and environmental uncertainty: the triumph of groundless hardware and technological communication as *sine-qua-non* conditions for modern habitation. In both of these projects, Ebeling rejects not only the foundational architectural myth, as romanticized by Laugier in his frontispiece for *Essai sur l'Architecture*, but also the generative elemental taxonomy of Semper's Caribbean hut as well as Le Corbusier's Cartesian skeleton in *Le Maison Dom-ino*. The alternative space one can sense in the *Raumkubus* is confirmed in the *Kugelhaus*: a new type emerging out of the new conditions. The technological pod is a refuge and retreat, a place for the subject to reunite with the self. From this point of view, the spherical house illustrates Sloterdijk's description of absolute islands: material entities where being-in-the-world is grounded in a fluid, mobile, and distant medium that nonetheless results in hyperconnectivity.[22]

If the illustration of the *Raumkubus* on the cover of *Der Raum als Membran* identifies the problem, the *Kugelhaus* constitutes the final response in a seminal epistemological turn. Ebeling decided to place his aerostatic project in a remote area, away from terrestrial turbulences. The *Kugelhaus* thus acts as a metaphor of his own detachment from his Bauhaus peers and, probably, of his own intellectual isolation. The sphere was equipped with multiple apparatus for detection, regulation, and communication, granting an individualized microclimate. It allowed the human habitation of an atomized, post-human future, artificially in command of his own immune architecture.[23] The entrenchment within his prosthetic membrane also becomes a trap.

We are witnessing the arrival of a new social paradigm in which identity relations will take place through the telecommunication of what Paolo Virno calls "the non-present or multitudes," a technological and anthropological twist to the social essence of mass communications.[24] This new subject encounters in Ebeling's sphere a form of representation but also a means for survival; it is a loophole for isolation and communication, floating above certainties that not so long ago had been grounded on solid terrain: postmodernity was around the corner. Ebeling's designs announced the arrival of the future cosmonauts and the more recent taikonauts. The conceptual seeds of the future *Cushicle* and *Suitaloon* by Mike Webb are already contained in his *Kugelhaus*, as are the mobile structures that Toyo Ito designed for the nomad women of Tokyo, artificially confined as prominent subjects of the Japanese megalopolis. All these singular structures materialize the isolation of the contemporary subject as the primary ontological condition of habitation within unstable political, social, cultural, economic, climatic, and psychological circumstances. But where in all of this can we locate political space as a social compact? In other words, where is the city?

22

Sloterdijk, "Absolute Inseln," *Schäume*, 317–337.

23

The description of the sketch goes as follows: "Technical Buildings of the future. Spherical house in the sea. Flight observation house, etc. The sphere's outer frame unites the production of light via the implementation of sun radiation, antennae and receivers of radio waves, heat regulation in the interior space; the floor based on rollers automatically compensates the sphere's fluctuations [so that it stays] horizontal." I have used the translation appearing in Spyros Papapetros, "Future Skins," *Space as Membrane by Siegfried Ebeling*, xviii.

24

Paolo Virno, *A Grammar of the Multitude: For an Analysis of Contemporary Forms of Life* (Los Angeles: Semiotext(e), 2003); *Multitude: Between Innovation and Negation*, trans. Isabella Bertoletti, James Cascaito, and Andrea Casson (Los Angeles: Semiotext(e), 2008).

Pep Avilés is an architect and historian, currently working as an assistant professor at The Cooper Union for the Advancement of Science and Art. He edited the Spanish edition of Siegfried Ebeling's *Der Raum als Membran* (Mudito, 2015). He is also a founding member of the experimental design platform The Fautory.

Dehydrated Architecture

CAROLINA GONZÁLEZ VIVES

Water is as "natural" a part of our climatic milieu as sunlight or oxygen, and yet within the fields of architecture and urbanism it is often viewed with either fear or fascination. Across the past century, an interesting dichotomy has been consolidated, introducing a clear spatial order. While water's value as an economic asset and major transformer of environmental conditions has led political, social, and cultural agents to mobilize it in anthropic management projects, constructing extensive hydrophilic landscapes that extract the maximum flow possible from ecosystems, simultaneously, building technology seems to have directed its efforts toward achieving a fully hydrophobic architecture—the creation of dry spaces.

Water vapor (the gaseous phase of the water cycle) is one of the most relevant factors affecting environments at all scales. Steam is the most prevalent greenhouse gas, making up between 1 percent and 4 percent of the atmosphere (as compared to carbon dioxide's 0.038 percent). Recent developments in atmospheric representation, notably satellite imaging, have vividly demonstrated links between massive deforestation and climatic consequences in developing countries such as loss of rainfall and rising temperatures.[1] At a smaller scale, fluid dynamics modeling has contributed to discussions about real-time relationships between water and urbanization.[2] Finally, the construction of infrastructures, buildings, and landscapes all point to the ways in which water is *cultural* as much as technical; any given project evinces a particular attitude toward the human body and its privileged position as a sensor of environmental quality.

Take, for example, the construction of modern hydrologic geographies in arid climates, such as those created in the nineteenth and twentieth centuries as part of the urbanization of landscapes like the American West or southern Europe. Such projects materialize their specific modernity through an understanding of geography as hydrologic engineering. The control of water—or, moreover, the elimination of dryness— becomes seen as a form of social transformation. This is not simply a matter of irrigation but also imagination. This cultural desire for the presence of water appears in countless literary works or paintings, documenting a clear association in the collective consciousness between aridity and misery,

1

Michal Kravčík, et al., "Water for the Recovery of the Climate— A New Water Paradigm," trans. David McLean and Jonathan Gresty (Žilina, Slovakia: Krupa Print, 2007). Available at www. waterparadigm. org/download/ Water_for_the_ Recovery_of_the_ Climate_A_New_ Water_Paradigm. pdf.

2

See, for instance, the interactive groundwater map developed in Portland, Oregon, to coordinate actions above and below ground toward improving infiltration as a flood control strategy, http:// or.water.usgs.gov/ projs_dir/puz/puz. html. The Arid Lands Institute's digital map of the L.A. region incorporates the behavior of soils, land uses, and the distribution of underground contamination to organize different hydrologic urban zones whose conditions recommend certain interventions, http://blogs.kcrw. com/dna/hadley- and-peter- arnold-envision- hydrological- zoning-as-a-way- out.

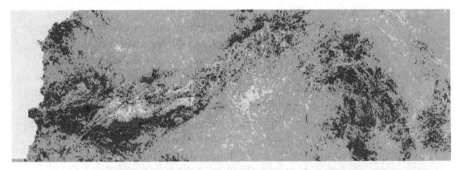

Annual average real and potential evapotranspiration in the Tajo River Watershed, Spain. Courtesy of SIA, Integrated Water Information System.

while the abundance of water takes on a utopian aspect.[3] This remaking of dry landscapes is a collective enterprise of gigantic proportions, one that concludes with the complete transformation of a previously arid ecosystem into a modern, efficient, productive hydrologic geography.[4]

The production of such landscapes involves deforestation and earth moving, compacting and leveling topographies to transform porous, absorptive ecosystems into smooth and impervious surfaces, thus extracting as much

3

See the work of the so-called Generación del '98, a group of Spanish writers including Antonio Machado, whose *Campos de Castilla* was published in 1912, or "la Escuela de Vallecas," a group of artists centered around Benjamin Palencia in the 1930s whose work deals obsessively with aridity as pictorial argument.

4

See Erik Swyngedouw's "Modernity and Hybridity: Nature, *Regeneracionismo*, and the Production of the Spanish Waterscape, 1890–1930," *Annals of the Association of American Geographers* 89 (1999): 443–465, or his recent book *Liquid Power: Contested Hydro-Modernities in Twentieth-Century Spain* (Cambridge, MA: MIT Press, 2015).

water as possible from them. These massive hydrological projects also involve constructing infrastructure whose enormous energy needs are ongoing, such as pumping stations, diversions, and desalination plants, each of which plays a role in blurring the existing topography. With their sublime and monumental character, they enact an idea of culture that portrays nature as an unlimited (and tamable) resource. But as these new hydrological landscapes extract water, desertification expands in kind. The absence of water in the soil (having been removed for human use) means no evaporation and thus no condensation. At a regional level, the conditions become extreme—greening for some corresponds with dryness for many. Soil dehydration is followed by

A golf oasis in the desert hills, Las Vegas, Nevada. Photograph by Alex Maclean.

Suburban backyard pools. Photograph by Alex Maclean.

the air dehydration, initiating anthropogenic cycles of landscape destruction over large areas of the planet. The cartography of moisture indexes economic power, even if the relationship of our daily lives to the vast implications of desertification is sometimes masked by the impressiveness of the infrastructure or the general invisibility of water in our experience of the city.

This landscape is not only built for productivity; it is shaped by cultural approaches, as well. The relocation of water and the expenditure of energy it entails are also motivated by aesthetic ideas about what qualifies as an inhabitable landscape (materialized most directly in large gardened areas, lawns, and golf courses, which account for a significant volume of the water consumed in the American West and Southern Europe).[5] The transfer of water is equated, in many situations, with a transfer of beauty—meaning that effecting change requires reimagining dryness as a source of identity with its own opportunities for expression and design, replacing the generic dream of the suburban lawn (oases in an otherwise grassless context) with another ecological imaginary.[6] The opacity of hydrological infrastructures make it such that the general public has little awareness of the huge volumes of water that invisibly cross urban space, not to mention the environmental impacts of that flow inside and outside the city, which include desertification and heat islands.

This fascination with harvesting water in arid landscapes has not been matched by an equal interest in the role of water within buildings themselves. On the contrary, architectural modernism has generally mistrusted water. Across the nineteenth century, the city came to be designed as a waterproofed machine, whose new networks defined the flow of water and organic waste while channeling a growing aversion to both. Earthworks and drainage systems de-watered the territory of the city, followed by the construction of an extensive and watertight carpet of pavement and architectural surfaces that hides the thickness of the soil and erases water from the urban map. This act of sealing the ground prohibits the exchange of water between land and air via infiltration, moisture retention, or evaporation. Dust, mud, and puddles disappear from the modern urban environment to make way for a more sophisticated atmosphere surrounding an aseptic surface.[7]

This newfound invisibility of water enables the mineralization of the world above ground, with the city becoming a terrain of minimum biodiversity. The metaphor of the mine, for Lewis Mumford, was a foretaste of the conditions of modern urban space—organic matter has disappeared, nothing edible remains, and all environmental conditions, including light and air, are artificially assisted. The construction of the modern city has been an effort to remove the body from stickiness, muddiness, and dustiness. To borrow a phrase from Julia Kristeva, this watertight urbanism is in a sense the "denial of pierced body."[8]

For architects, too often concerned with the realm of the visible, the undergrounding of water has removed it from their disciplinary focus for more than a century. The "hygiene" of modernism was perceptual (the production of white volumes) as well as atmospheric, in its watertight façades and frequently synthetic interiors. The modernist aversion to the messiness of the body (especially the female body) and organic matter, manifest in their glossy environments waterproofed at all scales, appears to continue today—think of the work of Kazuyo Sejima and Ryue Nishizawa of SANAA, whose laboratory-like spaces appear sealed with minimal thickness. Contemporary swimming pools further evince this conflict between water, buildings, and bodies. Only a tightly defined and polished pool with an almost radioactive

5

For example, in the Los Angeles region of California, measurements from the L.A. Department of Water and Power show that these areas consume 30 percent of the water supply.

6

Seminal texts such as *Scenes in America Deserta* by Reyner Banhan, or Gilles Clement's later works, have contributed to diversifying our references by linking notions of beauty to greater biodiversity. The development of a specific vocabulary for dry forms—a common lexicon for design and scientific disciplines—is another necessary project, creating a tool that can assign each component of the landscape a value on which to project its evolution.

7

See Mirko Zardini, "El suelo en la ciudad moderna y el predominio del asfalto," in *Naturaleza y artificio: el ideal pintoresco en la arquitectura y el paisajismo contemporáneos*, ed. Iñaki Ábalos (Barcelona: Gustavo Gili, 2009).

8

Julia Kristeva, *Powers of Horror: An Essay on Abjection* (New York: Columbia University Press, 1982).

Mussel Choir, The Living, Natalie Jeremijenko, and Mark Shepard, Postmasters Gallery, New York, 2014. Photograph by David Benjamin, courtesy of The Living.

color (demonstrating disinfection and sterilization) makes possible the social experience of diving with other bodies.

Proposals like The Living's *Mussel Choir* and Décosterd & Rahm's Omnisports Hall have synthesized with accuracy this ambition to rethink architecture's relationship with water and the organic, pointing out alternatives for managing urban metabolism.[9] The former employs mussels as living sensors for an urban water-quality-monitoring system. The movement of their shells reveals the presence of organic matter, much of it from the sewage network that combines fluids that have passed through and across other bodies. In the latter, the building is treated as a device that channels flows of matter and energy, establishing productive collaborations between species and biotopes. The indoor air is directed across soil to reduce its temperature before being breathed by athletes. The sweat of their moving bodies is extracted from the sports room by the mechanical ventilation system, where it is condensed and used to irrigate food crops grown in the bar courtyard. The architects' ironic conception is disturbing, transgressing invisible but rigid conventional hierarchies by revealing how human bodies, the living world, and inorganic matter are connected by the same flows. Both projects question our definition of waste and shift our conception of cleaning from disinfection to fertility, giving rise to new ways to think about urban metabolism, the position of water, the shape of architecture and the city, and the perception of biodiversity in the urban ecosystem.

Beyond their relationship with organic matter and hygiene, modernists' visceral aversion to wetness also played out in their increasing design engagement with the qualities of interior air. The advances in HVAC technology and easy access to energy that marked the twentieth century gave rise to an architecture whose form and materiality was no longer connected to climatic concerns. Greenhouse-like buildings with lightweight, sealed envelopes, and seemingly infinite interiors with mechanized climates, are part of a conceptual lineage connected to the environmental utopianism devised in the 1960s and 1970s by designers like Buckminster Fuller or the countercultural

9
Mussel Choir was an installation created by The Living with Mark Sheppard and Natalie Jeremijenko for the American Pavilion at the 2012 Venice Architecture Biennale. Décosterd & Rahm's Omnisport Hall was a competition entry submitted in June 1998 to the City of Neuchâtel, Switzerland; for project narrative and images, see www.philipperahm. com/data/projects/ sallcomnisports/ index.html.

Project for Floating Cloud Structures, Buckminster Fuller, 1960. Courtesy of the R. Buckminster Fuller Estate.

Restless Sphere, COOP HIMMELB(L)AU, Basel, Switzerland, 1971. Photograph © Katharina Vonow.

"Mobile Office," Hans Hollein, 1969.

atmospherics of Hans Hollein and Coop Himmelb(l)au. Conceived and located in wet geographies, their atmospheres are defined by brightness and completely dry air, in accordance with the modernist distrust of water, moisture, and shade. The substantial gadgetry of air extraction, filtration, and dehumidification is the hidden inverse of these transparent sealed envelopes and glossy finishes, ensuring the kind of ventilated and dehydrated environment that is considered optimal for health. Though designed to protect against the various forms of pollution seen by the twentieth-century city, such buildings have since revealed the inconvenience of inflexibility, their high dependence on energy, and above all, the clinical pathologies of sealed atmospheres. They have tabled the discussion about exterior environment by mitigating its effects and obscuring our awareness of whether these environmental effects come from inside or outside.

These typologies and constructive solutions meant for wet climates have been moved, unfiltered, to more arid latitudes, further drying already dry air. A paradoxical notion of comfort—one associated with sealed glass envelopes and air heating devices, which base their performance on dehydrating the air—has spread everywhere, beginning with office and commercial buildings and now applied to many other programs. Along with other factors such as the massive use of fossil fuels, these dehydrated and waterproofed architectures have proved to be little-recognized but effective tools in building new climates. The utopian projects' ambition to design the entire urban environment has been overwhelmed by this uncontrolled experiment, whose systematic addition of building-scale actions has resulted in large domes of altered, hot, and extremely dry air, floating above metropolitan areas as a particular form of landscape desertification.

During the last two decades, interest in environmental issues has brought this fascination with atmospherical densities into the realm of the quantifiable and useful. The Modernists' light, technified archetypes centered on air design have been treated as the starting point for sustainability in architecture, and architectural diagrams have been filled with vectors and arrows. But in working from these archetypes without interrogating their essential cultural assumptions, many of today's architects overlook powerful missing variables that could be integrated into a sustainable approach. Expanding the theory and practice of sustainability, including rethinking large-scale environmental projects in arid and desertified lands, requires assembling and examining references that go beyond sunny, dry atmospheres.

Matsys, Sietch Nevada Project, 2009.

Rather than aiming for absolute environmental control and seeking the thermal anesthesia of rationalized, positivist interiors, some architects have recently started to work with moisture and shadow—with denser, heavier environments like forest interiors or spaces dug out from the earth. This combination is essential in spatially defining an oasis, the traditional public space of arid lands. Oases have historically been linked with a mode of operation based on scarcity of resources and a mutually beneficial relationship with the natural environment. They constitute a necessary model for building interiors and exteriors in dry urban areas. Unlike those of previous sealed archetypes, the "envelope" of an oasis comprises discontinuous layers of plants and soil permeable to ecological flows, which together form an outdoor yet defined and thermally controlled space. Using little energy, oasis dwellers establish alliances with living systems and carefully manage local air and water cycles to create cooled exterior zones. Their approach and techniques should become part of architects' conceptual set of tools for designing in arid places on an environmental scale, complementing and supplementing technologically advanced solutions. The potential of such notions can be recognized, for instance, in Matsys' Sietch Nevada project for the exhibition *Out of Water*, which envisions colonizing the American West with excavated units, similar to the traditional courtyard wells of the Sahara desert.[10]

On the other hand, Modernist interest in the air has evolved into contemporary architects' engagement with the "cloud" and projects exploring plastic and sensorial celebrations of water, constituting water's return into contemporary architecture. The approach to climate design manifest in Prada Poole's El Palenque pavilion, shown at the 1992 International Exhibition of Sevilla, exemplifies this trend.[11] Its façades and roof are hybrid materials, simultaneously solid, liquid, and gaseous. Its thermodynamic behavior is regulated by the vaporized cloud surrounding its tentlike roof, its façade (formed by liquid water), and its radiant concrete foundation, which is cooled by an embedded water circuit that transfers heat to the ground. Its cloudy consistency exhibits the beauty that energy-efficient systems can create.

Subsequently, analytical studies of the behavior of the form in relation to moist air have been advanced through the development of agricultural greenhouse prototypes for arid landscapes, and through experiments like those carried out by research groups such as Watergy or the Blaustein Institutes for Desert Research.[12] Others go further in creating an image of hydrophilic

10

"Sietch Nevada" was exhibited at *Out of Water* in 2009; for project narrative and images, see http://matsysdesign.com/2009/06/25/sietch-nevada.

11

Precedents of the El Palenque pavilion include E.A.T.'s Pepsi Pavilion for Expo '70 in Osaka and more recent examples like Diller Scofidio's Blur Building, exhibited at the Swiss Expo in 2002.

12

For information on greenhouse research, see Martin Buchholz, ed., *Overcoming Drought*, a report published and funded by the European Union in 2009, http://www.user.tu-berlin.de/marbh/Implementation Guide.pdf. For information on Watergy, see www.Watergy.de. Projects conducted through the Blaustein Institutes that are particularly relevant include the work of Limor Shashua-Bar and David Pearlmutter.

Parque de los Pinos Urban Intervention competition entry, AMID (cero9), 2008.

architecture. For instance, AMID (cero9)'s 2008 proposal for Pinos Natural Park in Madrid, "Social Oxygen Balloons," takes a biotech approach, making the process of transpiration that takes place in its pavilion walls (the building cools itself by storing and evaporating water) visible at a macroscopic scale. The section describes a prickly body, like a hedgehog, whose "spikes" are elongated capsules embedded in the façade that feed plantation algae. These methacrylate blisters almost double the volume of the curved shell that encloses the interior space. They act as miniature greenhouses, connecting the pavilion's interior and exterior atmospheres through transfers of matter and solar energy—they collect and store rainwater, which, metabolized by the algae, becomes oxygen feeding the vegetation of the park. Visibly sweating, these cyborgs exhibit that they are active processors of water and air.[13]

13

For more on the project, see David Gissen, "The Material Transformations of AMID (cero9) Social Oxygen Balloons," *Architectural Design*, vol. 80, no. 3 (May–June 2010): 40–45.

There have been many propositions about how we might control global warming technologically. On the atmospheric scale, there has been speculation about seeding clouds for artificial rain or producing clouds of special gases to reflect solar radiation. But the local scale plays a vital role as well, and the composition of the air can be influenced by the hydrology of the ground—the systematic retention of water, the moisture of the soil, and evaporative cycles are crucial for the larger climate. Our technological culture exists in a complex exchange with our changing notion of the body, and water has played a crucial role in that techno-social relationship. The correspondences between urbanization and desertization, between cleanliness and mud, between bright, dry air and the moisture outside are deeply ingrained in our imaginations, and these correspondences offer implicit starting points from which to think about the design of landscapes and buildings.

These cultural constructions can evolve toward a coexistence with fluidity, toward a notion of sanitation closer to fertility than to sterilization, and toward a preference for hydrophilicity instead of drainage. This involves reconsidering things like urban metabolism and environmental quality, and it means recovering the place of water within urban structures and public space. Architecture has specific tools that can help intensify our perception of hydrological processes and give form to the climatic imaginations of other disciplines as well as our own.

Carolina González Vives holds a PhD in architecture from the Universidad Politécnica de Madrid and is an associate professor in the Architectural Design Department at the Universidad de Alcalá. During the research period for her dissertation, "Antidesertification Architecture," she collaborated with the Arid Lands Institute in Los Angeles, California. Since 2005 she has led her own office in Madrid.

The Crystal: Architecture Calibrating Climate Change for the Urban Future

MAY EE WONG

A view of the Crystal from above, WilkinsonEyre Architects, 2012. All photographs were taken by the author in the summer of 2015.

I arrived in London shortly after what was deemed the hottest day of the summer, with the city thrown into an unexpected swoon at 93 degrees. Coming from a place where 105 degrees was becoming the norm, 93 had seemed like it would be a breeze. But as I pushed myself onto the Heatherwick Studio–redesigned Routemaster buses and packed into the Tube, I began to experience the temperature's oppressive quality. With little ventilation and no air conditioning, these enclosed bubbles transporting the urban masses felt like microcosms of a suffocating planet: increasingly hot, carbon-dioxide-filled, over-teeming, and barely tolerable—exactly the sort of problems that the exhibition *Our Urban Future* would try to impress upon me and other visitors in the comfort of a mixed-use sustainable "intelligent building," the Crystal.[1]

Located at the Royal Victoria Dock in East London and opened in 2012, the Crystal was designed by WilkinsonEyre Architects in collaboration with interior architect Pringle Brandon Perkins+Will and engineering firm Arup. The building is owned and operated by Siemens, a multinational engineering conglomerate that has in recent years become a major global player in

1

Siemens, *The Crystal: One of the Most Sustainable Buildings in the World*, https://www.thecrystal.org/wp-content/uploads/2015/04/The-Crystal-Sustainability-Features.pdf.

renewable energy, and is home to Siemens' Center of Competence Cities.[2] As I would discover through my visit, the Crystal makes the company's implicit endorsement of various United Nations' agendas on climate change, sustainable development, and better urban futures under one roof. It does so largely by deploying architecture—in the form of edifice, model, and diagram—as a communicative infrastructure that constructs a climatic imaginary of abstract global risk for our contemporary times. This imaginary is further defined by the exhibition the building hosts, *Our Urban Future*, which presents to visitors a tightly woven narrative of energy-efficient infrastructural solutions and urban management policies that can be adapted across cities as the means to combat global warming. Mobilized in what anthropologist Anna Lowenhaupt Tsing terms a process of "scale-making," the Crystal acts as an "invisible technology," which, through analogy and experience, presents infrastructure as a coherent and scalable system, while positioning architectural authority as expertise on the specificity of a site or place.[3,4] Here architects are portrayed as collaborators in a smart ecology of urban sustainability, interpolated into a political economy of green development that naturalizes climate change as an extension of existing global capitalist imaginaries.

GAZING INTO THE FUTURE WITH THE CRYSTAL

Siemens' literature on the Crystal emphasizes superlatives. In addition to boasting that the Crystal is one of the planet's greenest buildings, executives assert that it houses the world's largest exhibition on urban sustainability, which provides visitors with an educational experience exploring "how we can create a better future for our cities."[5] This experience begins in a small auditorium where visitors are invited to watch a looped film that allows them to "immerse [themselves] in the forces of change." The film introduces these four "forces"—rising urbanization, scarcity of resources, growing populations, and the effects of climate change—with short montages of stock photography depicting the disappearance of geographically indeterminate forests and agricultural areas, a close-up shot of dry, cracked earth, and cars surrounded by pollution and smog. Then the screen turns black and an image of the globe emerges, juxtaposed with a graph showing the ever-climbing line of the earth's average temperature. Captions threaten the viewer in succession: "Drought." "Water Scarcity." "Desertification." "Storms." "Flooding." "Hurricanes." Then we are informed that "the frequency and intensity of extreme weather events are expected to rise, even with small temperature changes. Many governments have committed to action on climate change, but a global framework to fight climate change is not yet in place." The world that the film depicts is a post-apocalyptic one: these "forces of change" are crises. But the word "crisis" is never explicitly mentioned in the film, because at Siemens, every crisis is an opportunity.

Outside the auditorium, exhibits inform us that climate change matters because its various undesirable consequences are negatively affecting people around the world. The exhibits tell us that planetary carbon dioxide levels have risen as global growth has caused a significant increase in energy consumption, and this energy is primarily sourced through burning fossil fuels. After running through screens explaining how global warming is on the rise, I swiped one that brought me to a checklist of categories: "Problems. Opportunities. Actions." It informed me that I could click on a marker on the globe

2

According to a Siemens' press release, the Center of Competence Cities "is responsible for building up strategic partnerships between Siemens and city organizations. It is conducting urban research, promoting cross-Division and cross-Sector solutions for cities, bringing together internal and external urban experts and managing the worldwide network of Siemens' City Account Managers." See "The Crystal Celebrates Its First Anniversary with the Sustainable Cities Week," September 2, 2013, http://w5.siemens.com/greece/internet/en/press/2013/Pages/02092013.aspx.

3

Anna Lowenhaupt Tsing, *Friction: An Ethnography of Global Connection* (Princeton, NJ: Princeton University Press, 2005), 57. The term "scale-making" relates to the projection of global and local imaginaries that conjure or perform how a given economic project might succeed.

4

The phrase "invisible technology" is used by Daniel Libeskind in the exhibition's "Shaping Our Urban Future" video wall. I discuss Libeskind's mention of the term later in this essay.

5

Siemens, "The Crystal."

to select a particular city and read about its problems and their solutions. I
learned that close to 80 percent of the world's energy needs could be supplied
by renewables by 2050, and that 13 percent of the world's urban population
currently lives in low-elevation coastal zones susceptible to flooding.

Pitched as an experience that puts the layperson in the shoes of an urban
planner, a policymaker, and an architect, *Our Urban Future* simplifies
and categorizes the causes, effects, consequences, and conditions of climate
change into metrics, graphs, checklists, and operable actions for urban
sustainability across cities. The exhibition's translation of complex decision-
making processes into pleasurable interfaces distills an ideology of universal
quantification, standardization, and enforcement by which these urban professionals operate; one of the exhibits on city planning cites a quotation from
former New York City Mayor Michael Bloomberg—"If you can't measure it,
you can't manage it."

Our Urban Future presents us with a climatic imaginary commensurable
with what Ulrich Beck identified as "global risk," or the staging or manufacturing of hypothetical consequences on a global scale.[6] As communicated
by the exhibits, the exact effects and outcomes of climate change are unpredictable, but likely scenarios need to be registered and anticipated in order
to be properly controlled. In *Our Urban Future*, climate change indications
and risks are relayed through percentages, PPM (parts per million) statistics,
and their correlated temperature differentials drawn from climate science.
Resilience—as gleaned from the exhibition's "Safe and Sound" section—is
defined as disaster risk reduction from severe weather phenomena such as
rising sea levels, as well as terrorism.

FIGURING THE GLOBAL SUSTAINABLE CITY

The depictions of cities employed in *Our Urban Future* reveal insights into
the techno-centric scope and intent of the exhibition's curatorial vision. The
otherwise abstract nature of global climate change becomes legible through
the figure of the individual global city, which the curators extrapolate as case
studies for best practices. "Cities across the globe have launched initiatives to
combat climate change," intones "Forces of Change," as the illuminated globe
appears on-screen, marked with cities around the world that are labeled with
their 2025/2030 carbon dioxide emission target reductions. This is followed by
a segment that declares a universal commitment for "research and innovation"
and "developing technology" for "green living." This optimistic vision reflects
what Beck proposes as the "cosmopolitan moment of world risk society," in
which individuals, faced with the prospect of global disaster, could be forced
into simultaneous collective action in acceptance of our common vulnerability,
which exceeds the boundaries of the nation-state.[7] But while Beck proposes
that this "moment" entails the recognition of our plurality and common civic
responsibility, it seems that social and political questions play an ancillary role in
Siemens' cosmopolitan/climatic imaginary. Instead, their vision is constructed
mostly on technological grounds, delimited by the geographical reach of infrastructural solutions, with energy circulating and architecture communicating
various analogical and experiential connections between home, city, and world.

In her 2005 book, *Friction*, Tsing describes how climate models rely on
the global scale as precedent from which additional models are created.[8] The
climate-change-stricken city has similarly been projected on a global scale

6

Ulrich Beck,
World at Risk
(Cambridge:
Polity Press, 2008).
Reinhold Martin
describes how
official environ-
mental discourse
becomes inscribed
in Beck's socio-
technical terms
of risk, as well as
how "environment"
is abstracted into
organizational
systems of signs
in architectural
discourse in "Envi-
ronment, c. 1973,"
Grey Room 14
(2004): 78–101.

7

Beck, *World at
Risk*, 55.

8

Tsing, *Friction*,
104.

Forces of Change, a video that was part of the Crystal's exhibition *Our Urban Future*, showing carbon emission reduction targets of global cities.

and replicated at the Crystal as well as through a range of related platforms dedicated to the "global sustainable city"—international institutions such as the UN, projects such as 100 Resilient Cities, and media circulated by thinkers such as Stewart Brand. The notion of the global sustainable city models both problems that are complex in nature and solutions that have the capacity to sustainably address various aspects.[9]

As stated in its brochure, the Crystal's design concept "was driven by a vision of a multi-faceted urban world," with its geological shape meant to suggest "the many facets of sustainability and the complexity of urban life." Seemingly without irony, the building is also described in the same brochure as being inspired by the Crystal Palace, the iconic London exhibition hall whose architecture and exhibits celebrated the best in Victorian industrial engineering innovation. Drawing all of its inspirations together, its promoters summarize the building as "symbolizing the dawn of a new, sustainable age."[10]

As an "all electric" building, the Crystal showcases renewable energy, infrastructural technologies, and "smart grid applications" that constitute a model of the self-managing "smart" complex.[11,12] Its architecture articulates Siemens' definitions of sustainability and resilience in the age of climate change (i.e., the use of energy-efficient infrastructure that leads to a complex chain of beneficial implications) and demonstrates how they can be achieved. Within this vision, Siemens' marketers present architecture as an instrument that stabilizes the coordination between networks of infrastructural technologies—a recognizably coherent "smart" system sensitive to the immediate environment and adaptable across cities. This systemic notion of infrastructure runs counter to its empirical reality, for as historian of technology Paul Edwards notes, "*infrastructures are not systems* but networks or webs."[13] There can be some degree of coordination or regulation to infrastructure, but it can never be designed as a single system. While the Crystal's "innovative building management system

9

Examinations of various articulations of the global sustainable city reveal how "sustainability" reflects different meanings and is employed toward myriad ends.

10

Siemens, "The Crystal," 17.

11

Siemens, "The Crystal," 13.

12

Siemens, "The Crystal," 20.

13

Paul N. Edwards, *A Vast Machine: Computer Models, Climate Data, and the Politics of Global Warming* (Cambridge, MA: MIT Press, 2013), 12. Italics his.

that automates and manages energy, building operations and infrastructure" promises the user complete control over the building's internal environment, this sense of total control is at most an impression mediated by the user's encounter with the architecture.[14]

Siemens also mobilizes architecture as an epistemic infrastructure in the form of diagrams. The installations in *Our Urban Future* serve as models that mediate between the abstractions of urban issues, and the particularity of applied technologies.[15] Its statistical charts and videos follow the methodological precedent established at the UN "Habitat I" conference held in Vancouver in 1976, where filmmakers captured and shared human settlement problems and solutions as audiovisual statements that created a forum for global management between a diverse group of experts.[16] Thirty-six years later, as the venue for UN-Habitat's 2012 conference, *Urban Planning for City Leaders*, the Crystal and *Our Urban Future* provided a similar forum for various groups of international experts to gather and discuss issues of urban sustainability.

SUSTAINABILITY FROM THE BUILDING UP

At more than 6,300 square meters, the Crystal stands as an impressively large, sleek, but underwhelming edifice. Its shaded glass walls are sharply angled to resemble the edges of a crystal, but the complexities of its geometry are only fully visible from above. With its sculptural, iconic form it embodies Hal Foster's "banal cosmopolitanism," connecting regenerated, post-industrial East London to other global cities where similar landmarks, such as Norman Foster's Hearst Tower in Manhattan, connote the sustainable future.[17]

My technical tour guide informed me that the all-glass shell designed by Wilkinson-Eyre is but one part of an infrastructural ecology embedded within the building and its immediate surroundings. The entire complex is designed as an integrated system that generates its own power (through photovoltaic cells and geothermal wells) and circulates it through the building in the most efficient

14
Siemens, "The Crystal," 12

15
Simon Sadler highlights architecture as a hermeneutic in his reflection upon countercultural architectural diagrams in "Diagrams of Countercultural Design," *Design and Culture*, vol. 4, no. 3 (November 2012): 345–367. Orit Halpern also discusses the diagram in the context of design and "communicative objectivity" in her book, *Beautiful Data: A History of Vision and Reason Since 1945* (Durham: Duke University Press Books, 2015).

16
Felicity D. Scott, "Carrying On Talking," *Texte Zur Kunst* 92 (2013): 108–33.

17
Hal Foster, *The Art-Architecture Complex* (New York: Verso, 2011).

Our Urban Future's exhibits present cities through metrics communicated in diagrams.

Interactive kiosks allow visitors to *Our Urban Futures* to construct their own sustainable buildings.

manner to minimize energy loss. He told me that the use
of the glass shell signals a move away from passive design
strategies in favor of high-tech solutions, as the shell is wired
to an energy management system that monitors thirty-five
hundred data points and controls the entire building—the
Crystal's hundred and fifty ventilation panels, for example,
can be controlled from a single console or set to automatically adjust to external temperature conditions.[18,19] As the
day took a fickle turn toward more typically cool English
weather, I could not experience the building's full ventilation
capacities, and I wondered about the amount of solar energy
customarily captured by its photovoltaic cells. (My guide did
explain to me that the building's energy needs are supplemented by the grid.)

Out in the parking lot, he introduced me to the Tesla
Supercharger, which was, as far as he knew, the only one in
London, and which serves as a central charging station for a
local fleet of taxis. He also pointed out a rainwater catchment
system that supplies the building's toilets and irrigates its
adjacent "biodiversity garden." Back in the exhibition hall I
was reintroduced to the Crystal's architecture—this time in
diagram form, complete with statistics on the various quantities of energy the building generates and consumes. This area
included a simulation in which visitors can pick and choose
tools "to create a smart, sustainable building that runs efficiently and also keeps occupants comfortable and happy."

My guide delimited the architecture of the complex to
the façade alone, in opposition to its engineering technology. But the Crystal's architecture could also be understood
as the communicative interface that mediates between the
abstraction of energy usage and the ecology of the building's site. Daniel Libeskind is featured in the exhibition's
"Shaping Our Urban Future" video wall as the architect
interviewed among mayors and politicians such as Boris
Johnson and Frank Jensen, as well as UN-Habitat's executive director, Joan Clos.[20] Libeskind promotes the architect's
expertise as the ability to understand place and environmental orientation—or in his words, "knowing where to
put something and how to put it." Sustainability, he says,
is not a new notion, and architects have understood how to
design with it in mind since Vitruvius, who recommended
they observe the flight patterns of birds at a building site.[21]
Further linking environmental stewardship with a sense of
architectural tradition, Libeskind describes architects as
bearers of memory and care, providing much needed stability
in times of change. He also asserts that there is no contradiction between aesthetics, the inner structure of a building,
and the environment—architecture functions as "invisible
technology." This serves as an endorsement for the smart
complex whose design concept relies on identifying natural
conditions and other aspects of the building's context as
abstract, calculable resources.

18

Siemens, The
Crystal: One of the
Most Sustainable
Buildings in the
World, 2.

19

Siemens, The
Crystal: One of the
Most Sustainable
Buildings in the
World, 5.

20

UN-Habitat, or
the United Nations
Human Settle-
ments Program,
focuses on tackling
urbanization
problems and
supports "socially
and environmen-
tally sustainable
human settlements
development and
the achievement of
adequate shelter
for all" ("UN-
Habitat at a
Glance," http://
unhabitat.org/
about-us/un-
habitat-at-a-glance).
Mandated in 1978
and established in
its current form in
2002, UN-Habitat
provides policy
expertise on issues
of urban growth
such as housing
and infrastructure,
especially in devel-
oping countries.

21

It is likely that
Libeskind was
referring to Vitru-
vius's suggestion
to "consider and
observe the natures
of birds, fishes
and land animals
… to reflect upon
distinctions of tem-
perament" in his
chapter on "The
Site of a City"
(Book I, Chapter
IV), a section
that beseeches

the architect to
be sensitive to the
specific climatic
conditions and
ecological relations
of a potential city
site, especially in
this case, observing
patterns of how
different living
beings respond
to the natural
conditions of the
site. See Vitruvius
Pollio, Vitruvius:
The Ten Books
on Architecture
(Cambridge, MA:
Harvard University
Press, 1914), 19.

22

Neil Brenner and
Christian Schmid,
"The 'Urban Age'
in Question,"
International Jour-
nal of Urban and
Regional Research
38 (2014): 731–
755.

23

"London Launches
Low Carbon
Green Enterprise
District," BBC
News, May 27,
2010, http://
www.bbc.com/
news/10171031.

24

The London Dock-
lands Development
Corporation
(LDDC) was an
Urban Develop-
ment Corporation
set up by the Brit-
ish government
to regenerate the
London Docklands
area in 1981. In
the 1990s, their
redevelopment
plans were stalled
during a deep
recession—espe-
cially after the
international
property develop-
ment firm Olympia
and York, which
had undertaken
the development
of Canary Wharf,
went bankrupt in
1992.

View from the Royal Victoria Dock, with the Emirates Air Line cable car system (WilkinsonEyre Architects, 2012) seen in the background and the Crystal at right.

URBAN AGE ECOLOGIES

While the Crystal's depictions of climate in the age of sustainability might evoke far-reaching temporalities, my visit also reminded me that the scales of climate are extensions of commercially driven global imaginaries in the here and now. These global imaginaries are centered on the city as an economic node, and now include the ideology of burgeoning urbanization driven by UN-Habitat and perpetuated by organizations, journalists, architects, politicians, and urbanists from around the world. Termed the "Urban Age Thesis" by Neil Brenner and Christian Schmid, the central claim of this ideology—that more than half of the world's population now lives in cities—justifies a centripetal focus upon the city as the locus of global research on urban policy and innovation, even as this thesis, as Brenner and Schmid demonstrate, is empirically suspect.[22]

In an effort to situate the project in its urban context, taking a leaf out of Vitruvius's book, I walked around the Crystal. The building is located in the London Docklands—more specifically, at the Royal Victoria Dock in the Green Enterprise District. This district is a regeneration project of Boris Johnson's administration, which since 2010 has aimed to make London "a global leader in the low carbon economy."[23] Formerly the site of the Port of London (from 1960 to 1980), various parts of the London Docklands have seen uneven attempts at redevelopment, including a period of decline under the London Docklands Development Corporation (LDDC) in the 1990s and the failure of the Millennium Dome (since rebranded as the O2 Arena) in 2000.[24,25]

25

Opened in 2000, the Millennium Dome was the site of an exhibition, the *Millennium Experience*. The Dome was widely acknowledged to be a failure due to poor crowd attendance and financial and management problems. In 2005, it was rebranded as an entertainment venue and renamed the O2 Arena.

View of the Crystal from the Emirates Air Line, with the O2 Arena (Richard Rogers, 1999) in the background.

Since the London Docklands is the setting of J. G. Ballard's 1975 novel *High-Rise*, I initially saw the district through Ballardian-colored glasses. But at the Royal Victoria Dock, I realized the area was far more mundane than it was dystopian. Rather than a blighted post-industrial landscape waiting to spark into nihilistic implosion, this was an animated place full of everyday activity. On my way to the Crystal, walking by cranes and scaffoldings encircling a half-built apartment block that was to be called the Pump Tower, I noticed that the construction area was mostly located at the back end of the building facing the Greenwich Peninsula, the other prime area that was targeted for urban regeneration. Behind the Crystal was a showcase for yet another upcoming condominium project. This was Hoola London, which featured "smart" technology by Siemens and promised panoramic evening views of the neighboring Richard Rogers–designed O2 across the River Thames.

But judging from my walk along the waterfront, the area didn't yet live up to the "dazzling, exhilarating and growing" Green Enterprise District described on the Pump Tower's website.[26] It was a modest space of leisure; the river lined with cranes and warehouses that were remnants of the dock's steamship-handling heyday that had begun at the end of the nineteenth century. Pedestrians ambled down the concrete sidewalks that flanked the river. A Spanish-speaking family lay on a white beach towel in their bathing suits. The other main attraction in the area, besides the water, was the Emirates Air Line, a cable car traversing the River Thames between the Royal Victoria Dock and the Greenwich Peninsula. Also designed by WilkinsonEyre, the

26

City & Docklands Property Group, "Pump Tower— Royal Victoria Dock, London E16," http://www. cityanddocklands. com/PT- Information.html.

Air Line is the source of most of the Crystal's visitors, who are drawn by the dramatic aerial views of the building available from their elevated cabins. Intended to model an Emirates flight, the Air Line experience begins by greeting passengers with a picture of the airline's flight attendants at the ticket counter. The cabin exteriors are decorated with images of other cities in the world that can be visited via Emirates along with captions that suggest people can view the city "from a different perspective" or "with fresh eyes." Once Air Line passengers touch down at North Greenwich station, they can attend the Emirates Experience, an interactive exhibit that simulates a thirty-minute flight in an Airbus A380 or Boeing B777.

There is a distinct irony in disembarking from a simulated experience of flight made possible by fossil fuel into an exhibition that promises a future without it. The O2, the Emirates Air Line, and the Crystal form a chain of spaces that direct visitors through various scales of projects aimed at cementing London's position as the preeminent financial capital in the circuit of global cities. Anna Tsing defines scale as the spatial dimensionality necessary for a particular kind of view, but I would like to propose that scale is the spatial dimensionality necessary for a particular understanding of time, as well.[27] The coexistence of the shipping cranes, the Air Line, and the Crystal at the Royal Victoria Docks reveals how cities are constantly remade through capitalist urban projects that exceed existing temporal and geographical boundaries. These "gigantic forgetting machines" leave behind just enough of the past and inscribe just enough of the future to allow people to understand their inevitable destruction and reconstruction under capitalist development.[28] This phenomenon is singularly inflected at the Crystal, where architecture calibrates the scale of climate change amid the forces of green capitalism so that visitors can experience change without changing.

27

Tsing, *Friction*, 58.

28

Jerry Herron, "Detroit Borderama: Part 2," *Places Journal*, July 7, 2010, https://placesjournal.org/article/borderama-detroit-2.

May Ee Wong is a PhD candidate in the Cultural Studies Graduate Group at the University of California, Davis, with a designated emphasis in critical theory. Her dissertation research examines epistemologies and ideologies of contemporary ecological and complex systemic discourses pertaining to the urban laboratory and the global sustainable city.

Risk: Excerpts from the Environmental Division of Labor

REINHOLD MARTIN

This paper was first presented at the conference "After the Spectacular Image: Art, Architecture, and the Medium of Climate Change," organized by Daniel Barber and hosted by the Princeton Environmental Institute. With thanks to Daniel Barber and Alexandra Quantrill for conversation on related matters.

1

Ulrich Beck, *Risk Society: Towards a New Modernity*, trans. Mark Ritter (London: Sage Publications, 1992), 19–50.

2, 3

Beck, *Risk Society*, 21.

4

Brian W. Edwards and Emanuele Naboni, *Green Buildings Pay: Design, Productivity, and Ecology* (New York: Routledge, 2013), Sect. 5.2, "Bank of America Tower, New York, by Cook + Fox Architects."

Thirty years ago, the sociologist Ulrich Beck argued that as risk distribution overlaps with and ultimately merges with wealth distribution, a reflexive "risk society" emerges, overlaps with, and ultimately replaces the "class society" engineered by industrial modernity.[1] A contemporary monument to this "risk society" is the Bank of America Tower, also known as One Bryant Park, in Midtown Manhattan, designed by Cook + Fox Architects for the Durst Organization and completed in 2010. This tower was the first commercial skyscraper to achieve LEED Platinum certification from the US Green Building Council, which, despite its official-sounding name, is a private organization. Hundreds of pages of documentation were no doubt assembled to achieve this result, among them, one imagines, a LEED pre-certification application for environmental air quality. On such pages—which likely reported the presence of a 4.6-megawatt cogeneration plant, a thermal ice storage system, fritted glass, and waterless urinals—would surely have been text documenting the MERV 15 air-filtration system installed on all of the building's air-handling units. MERV stands for "Minimum Efficiency Reporting Value," and a MERV 15 rating, which is normally achieved with 12 to 36 inches of microfine fiberglass and removes 95 percent of all particulate matter from the air, including bacteria, is standard for hospital inpatient care but not for office buildings.

In 1986, Beck defined risk as "a systematic way of dealing with hazards and insecurities induced and introduced by modernization itself."[2] Always present, these risks came to be widely recognized during the 1960s, well before anything like climate change had been verified. But even then, when puffing smokestacks still punctuated Western skylines and workers occasionally still went on strike, claiming those factories as their own, risk was not limited to the visible dangers of industrialization. Instead, says Beck, "the risks of civilization today typically escape perception and are localized in the sphere of physical and chemical formulas."[3] In keeping with this axiom, the risks managed by the Bank of America Tower are largely invisible. They are measured by physical and chemical formulas that describe the behavior of greenhouse gases, as well as by MERV ratings and the dimensions of fiberglass filters. As in Beck's thesis, risk society and class society have indeed met in the particle-free air inside this building, which is filtered, heated, cooled, and lit to support the activities inside. But when we look more closely, we do not see one regime replacing the other; rather, we see a striking interdependence.

One Bryant Park was designed as a speculative core-and-shell office building and was LEED rated prior to acquiring its main tenants, for whom the building's "high environmental standards" were reportedly an important factor in securing tenancy.[4] The Bank of America occupies 75 percent of

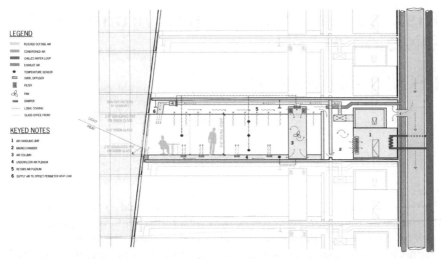

LEGEND

FILTERED OUTSIDE AIR
CONDITIONED AIR
CHILLED WATER LOOP
EXHAUST AIR
● TEMPERATURE SENSOR
SWIRL DIFFUSER
FILTER
FAN
DAMPER
LOW-E COATING
GLASS OFFICE FRONT

KEYED NOTES

1 AIR HANDLING UNIT
2 MIXING CHAMBER
3 AIR COLUMN
4 UNDERFLOOR AIR PLENUM
5 RETURN AIR PLENUM
6 SUPPLY AIR TO OFFSET PERIMETER HEAT LOAD

Section of a typical office floor at the Bank of America Tower (One Bryant Park), COOKFOX Architects, New York, 2010. Courtesy of COOKFOX Architects.

the overall square footage, about one-third of which comprises the bank's trading floors. As a result, and despite the platinum rating, it has been widely reported that the building emits more greenhouse gases than any similarly sized tower in the city, largely due to the energy loads generated by the 24/7 activities of those floors, with their screens, servers, and sleep-deprived bankers.[5] All of which is, of course, in support of the bank's ultimate purpose: to manage and profit from financial risk. At One Bryant Park, the Bank of America does so at fair expense to the external thermal environment but at relatively minimal environmental risk to the material comfort and well-being of bank employees, who breathe particle-free conditioned air as they speculate financially on the material risks taken by nations floating their currencies or homeowners mortgaging their houses. Of special interest to architects and architectural historians, then, is the fact that the coincidence of these two forms of risk, environmental and financial, is underwritten by the partition of what modernism used to call space and what we must now call *air*, and a corresponding division of labor.

That division, which is orchestrated above all by drawings and other visual documents, begins with the consultants. The firm of e4, Inc., a green building consultancy who compiled the documents that secured One Bryant Park's LEED rating, was only one of the building's approximately two dozen authors. Along with the architects, Cook + Fox and Adamson Associates, the list includes the MEP engineers Jaros, Baum & Bolles, the environment consultant Viridian Energy, the façade consultant Israel Berger, the geotechnical engineers Mueser Rutledge, and a whole host of others, to say nothing of the hundreds working under the general contractor, Tishman Construction, or those in the factories who produced its recyclable steel, or the engineers and line workers at Carrier who designed, produced, and installed the air filters.

This is nothing new. But I want to emphasize the obvious—that the technical performance on which the building's LEED certification rests belongs to the same class-based, gendered, and racialized transnational division of labor that organizes the economic processes on which the Bank of America's traders speculate. The issue is not that LEED certification uses spurious criteria or that a building's life after occupancy may negate those promises of

5

On the controversy over the legitimacy of the Bank of America Tower's LEED rating, see:

Sam Roudman, "Bank of America's Toxic Tower," *New Republic*, July 28, 2013, https://newrepublic.com/article/113942/bank-america-tower-and-leed-ratings-racket.

Martin C. Pedersen, "Broken Promises," *Metropolis*, July 30, 2013, http://www.metropolismag.com/Point-of-View/July-2013/Broken-Promises.

Sam Roudman, "The Debate on the Bank of America Tower Continues," *Metropolis*, August 7, 2013, http://www.metropolismag.com/Point-of-View/August-2013/The-Debate-on-the-Bank-of-America-Tower-Continues.

Roy Alter, "LEED-Bashing: Is the Bank of America Tower Really a 'Toxic Tower'?" *Treehugger*, July 31, 2013, http://www.treehugger.com/green-architecture/leed-bashing-bank-america-building-really-toxic-tower.html.

Koben Calhoun and Roy Torbert, "The Bank of America Tower—What Is and What Could Have Been," *Rocky Mountain Institute Outlet*, September 5, 2013, http://blog.rmi.org/blog_2013_09_05_the_bank_of_america_tower.

performance. This certification and the documents that subtend it have many functions, such as attracting tenants (including, in this case, Al Gore). But principally, they serve to bracket this other, more diffuse distribution of risk and its material support.

Take an earlier and simpler architectural example, which also has to do with the management of light and air at a building's outer surface, though with a much different, much more visible sort of filter—the *brise-soleil,* or sun breaker, that Le Corbusier designed for the Secretariat building at Chandigarh, capital of the Punjab, in 1952. Among the canonical images of this *brise-soleil* are the construction photographs of the building published in Le Corbusier's *Oeuvre Complète*. In his essential account of Chandigarh's design and construction, Vikramaditya Prakash calls our attention to the rural woman

Laborer working in front of the under-construction *brise-soleil* of the Secretariat building at Chandigarh, Le Corbusier, c. 1956. Photograph by Lucien Hervé, the Getty Research Institute, Los Angeles. © J. Paul Getty Trust.

carrying construction materials on her head in the foreground of one of these images. We can infer that she most likely lives on-site, and that her children or those of her extended family are most likely playing in the dirt just outside the frame. Most likely, too, she cannot read—in contrast to the nearly all male, multicultural though probably all English-speaking, and certainly all literate design team—much less interpret the technical drawings passing from the architect's office to the construction site. There is probably a string of supervisors and interpreters, all male, who translate the instructions written in English on the drawings into oral instructions, probably shouted in Punjabi or Hindi or in a village dialect, that inform her of what to do with the materials she carries.

If the *brise-soleil* at Chandigarh belongs to a tradition that paid detailed attention to environmental matters in the form of passive, "tropical" architectural elements, we recognize this construction laborer, and the risks she bears, as a precursor to those who, wherever they are, fabricated and installed the filters separating the dust and bacteria from the Bank of America's air—subordinate members of a "class society" on which the allegedly immaterial labor of bankers places its risky bets. Class- and caste-based as its plan is, a city like Chandigarh and everything that it stands for originally accommodated a relatively wide spectrum of modern subjects in its architecture and planning— from peons to members of parliament—even as it excluded others, like the manual construction workers. LEED certification recalibrates life in the far less visible terms of gases emitted and particles filtered, by building in a rather more stark, if more geographically distributed, arrangement. In short, the partitioning of the Bank of America's air is far more absolute, sociotechnologically speaking, than the partitioning of space visible in Chandigarh. At the Bank of America, either you are in or you are out. Financialization depends on such partitions, which are produced and repeated at countless scales.

How, then, did this happen? How did the sociotechnical imagination by which we still gauge our modernity come to turn on a materially real division of air as well as of labor, and the increasingly stark filtering out of one world from another—a precarious world outside that assumes the risks, and a securitized one inside that speculates on and profits from them? The question is far too vast for an essay like this one. What follows, in summary form, are some of the well-known architectural ways of drawing (and thereby partitioning) air, followed by a slightly more detailed example that lies in the deep background of the Bank of America Tower's contradictions. Revisiting this genealogy is a way of showing that the history of architecture is not a matter of aesthetics doing one thing and technics doing another. Rather, this series of enclosures remind us that architecture is a matter of filters of various kinds, filters that quite literally draw air through themselves and in the process, divide populations, in the imagination and on the ground.

In 1960, Buckminster Fuller captured the cognitive landscape of environmental risk in his dramatic photomontage of a dome over Manhattan, which builds an image of inside and outside that anticipates the actual partition achieved by the Bank of America's particle filters. Inside is clean air; outside is polluted air. Inside the dome—one among innumerable inflatables projected and occasionally realized during this period—the pressurized, conditioned air also acts as structural support. Fuller calculated that the dome's surface area would measure 1/85th of that of the buildings it covered, thus reducing energy loads required to heat and cool that air to 1/85th of existing levels. Add to this François Dallegret and Reyner Banham's notorious "Environment Bubble" photomontage/drawing of 1965, which juxtaposes the ludic,

Project for a geodesic dome over Manhattan, R. Buckminster Fuller, 1960. Courtesy of the Estate of R. Buckminster Fuller.

psycho-sexual freedoms of a technologically mediated, air-conditioned interior with an indeterminate and possibly hostile exterior. Mix these with Archizoom's laconic diagrams for "No-Stop City," a "paper" project drawn with the aid of a typewriter, its subroutines transcribed into a repetitive, algorithmic loop. Add Superstudio's "Supersurface," which transcribes the statistical survival of the species into the repetitive flatness of a "life without objects" (i.e., a life after consumerism) in, or rather on, a limitless, gridded technological desert rather than inside an environmental bubble. Finally, add Cedric Price's proposal for a Fun Palace for a series of working-class London neighborhoods in the early 1960s. Instead of inflatables, a movable steel frame regulated by a cybernetic command and control system accommodates indeterminate, unpredictable leisure activities. How is it that the residents of these neighborhoods had so much leisure time? Many of them were unemployed. This area of London was already experiencing deindustrialization and the consequent movement

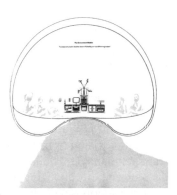

The Environment Bubble, photomontage drawing by François Dallegret for Reyner Banham's "A Home Is Not a House," 1965. © Collection FRAC Centre, Orléans. Photographed by François Lauginie.

Supersurface, The Happy Island, Superstudio, 1971. © The Museum of Modern Art/Licensed by SCALA/
Art Resource, NY.

of jobs to less expensive locales that belonged to the risk/reward calculus of
creative destruction.[6] Price's privately funded people's palace offered fun, as
compensation.

It would seem nevertheless that risk brings rewards when it supposedly
displaces class on modernity's battlefield. Production is exchanged for pleasure, work is exchanged for leisure, and the museum replaces the factory,
most vividly, of course, in Piano + Rogers Pompidou Centre. Most telling in
their early drawings of the project is the conditioned air, rendered in section
as a neutral white set off against the improbably blue Parisian sky, awaiting
one event after the other, one exhibition after the other, one film after the
other, superimposed interchangeably in a string of serial pleasures without
end. As Marx and Engels predicted, all that is solid does indeed seem to
have melted into air, even as the Pompidou's mass-produced structure is
fetishized in self-consciously technical drawings abstracted from the steel they
described—to say nothing of the workers who made and assembled that steel.

Nevertheless, architecture did continue to manage other risks belonging to
the supposedly obsolete class society. Alongside the rise of the multinational
construction firm and other changes in the erection of buildings globally, a
changing division of labor among architects, engineers, consultants, contractors, managers, builders, and workers spanned the planet in a newly
decolonized, decolonizing, or otherwise networked sphere. As the complexity
of construction rapidly increased, as its legal and financial regimes crossed
many more borders, and as computerization took command, the construction
industry was increasingly awash in drawings, models, technical specifications,
feasibility studies, legal documents, insurance applications, environmental

6

On the East Lon-
don sites for which
the Fun Palace
was proposed, see
Stanley Mathews,
*From Agit-Prop to
Free Space: The
Architecture of
Cedric Price* (Lon-
don: Black Dog
Publishing, 2007),
143–191.

Interior perspective of the Fun Palace project, Cedric Price, 1964. © Cedric Price fonds, Collection Centre Canadien d'Architecture / Canadian Centre for Architecture, Montréal.

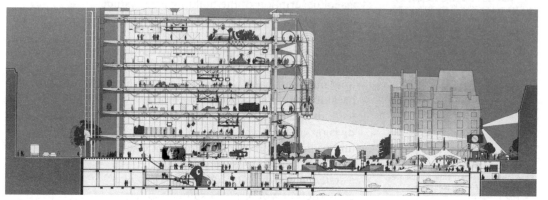

Revised competition scheme for the Centre Georges Pompidou, Renzo Piano and Richard Rogers, Paris, 1972–76; section. © Studio Piano & Rogers, architects.

impact reports, code reviews, client presentations, and other precursors to today's LEED certificates.

   In the midst of all of this, the bank reemerged (along with the museum as an archetype. But rather than the purpose-built monuments of nineteenth-century banking, this period saw the speculative office building that invariably houses a bank—or seems like it should—become paradigmatic, with its sheer iterability, its systematic adherence to the socioeconomic norms that are now ratified by the LEED certification system and other, correlate regimes. Since the 1960s, many monuments to banking and finance have forecast this normalization, this quiet establishment of platinum standards by which conformity to the neoliberal hegemony is enforced via a metaphysics of the Anthropocene. But none more so —and despite its theatricality—than Norman Foster's purpose-built headquarters building for the Hongkong and Shanghai Banking Corporation (HSBC), which was begun in 1979 and completed in 1985 in what was still the British colony of Hong Kong.

   The list of the consultants involved in the design and construction of the project was extensive: structural engineering by Ove Arup & Partners, quantity surveying by Northcroft Neighbour & Nicolson with Levett & Bailey, mechanical and electrical engineering by Roger Preston & Partners, landscape by Technical Landscapes Ltd., lighting engineering by Claude and Danielle

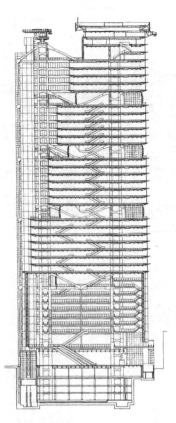

Hongkong and Shanghai Banking Corporation Headquarters, Foster Associates, Hong Kong, 1979–85; elevation. © Foster + Partners.

Engle Lighting and Bartenback Wagner Lichttechnische Planung, alongside further consultants for project planning, maintenance, wind testing, and so on.[7] Managed by this network, the design of the building as an integrated system assembled from mainly prefabricated parts likewise allowed these parts to be manufactured in disparate places, and made both design and construction a sophisticated logistics problem. (Alexandra Quantrill is writing a PhD dissertation on precision that treats the technical aspects of the building in greater depth.)[8] A very tight site favored prefabrication, as did the limited availability in Hong Kong of technically skilled construction workers. The visual description of the building as a system began with the client presentation with which Foster Associates won the commission, where annotated sketches are transformed into slightly less sketchy diagrams explaining how the kit of parts will work. In addition to breaking down a complex system into its constituent parts and a corresponding division of labor, these drawings did the important work of verifying to bank executives that the overall system is just that— a *system*—in which technical variables like construction workers, consultants, office space, and conditioned air are input, and optimal solutions are output. Behind the seeming realism of the project's design imagery are uncounted calculations and uncounted abstractions, in correspondence with the project's division and distribution of labor, of responsibility, and of risk: the diagrams, details, charts, models, tests, and specifications that circulated through overlapping channels occupied by architecture, engineering, construction, and finance.

Optimizing for specific variables is the basis of risk assessment. And though money had long circulated as mathematics, the sort of banking done by HSBC in the 1980s needed a vault. In Hong Kong, the vaults were to be on-site and in the basement, which had to be excavated from unstable soil near the city's waterfront without disturbing the foundations of neighboring buildings, or settling to a degree that would introduce too much uncertainty into the calculations for the new building's structural system. Hence,

7

Stephanie Williams, *Hong Kong Bank: The Building of Norman Foster's Masterpiece* (London: Cape, 1989). For technical details on the design and construction of the building, see in particular the special issue, "The Hongkong Bank: The new headquarters, Architects: Foster Associates," *The Arup Journal*, vol. 20, no. 4 (Winter 1985).

8

Alexandra Quantrill, "The Aesthetics of Precision: Research and Technique in the Architecture of Sealed Space, 1946–1986," PhD dissertation in progress, and in particular the chapter "Containment and Value at the Hongkong and Shanghai Banking Corporation Headquarters, 1979–1986."

the architects, engineers, builders, insurers,
and client had to be absolutely sure that the
excavated ground would only settle within
tolerable limits. This, of course, is a common
situation with large buildings. But here, given
the site conditions and the requirements
of the vault, the challenge was particularly
acute. Ove Arup & Associates studied the
contours of the predicted settlement as well
as those of the actual settlement as measured on-site, comparing the predictive
calculations with actual measurements made
on-site to verify their models. All of these
architectural visualizations were made with
mathematics, via an instrument field comprising piezometers, inclinometers, and other
sensors collecting data on site, and a large
mainframe computer in Arup's office.

Arup also used their mainframe
computer to calculate the building's superstructure under both static and wind loads.
Their structural model comprised some
3,200 nodes and 3,000 elements—minuscule by today's computational standards.
The model was run through finite element
analysis to determine east-west and north-
south wind deflection. Calculations were
also made to test the frame's ability to
withstand what Arup's published description called a "malicious" event, meaning

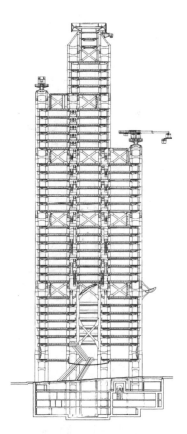

Hongkong and Shanghai Banking Corpo-
ration Headquarters, Foster Associates,
Hong Kong, 1979–85; section.
© Foster + Partners.

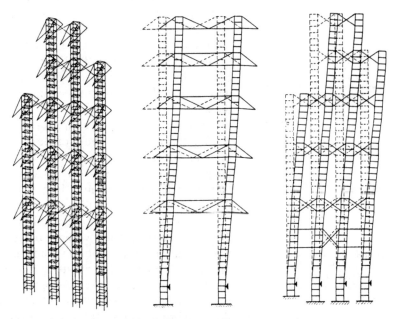

Superstructure analysis computer model for the Hongkong and Shanghai Banking Corporation Headquarters,
Ove Arup & Partners and Foster Associates. Courtesy of Arup.

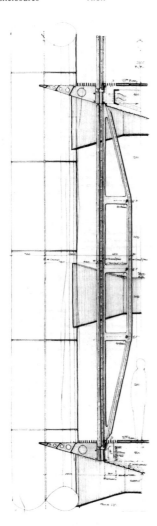

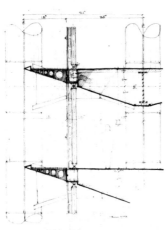

Sketches for the wall components of
the Hongkong and Shanghai Banking
Corporation Headquarters, Cupples
Products with Foster Associates.
© Foster + Partners.

an explosion or a terrorist attack. In other
words: calculated risk.

Wind testing for the building was probably the most extensive ever done to that
point. Data were gathered on wind direction
and velocity in the Hong Kong area, which
experiences typhoons on a regular basis.
Wind profiles were drawn and topographic
models were made for wind tunnel testing,
and prototype structural elements were
constructed and were themselves measured
under projected wind loads. Both the structural frame and the cladding were modeled
and tested against the different types of wind
data. The exterior wall components that
had to withstand these loads were detailed
by Cupples Products in St. Louis. Technical sketches, design development studies,
construction documents, specifications, and
shop drawings had to travel or be otherwise
translated from St. Louis to London to
Hong Kong and back. As Hong Kong was
still a British colony, most of the contractors
working on the project were either British
or French. So the likelihood that the documents in circulation were mostly annotated
in English may not have met with too much
difficulty. But it also registered another,
related hegemony. As at Chandigarh, it is
equally certain that not every worker actually building the building and reading the
drawings was able to read English. At some
point along the way, translation would be
required. This was especially likely given the
technical sophistication of the building, to
ensure that workers were obeying the drawings' thoroughly risk-managed commands to
the letter.

Meanwhile, back in the highly regulated
office interior, the Hamburg office of
the Quickborner Team, a prominent space-
planning firm, tabulated the results of a
handwritten survey (again in English) among
the future users of the building and entered
these into a computer to optimize fittings.
This calculation of value was not based
not on generic norms but on specific user
priorities. Variables were input, and optimal
solutions were output. (When applied to
the social sphere, Michel Foucault called
this "environmentality.")[9] To read a system
like this involves seeing all of the system's

9
Foucault uses the
term in his lecture
notes from 1978
to 1979, published
in *Naissance de
la biopolitique:
cours au Collège de
France 1978–1979*,
ed. Michel
Senellart (Paris:
Gallimard, 2004),
266: "Non pas un
individualization
uniformisante,
identificatoire,
hiérarchisante,
mais une *environ-
mentalité* ouverte
aux aléas at aux
phénomès transv-
ersaux. Latéralité."
Emphasis added.
Though the English
translation renders
"environmentalité"
as "environmen-
talism," the more
literal translation
resonates with
Foucault's notion
of governmentality.
See Foucault, *The
Birth of Biopolitics:
Lectures at the
Collège de France
1978–1979*, trans.
Graham Burchell
(New York: Pal-
grave Macmillan,
2008), 261.

FLEXIBILITY	20, 25, 15, 10, 9	(15)	15, 16
COMMUNICATION	10, 7, 8, 15, 14, 12, 5	(10)	10, 11
SPACE EFFICIENCY	15, 25, 18, 20, 17	(17)	18, 17
SECURITY + CONTROL	5 (M = for banking)	(5)	5
ENERGY / RUNNING COST	5, 10, 15, (58)	(9)	10, 9
EXTERNAL IMAGE incl. Banking Hall.	I	(20)	20
CUSTOMER BENEFITS	5, 7, 6	(5)	6
EMPLOYEE BENEFITS	12, 9, 10, 7, 5	(7)	8, 7
URBAN BENEFITS	3, 5	(4)	5

Chart of survey criteria assembled by the Quickborner Team, the space-planning consultant for the Hongkong and Shanghai Banking Corporation Headquarters. © Foster + Partners.

components assembled around a kind of absolute interior, a No-Stop City, or a continuous, gridded Supersurface into which could be plugged the results of user-generated surveys tabulated by space-planning consultants half a world away. Looking closer reveals the invisible air of capital, which has acquired distinct material properties of its own. Looking closer still makes visible the workers whose risk-laden lives made that air possible, and the financial calculations that paid for it.

The stated aim of all of this—like the aim of the Bank of America Tower twenty-five years later—was to maximize efficiency and flexibility. The actual result was a new kind of security matched to the imagined (and economically performative) financial security of the bank itself. This sense of security was not based on knowing that there was money in the basement vaults; rather, it was based on the probabilistic calculations of the risk-managed system swaying gently in the wind above. Today, in the LEED platinum Bank of America Tower and many others like it, a corresponding environmental, financial, and sociotechnical regime, and an even more complete division of labor, converts that wind into filtered air, in order to manage and exploit the risks inherent to a system built from, on, and in it.

Reinhold Martin is a professor of architecture in the Graduate School of Architecture, Planning and Preservation at Columbia University, where he directs the Temple Hoyne Buell Center for the Study of American Architecture. This text draws on lectures from his course, "Architectural Visualization since 1900."

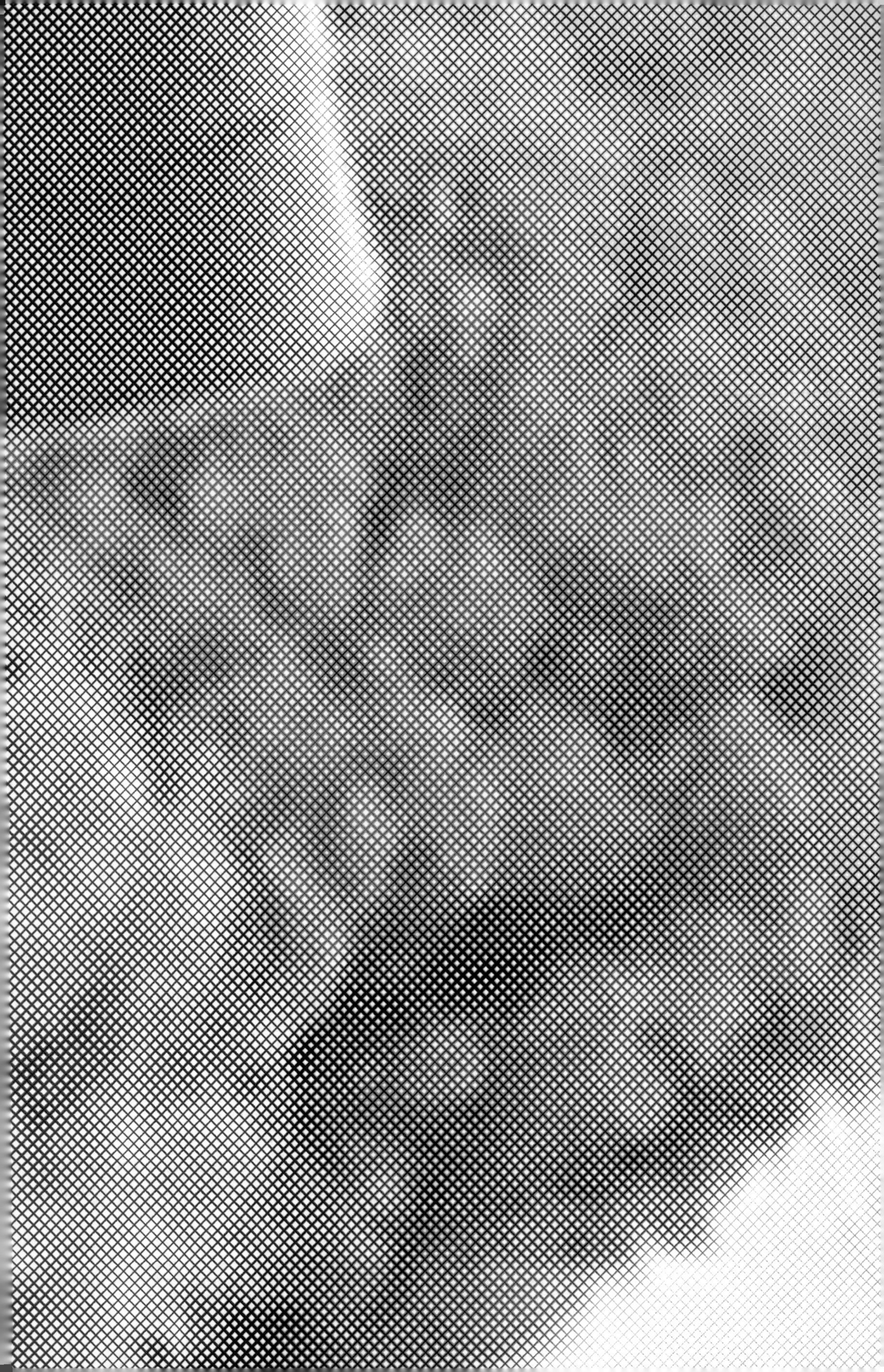

Inhabiting the Spaceship:
The Connected Isolation of Masdar City

GÖKÇE GÜNEL

Masdar Institute, a graduate-level research center that focuses on renewable energy and clean technology, was designed by Foster + Partners. Photograph by Gökçe Günel, 2014.

"The first night of living in a Masdar apartment was hilarious. I didn't understand how anything worked: the stove, the lights, the bathroom faucet, the cabinets, and I couldn't figure out how to turn off the AC," wrote Laura Stupin on her blog in September 2010, just after moving into the new Masdar Institute campus. Her studio apartment was situated at the center of Masdar City, inside a dormitory building she shared with her fellow students. "The Masdar Institute is the first part of the city to be completed, it includes the library, laboratory buildings, and the student residences," Laura continued, "and all these buildings fit together in a cube. And this cube is located in the middle of what is still a giant, flat, dusty, deserty construction site as progress on other phases of the city continues. It's quite a mind flip to be in such a strangely beautiful environment, then look [out] a window and see flat dusty landscape stretching out to the horizon. It really feels like I'm living in a spaceship in the middle of the desert."[1]

Laura was in her mid-twenties, and had moved to Abu Dhabi from the United States, after receiving her bachelor's degree from a private undergraduate engineering college in Massachusetts. Her ambition was to learn

1
Laura Stupin, "I Live in a Spaceship in the Middle of the Desert," September 25, 2010, http://squidskin.blogspot.com/2010/09/i-live-in-spaceship-in-middle-of-desert.html.

A rendering of the Masdar City master plan, which was circulated in the media between 2007 and 2010.
© Foster + Partners.

about renewable energy and clean technology at Masdar. In September 2010, when she posted her entry titled "I Live in a Spaceship in the Middle of the Desert," she received unexpected attention from journalists and researchers around the world. Major media outlets, such as the *Guardian* newspaper, reviewed her comments.[2] Like the other students who had moved to Masdar, she was trying to make sense of her experience with Abu Dhabi's emergent renewable energy and clean technology infrastructures.

Masdar, meaning "source" in Arabic, was founded in May 2006 as a multi-faceted renewable energy and clean technology company. It is widely known

2

John Vidal, "Masdar City: A Glimpse of the Future in the Desert," the *Guardian*, April 26, 2011, http://www. guardian.co.uk/ environment/2011/ apr/26/masdar-city-desert-future.

for Masdar City, the "futuristic" eco-city master-planned to rely entirely on renewable energies by the London-based architects Foster + Partners. While the eco-city and its multiple infrastructures were central to Masdar's development, Masdar has also been investing in renewable energy through its other operations—Masdar Power, Masdar Carbon, and Masdar Capital— in an attempt to ensure Abu Dhabi will remain a significant player in the global energy industry well after its oil reserves run dry. Masdar Institute, the energy-focused research center, set up and supervised by MIT's Technology and Development Program, operates on a growing campus within the eco-city site. This campus was Laura's "spaceship in the desert."

Since the 1960s, space technologies have inspired ecologically sensitive architecture, producing a blueprint for survival in a context of rising environmental concerns. As historians of science such as Peder Anker and Sabine Höhler note in their overviews of ecological design, the space program of the 1960s had considerable impact on the ways in which designers imagined and planned eco-friendly life on earth.[3] Buildings, perhaps best symbolized by the well-known Biosphere 2 project, would constitute self-regulating and decentralized systems with comfortable climatic conditions for humans, provide enclosed shelters for an impending ecological disaster, and perhaps serve as a means of escape from possible destruction on earth. Occupying buildings inspired by space technologies, humanity would behave like astronauts with clear outer space missions.

In these histories, the spaceship is a finite, technically sophisticated, and insular habitat for an exclusive group of beings facing an outside world of crises. In his book *Shipwreck with Spectator*, Hans Blumenthal explains how humans "prefer in their imagination, to represent their overall condition

3
Peder Anker, *From Bauhaus to Eco-house: A History of Ecological Design* (Baton Rouge: Louisiana State University Press, 2010). Also see Sabine Höhler, *Spaceship Earth in the Environmental Age, 1960–1990* (London: Pickering and Chatto, 2015).

The Masdar Institute campus includes dormitories, a knowledge center, laboratories, and a sports facility.
Photograph by Gökçe Günel, March 2014.

in the world in terms of a sea voyage."[4]
The idea of the spaceship (much like the
submarine that preceded it) then serves as
an extension of the ship metaphor, demonstrating the inevitable boundaries of human
activities, vilifying the space beyond human
habitability, and producing the outside as
a vacuum that should not be inhabited. As
seas full of mythical monsters surround the
livable environments on earth, the ship
provides a safe interior space. Thanks to
its strict boundaries, it acts as an ark, or
as the German philosopher Peter Sloterdijk
suggests, an "autonomous, absolute,
context-free house, the building with no
neighborhood."[5] This way, the ship puts
forward an alternative environment of peace
and rationality, standing in opposition to the
destructive and irrational crises of earth.

In prioritizing enclosure for some
over collective survival—the tension that
underpins most space-faring movies—the
spaceship also advances the principles of
selection and endorses what Sloterdijk calls
"exclusivity dressed up as universalism."
Despite saving only a very small number
of those who suffer a metaphorical shipwreck, the spaceship insists on addressing
the planetary-scale questions of survival in
the unknown, the sustenance of the species
beyond ecological disasters, and the preservation of an existing civilization albeit in
highly limited and confined form.

THE MOON LANDING

Inspired by this history of ecological architecture, Masdar City is intended to maintain
the lives (and livelihoods) of its residents
by relying on renewable energy and clean
technologies, and performs the role of
"a spaceship in the desert." This ecological
mandate would assist Norman Foster,
founder and chairman of Foster + Partners,
in producing a legacy for himself. According
to one of the on-site Foster + Partners
architects, "Norman wants to be the Bucky
Fuller of this century."

Buckminster Fuller conceived of the Earth
as a beautifully designed spaceship that lacks
comprehensible instructions. To satisfy this

4

Hans Blumen-
thal, *Shipwreck
with Spectator:
Paradigm of a
Metaphor for Exis-
tence* (Cambridge,
MA: MIT Press,
1996), 8.

5

Peter Sloterdijk,
Spheres 2: Globes
(Los Angeles:
Semiotext(e),
2014).

R. Buckminster Fuller's *Operating Manual
for Spaceship Earth*, 1968.

6
Buckminster
Fuller, *Operating
Manual for Space-
ship Earth* (1969;
repr. Baden: Lars
Müller, 2008).
See also Anker,
*From Bauhaus to
Ecohouse.*

7
Fuller, *Operat-
ing Manual for
Spaceship Earth,*
52–54.

need, he wrote *Operating Manual for Spaceship Earth.*[6] "We are all astronauts," Fuller asserted. "We have not been seeing our Spaceship Earth as
an integrally-designed machine which to be persistently successful must be
comprehended and serviced in total."[7] Since "no instruction book came with
it," humankind was confronted with the challenge of self-instruction in order
to successfully operate Spaceship Earth and "its complex life-supporting
and regenerating systems." Earth was an operable technological object, fully
accessible to humankind. Fuller not only wrote about his technocratic understandings of earth but also conceived many design and engineering projects
illustrating his philosophy, such as the geodesic dome.

   As a young architect, Norman Foster met Buckminster Fuller in 1971 to
collaborate on the construction of the Samuel Beckett Theater in Oxford.
The theater, which marked the beginning of their twelve-year relationship,
was a subterranean building intended to be used as classrooms and
exhibitions space for St. Peter's College. It benefited from the geodesic,
lightweight structures that made Fuller famous. Although it was never built,
Foster claims this building had a significant impact on the later stages of
his career, not only because it initiated his relationship with Fuller but also
in more formal ways: "I remember that Bucky made the comparison with

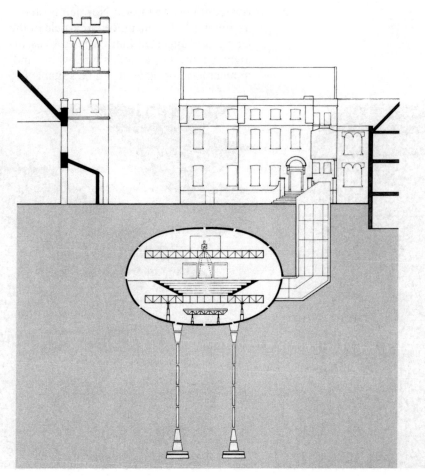

Norman Foster met Buckminster Fuller in 1971 to collaborate on the construction of the Samuel Beckett Theater in
Oxford, a subterranean building intended to be used as classrooms and exhibition space for St. Peter's College.

Astronaut Gene Cernan with the lunar module of Apollo 17.

a submarine because the structure of the building had to be resistant to water, like a seaworthy vessel. The building had to stand up to the ground water and other natural underground forces. So it's no coincidence that my later underground projects also take the form of ships and submarines."[8] Although none of their collaborative projects were built, in his Pritzker Prize biography, Foster says, "The thing about Bucky was that he made you believe anything is possible." Foster adds, "But perhaps the themes of shelter, energy and environment—which go to the heart of contemporary architecture—best reflect Bucky's inheritance... For me Bucky was the very essence of a moral conscience, forever warning about the fragility of the planet and man's responsibility to protect it."[9] For Foster, Fuller was what he termed "a green guru."

The legacy of the ship and the submarine continues to inform Norman Foster's design work. For instance, an article in the *Guardian* suggested that Foster's "reassuringly technical, graceful, silver, white, and immaculate" designs would be suitable for

8

Ernst von Meijenfeldt and Marit Geluk, *Below Ground Level: Creating New Spaces for Contemporary Architecture* (Basel: Birkhäuser, 2003), 130.

9

Thomas Tse Kwai Zung, *Buckminster Fuller: Anthology for a New Millennium* (New York: St. Martin's Press, 2002), 2.

The laboratory façades at Masdar Institute are composed of insulating cushions that shade the interiors of the building and remain cool to the touch under the desert sun. Photograph by Gökçe Günel, April 2011.

architecture on the moon.[10] More recently, Foster + Partners publicized renderings for a settlement on Mars, constructed by robots prior to the arrival of humans. "Designing for extra-terrestrial environments provides an exciting platform for experimentation that is at the front line of innovative technology," one of Foster's partners commented.[11]

In conversation with this lineage of outer space designs, Norman Foster also proposed that he understood practicing architecture in the Arabian Gulf to be similar to lunar exploration.[12] News commentaries, rather fascinated with the idea of constructing an eco-friendly city in the desert accordingly state, "The inhospitable terrain suggests that the only way to survive here is with the maximum of technological support, a bit like living on the moon."[13] In his autobiography, the leading Emirati entrepreneur and businessman Easa Al-Gurg also demonstrates that Emirati rulers understood the desert as a moonscape by describing how Sheikh Rashid bin Saeed Al Maktoum of Dubai dismissed the moon landing as a hoax, arguing that the landscape looked like the empty terrain in Ras Al-Khaimah, one of the seven emirates that make up the UAE. "Maybe it was filmed there," Sheikh Rashid said.[14]

Quite appropriately, the on-site architecture team at Masdar presented a slide show that included a lunar image, juxtaposing a Lunar Module with the gray, lightweight cladding of the laboratory buildings on the Masdar Institute campus. The laboratory façades were composed of insulating cushions, the architects explained, which shaded the interiors of the building and remained cool to the touch under the desert sun. In December 2010, Fred Moavenzadeh, then president of Masdar Institute, spoke on Richard Quest's CNN documentary about Masdar City and explained that when the United States wanted to send a man to the moon, it produced NASA. Now, when the United Arab Emirates is transforming and diversifying its economy, it is building Masdar City.[15]

10

Jonathan Glancey, "Man on the Moon: Norman Foster Prepares for Architecture's Lift Off," the Guardian, September 9, 2009, http://www.guardian.co.uk/artanddesign/2009/sep/22/moon-norman-foster-architecture.

11

Norman Foster takes on Mars: Rachel Reilly, "Architect's Firm Reveals their Award-Winning Vision for an Astronaut's Life on the Red Planet," Daily Mail, September 29, 2015, http://www.dailymail.co.uk/sciencetech/article-3253478/Norman-Foster-takes-Mars-Architect-s-firm-reveals-award-winning-vision-astronaut-s-life-red-planet.html. See also Amy Frearson, "Foster + Partners Reveals Concept for 3-D-Printed Mars Habitat Built by Robots," Dezeen, September 25, 2015, http://www.dezeen.com/2015/09/25/foster-partners-concept-3d-printed-mars-habitat-robots-regolith.

12

Timur Moon, "Norman Foster: Building an Oasis," the National, November 28, 2010, http://www.thenational.ae/arts-culture/norman-foster-building-an-oasis.

13

Tom Heap, "Masdar: Abu Dhabi's Carbon Neutral City," BBC News, March 28, 2010, http://news.bbc.co.uk/2/hi/middle_east/8586046.stm.

14

Al-Gurg, Easa Saleh, The Wells of Memory: An Autobiography (London: J. Murray, 1998). Easa Al-Gurg started one of the UAE's most prominent family businesses in 1960, after gaining considerable experience in the fields of banking and finance. He was an adviser to Sheikh Rashid, the late ruler of Dubai, and was also one of the people who attended the meetings in which the UAE was formed on December 2, 1971. Easa Al-Gurg also served as UAE ambassador to the UK and the Republic of Ireland for almost two decades, starting in 1991.

15

Richard Quest, "World's Most Futuristic City" CNN, December 6, 2010, http://www.youtube.com/watch?v=IJjbqDq9_QE. Fred Moavenzadeh repeated this claim in an interview with Wired magazine in 2013, suggesting, "When you look at the whole concept of Masdar in the 21st century, it's the same as NASA in the 20th century ... It has the same rationale, the same philosophy. NASA put a man on the Moon to show the strengths of the United States in that area of technology. And Masdar is being developed to show Abu Dhabi's commitment to clean air and technology." See Patrick Kingsley, "Masdar: the Shifting Goalposts of Abu Dhabi's Ambitious Eco-city," Wired, December 17, 2013, http://www.wired.co.uk/magazine/archive/2013/12/features/reality-hits-masdar. Interestingly, in 2015 the Dubai government set up the Mohammed bin Rashid Space Centre in an attempt to prompt economic diversification, scientific research, and technological complexity in Dubai and the wider UAE. So far the center has conducted projects in collaboration with the South Korean space program and aims to send a probe to Mars. For more information, see http://mbrsc.ae/en/page/introduction. Also see the Guardian article about the space center: Kareem Shaheen, "Emirates Space Mission Hopes to Launch New Era in Middle East," the Guardian, July 19, 2015, http://www.theguardian.com/world/2015/jul/19/emirates-space-mars-mission-middle-east.

THE FRONTIER

The spaceship analogy reconfigured the desert as an undiscovered frontier from which a novel means of livelihood could emerge.[16] In this voyage, the frontierspeople of Masdar City would be in control, both abiding by the principles of the Abu Dhabi government, and taking initiative to trigger a future of innovations in renewable energy and clean technology. In an ever-expanding geography, the students would act as astronauts—steering the spaceship and managing the successful institution of a new resource economy within oil-exporting Abu Dhabi. Perhaps this could be the reason why Laura's blog featured an optimistic comment from a former Masdar employee: "Brave people living on the island, it will get better and better…"

In his book *Carbon Democracy*, Timothy Mitchell shows how conceptions of endless oil supplies enabled progress to be conceived as infinitely expandable and without any material constraints. In the mid-twentieth century, the cost of energy did not present a limit to economic growth, as oil prices continuously declined. Given how simple it was to ship oil across the world, this resource could easily be treated as inexhaustible. This belief in the infinity of oil also played a key role in producing "the economy" as an object, which could likewise expand without limit.[17]

In contrast to this history of oil, the Masdar City project acknowledges the fact that fossil fuels may eventually disappear. Nevertheless, the idea of resource infinity, this time enabled by renewable energy and clean technology, still characterizes the ways in which producers of the city imagine the future. In response to depleting oil resources, the eco-city promotes the infinity of sunlight and wind. The spaceship in the desert has the capacity to journey through endless space and confirm the vision of a boundless frontier where new types of resources await discovery.[18] As an exploratory vehicle, the Masdar City project is intended to challenge and resolve the problem of finitude. The frontier narrative also led the producers of Masdar City to conceptualize the present as a moment of potential while concentrating their efforts on constructing a future that would be incubated inside this enclosed space within the Abu Dhabi desert. Masdar City could give rise to a new generation of resource pioneers, who would hurdle through unbounded territory.[19] Accordingly, it was not only the space of the Abu Dhabi desert that awaited another discovery but also its temporality. The spaceship analogy spoke to a future of technical adjustments that would potentially emerge from this enclosed space, possibly allowing the eco-city to be replicated in other settings around the world.

16
Looking back at the Apollo space program (which the president of Masdar Institute touched upon in Richard Quest's CNN documentary), David Mindell explains how President Kennedy had seized and mobilized the powerful mythology of the frontier in aiming for the moon. The term *frontier*, originally meaning "border" or "borderline," obtained new meaning during the settlement of the American West in the eighteenth and nineteenth centuries. In this narrative, the heroic pioneers were headed to an unknown geography full of unpredictable dangers as well as antagonistic competitors, yet they would make use of self-control, self-reliance, and humility to open up this new frontier. David Mindell, *Digital Apollo: Human and Machine in the First Six Lunar Landings* (Cambridge, MA: MIT Press, 2008).

17
Timothy Mitchell, *Carbon Democracy* (New York: Verso, 2012).

18
For a longer exploration of this theme, see Rosalind Williams, *Notes on the Underground: An Essay on Technology, Society, and the Imagination* (1990; repr. Cambridge, MA: MIT Press, 2008), 7.

19
As Sabine Höhler reminds her readers, "Appropriating space by compiling, registering, and neatly arranging the elements within it is a strategy not limited to the modern era of scientific collecting, archiving, and interpreting of the world. The procedure recalls the primal ship representing the inventory of the world, the biblical ark." Sabine Höhler, "The Environment as a Life Support System: the Case of Biosphere 2," *History and Technology*, vol. 26, no. 1: 39–58. Also see Sabine Höhler, *Spaceship Earth in the Environmental Age, 1960–1990* (London: Pickering and Chatto, 2015).

EXPORTING THE SPACESHIP

Nevertheless, Masdar Institute students, the frontierspeople of Abu Dhabi's emergent eco-city experiment, remained unsure about the translatability of Masdar City into other settings. As such, on February 1, 2011, they gathered in the Masdar Institute auditorium to stage a debate on whether "Masdar City is an elite enclave of sustainability, unsuitable for the rest of the world" or not. The graduate students, who came to Masdar City from countries like the United States, China, India, Egypt, Jordan, Iran, Turkey, and Iceland, were struggling with such questions, and chose the context of a debate club performance as a way of thinking about them.

The team that perceived Masdar City as "an elite enclave of sustainability" argued how Masdar is "too unique" to be applied elsewhere. First, Masdar was very expensive. Which other country, other than the oil-rich UAE, would be able to devote $22 billion for an eco-city?[20] Second, they recalled how this project had been put together to contribute to the economic diversification of Abu Dhabi, and perhaps would not be financially feasible or meaningful for other countries with different economies. Masdar City was expected to help the UAE transform its brand image from oil producer to technology developer, and induce a perception shift, perhaps attracting foreign investments or facilitating the creation of local start-up companies focusing on renewable energy and clean technology. Third, the political climate of Abu Dhabi was working in favor of Masdar City by providing prolonged commitment and stability—the government often served as a steady source of financing for the project. "Well, other than that," the pro-team reiterated, "the concept of a green city has existed for a long time." In this understanding, Masdar City no longer comprised a vision that would unfold into the future. Rather, it remained an island contingent on a specific set of circumstances, only available within the United Arab Emirates. Abu Dhabi's oil capital, its future economic vision, and its political environment were thus perceived as preconditions for launching the spaceship.

In response, the team that defended the global applicability of Masdar City proposed that the eco-city should rather be understood and framed as a prototype: Abu Dhabi would shoulder the burdens of building the eco-city, and others would benefit. "Every new idea is expensive," one of the students underlined. "Think about the car: first rich people had it and now it has spread around the world." Masdar City could become less expensive in an undefined future. It could be exported to other countries as a whole, in the same way that the car and its infrastructures had been exported. In the meantime, the experiments taking place at Masdar would be learning experiences for the students, researchers, and faculty, opening up global horizons for research on renewable energy and clean technology and eventually adapting them to other regions in bits and pieces.

At the end of the meeting, one student approached me to express his dissatisfaction at how none of the students in the debate teams had actually defined what Masdar City *was* or what exactly they expected to spread around the world: "No one talked about the personal rapid transit units or the motion sensors," he specified, pointing to the technological artifacts that seemingly defined the eco-city for him. In which of its materializations did Masdar inhere, and what would it pass on to the rest of the world, he wondered, and what exactly was the future that the spaceship promised?

20
For a review of these figures, please see "Work Starts on Gulf 'Green City,'" *BBC News*, February 10, 2008, http:// news.bbc.co.uk/2/ hi/science/ nature/7237672. stm.

"Man with a brush" clears the thick coatings on solar panels, which result from a mixture of dust and humidity, and ensures their efficacy. Photograph by Gökçe Günel, April 2011.

MAN WITH A BRUSH

Abu Dhabi is perceived to be a perfect location for harnessing solar energy. However, according to an engineer at Masdar—let's call him Mahmood—this perception was not completely accurate. Mahmood was an Egyptian man in his mid-thirties, and had recently finished his PhD at an American university. Wishing to be closer to home, he accepted a position at Masdar as his first job. As we chatted outside the solar power stations, he stated that high levels of dust and humidity not only blocked direct solar rays but also resulted in thick coatings on solar panels, diminishing their effective functioning. "Although we can't fix the first problem that easily, we have found a solution for the second problem." He continued, "We call it 'man with a brush.'"

While there was extensive research toward a solution for removing dust, humidity, and mud from solar panels in the UAE, during the time of our conversation, no technical solution had proved as effective as the use of labor. In Mahmood's narrative, the "man with a brush," a worker dedicated to gently wiping away dust and mud from the solar panels, became part of the picture mainly because he exposed the potential that is embedded in the solar panels. In some ways, "man with a brush" would allow for the proper functioning of not only the solar panels, but also the economic diversification project of Abu Dhabi. In this context, the "man with a brush," who was essential to the construction and maintenance of the spaceship, was framed as a disposable tool and abandoned outside the technologically complex vessel.

CONNECTED ISOLATION

Released in 2014, the science fiction film *Interstellar* portrays a nearly unin-habitable world consumed by dust storms. Life becomes increasingly difficult for the film's protagonists as they lack breathable air and nutritious crops. Fast-forward to the final scenes of the movie, however, and the protagonists have arrived in a peaceful spaceship floating far away from earth, one that contains the same house, town, and lifestyle they occupied on earth. In pre-serving the status quo the residents of the spaceship have left the rest of the earth to its collapse, celebrating the fact that they can inhabit technological dreams without attending to problems on earth in a collective manner.

In *Interstellar*, the American house, the baseball fields, and the beer bottles stand in as manifestations of what Sloterdijk calls "connected isolation."[21] To construct an artificial lifeworld inside the spaceship, materials, designs, and building techniques have been imported, allowing the frontierspeople to fall back on their prior social lives and political conceptions. In a context of complete isolation, the absence of these attachments could perhaps facilitate the production of more radical transformations. But as Sloterdijk argues, "All current and future space-insulators will remain imitators, like their distant forefather Robinson Crusoe… [and] create things exclusively out of what they brought with them, just as he did. It goes without saying that conven-tional astronauts are far removed from being the pure consciousness of their island."[22] The inhabitants of the spaceship are bound by imported ideas that preclude the necessary shifts in perspective.

Despite the numerous technical adjustments they offer, contemporary forms of urban development, such as Masdar City, produce future scenarios that are akin to *Interstellar*, inviting future residents to both preserve existing social relations and survive ecological disasters inside serene and optimistic spaceships—so caringly cultivated, and as the Richard Brautigan poem goes, "all watched over by machines of loving grace."[23]

21

Peter Sloterdijk, "The Absolute Island," in *Olafur Eliasson: Minding the World*, ed. Olafur Eliasson and Gitte Orskou (Denmark: ARoS Aarhus Kunstmu-seum, 2004), 161.

22

Sloterdijk, "The Absolute Island," 162.

23

Richard Brautigan, *All Watched Over by Machines of Loving Grace* (San Francisco: Communication Company, 1967). Available at: http://allpoetry.com/All-Watched-Over-By-Machines-Of-Loving-Grace.

Gökçe Günel is a lecturer in anthropology at Columbia University. She finished her PhD in anthropology at Cornell University in 2012. Her forthcoming book, titled *Spaceship in the Desert: Energy, Climate Change, and Green Business in Abu Dhabi*, focuses on the construc-tion of renewable energy and clean technology infrastructures in the United Arab Emirates, more specifically concentrating on the Masdar City project.

The Appearances of the Letters of the Hollywood Sign in Increasing Amounts of Smog and at a Distance

HOLLYWOOD	HOLLVWODD	KOLLYWCOD	KDLLYWDOD	HOLLYWOBD
NOLLYWOQD	HDLLVWOOP	HDLLYWOQD	HCLLYWOOO	NCLLYWDOD
NCLLVWOOD	HOLIYWOCD	HOLLVWCCD	HBLLYWDOD	NDLLVWODD
HDLLVWOGD	ROLLYWDOP	NCLLYWCOD	HOLLVWOBD	BQLLYWOOD
KOLLVWOOB	KDLLYWOGD	KOLLYWCCD	BDLLYWOOP	HGLLYWOOO
NOLLVWDCD	HOLLYWCBD	NDLLVWOCD	KCLLYWDDD	BOLLVWOCD
HCLLVWOOO	KOLLYWOCB	HDLLYWDCP	BDLLYWCOD	NOLLVWOBD
KCLLYWDCD	HRLLYWOOB	HBLLYWOQD	HDLLVWOOR	HPLLYWODP
HQLLYWOPD	NDLLYWDOO	NOLLVWOPD	HQLLYWDQD	HOLLVWGDP
HOLLVWPDD	HOLLVWRDD	KOLLYWBOP	KOLLVWDQD	HDLLVWQOP
KCLLVWCOD	RPLLYWOOD	HDLLYWQDP	HQLLYWGOP	HQLLVWGOD
RQLLYWDOD	HQLLYWDDP	ROLLVWOOO	ROLLYWCCD	KDLLVWQOD
HDLLVWOQP	HDILVWOOP	FOLLYWDOP	KGLLYWDOP	NOLLTWDOD
NOILVWDOD	RCLLYWOQD	NOLLVWOCB	KCLLYWODB	KDLLVWCDD
NQLLYWOCP	HCLLYWDBD	HDLLVWCQD	KCLLVWDDD	NCLLYWOOR
KCLLVWOOO	HCLLYWQGD	HDLLYWCOR	KCLLYWOBD	KOLLVWCOB
HDLLVWDCP	HOLLYWCPP	KOLLYWCBD	RCLLVWOOP	HOILVWCOP
EOLLVWOCD	HCLIYWDOP	NCLIYWDOD	NOLLVWDDO	NQLLYWGDD

```
NDLIYWQDP   EQLLVWOQD   EQLLVWDDD   RDLIYWROD   RDLLVWCCP

NGLLVWOQO   HQLLYWCCU   HDLLYVDOR   EOLLYWORO   ROILYWDDB

NOLLYVDQP   NOLIVWGDP   NDLLYWCCU   RDLLVWQOO   HBLIVWDOP

HDLLYWQGU   KGLLVWBOP   NDLLYVOQP   NDILVWDDP   HDILVWDDB

HDLLYWBBP   BOLLYWDQU   KDLIYWODR   FBLLYWDOP   NGLLYWORO

ROLLVVDDD   NGLIYWODO   HGLLVWOBO   HOLLVWDBU   BOLIYWODR

NDLLTWODB   EQLLYWOBD   HDILYWDBP   RDLIYWBOD   NOLLYVBDD

BDLLVWDOU   RDLIYWDQD   NOLIVWBDD   NGLLYWDPP   HGLLVWQDO

HOLLTWRDP   NQLLVWCCP   NDILVWDOB   NCLIVWDCD   FCLLYWCOB

NRLLTWDOD   ROLLVVODP   RDLLTWODP   NOLIVWQDP   BDLIVWOOB

KOLLVWRGP   HDLLTWQDP   BCLIYWCOP   HGLIYWQOO   NPILVWDOD

HQLIVWOGP   NOLIYWOBO   BDLIYWOOU   HGLLVWOGU   EDLLYWOGB

HQLLYWOPU   NOLLTWQQD   HCLLYWGRO   HQLLVWCRP   NDLLVWDCR

KOLLYWPCR   HBLLVWCDO   NCLLVWOPO   NCLLYWGQB   HBLLVWCBD

KCLLYWBRD   HCLLVWBDB   HDLLVWCQR   HOLLVWPCR   KCLLVWQBD

BQLLVWCOB   HBLLVWCRD   BCLLVWBDD   KDLLYWQCR   RQLLVWDCP

NQLLVWOCR   RCLLYWOPO   BOLLYWRCO   KBLLVWCOB   RQLLVWCDP

NCLLYWPPD   KCLLVWPQD   KPLLVWQCD   HCLLVWDBB   HDLLVWCBB

HOLLVWCPR   RPLLYWQCD   NRLLYWCOR   KCLLVWRQD   BBLLYWCOO

BQLLVWCOO   KQLLYWCGB   BDLLYWDCR   RDLLYWCQO   HCLLYWGPO
```

BQLLVWCDO NQLLVWQCB BQLLVWGCP KRLLYWRCP HBLLVWCPP

NCLLYWPQB BPLLVWOCO HCLLVWRGO NDLLTWOCR HQLLTWCDO

NDILVWCDO BCLLVWRQD NDLIYWQCO NDILYWCDR NBLLVWGCP

BDLLVWRCP NGLLYWCRO NOLIVWCQO NDLIVWDCO EOLLYWCRO

FCLLVWDOU HPILYWDCO BOLLYWCPR KGLLTWCDP NCILYWBDP

KQILVWGCD NCLIVWCCP EOLLYWCPB BDLLVWCDU HCILVWQPD

EOLLYWCRB RDLLYVQCD NOLIYVDCP EOLLYWPCO RDILYWQCP

ROILVWOCR HRLIYWCDB HPILYWCDO FDLLVWCOR BDLIYWCDO

BCLLVWBGD NOILYWGCR BBILYWDCD RCLLYWGDR BQLLVWOCU

EOLLTWCQD RDILYWCOR KCILVWDOR NCLLTWGDP BDLLYVCDP

BDILVWGCD BCLLYWDQU HDLIVWGCO HCLLVWBGB FOILYWCQP

HDILTWCOB NOILVWCDR FQLLYWCDO BGLLVWCDO HPLIYWQCP

KOLIVWCGB EOILYWDCO FOILVWDCP NGLLVWCPP NOLLTWDCU

FDLLVWDCB KCLLVWQBP NOILYWPCB FDILYWDCP RDILVWCOO

NOLIYWQCU ECLLYWRQD HQLLYWCPU KRLLYVCDD HDILVWCGO

KDILYWCBP HCILYVDDP FDLLYVCDD NCILVWDPD BGILYWOCO

NPLIVWDCD FCLLVWDBD NGLLYWBCB BCLLTWDOB ECLLYVDDD

EDILYWCGD NCLIYWGRD ECLLYWOPO FGILVWCOD NRLLVWCQP

RDILYWCGP NCLLVVDGD KDLLYWBBB ROLLYWBBB RRLLYWDQO

NBLLVWROB NPLLYWDBO RRLLVWQGD NRLLVWQDO BQLLYWDRO

```
RCLLVWOBU   NGLLTWCDO   FPLIYWOCP   KDILTWCOB   NQLLTWDCO

RGLLTWCDP   NCILYWCCU   NOLIVWBCO   NRLLYWCGU   NCLIYWGQO

EDLLYWGCU   NCLIVWQGP   HCILTWCCP   FCLIYWDGP   EPLLVWOCB

NDILVWQCB   BQLIYWDCB   BQLLTWOCB   KGLIYWCQO   NCLLYVPOO

NPLIYWCPD   NCLLYVCCO   EGLIYWOCB   KDLLYVCPP   BGLLVWCGB

FOLLVWCRO   FCILVWOGP   HCLLTWDQR   HCLLYVBDO   BOILYVCOO

BCLLVWGBP   KPLIVWCDP   BRLLYWOPR   HBLLVWDBR   NBLLYWBBP

BRLLYWCCR   KDLLVWBPO   RDLLVWRGB   BDLLVWGBB   NDLLVWPBO

HRLLVWBBP   RPLLVWOQR   RBLLVWGOR   KBLLVWDBB   BBLLVWPOB

KQLLVWDPR   NPLLYWRQO   NPLLVWBPD   KRLLVWDQR   BOLLVWRPO

BCLLVWCPB   RRLLVWGPD   RBLLYWQQO   KQLLYWPBB   KPLLVWPRD

BQLLYWDRR   RBLLVWBOB   NQLLVWGRO   NPLLVWBDB   NDLLYWRRR

KBLLYWPBP   NQLLYWPRO   BQLLYWQBB   NBLLVWDBO   NDLLVWRQR

KBLLVWBDB   RBLLYWGQB   BPLLYWROR   BRLLYWRRD   BCLLYWPCR

HPLLVWGQR   HBLLVWQRO   RPLLVWPOO   BDLLVWRRP   KRLLYWGBO

NPLLYWPGO   KDLLVWRQR   HBLLVWGPO   ROLLVWPPB   RRLLVWDRP

RBLLVWRDP   HBLLYWQBR   NPLLVWDPB   NPLLVWGRP   BPLLVWBOO

HPLLVWQRB   HRLLYWQRR   RBLLYWBGP   NBLLYVDPD   RDILYWPGP

NDILVWPOR   FDLIYWCCB   KCILYWCBO   RDLIYWRDO   HQLIYWGGR

RDLIYWGDR   HBLIYWROU   FPLLYWDDR   BPLLYWDQR   KGLIVWPDP
```

HBLLVWQPR	BPLLYWRRP	RDLLYWPRR	RRLLVWOBR	RDLLVWGRR
NQLLYVRDB	HCLITWCQB	EBLLVWDDU	NBLIVWQOR	EOLIVWGDR
NDLIYWRPB	FDLLYWRPO	EBLLVWDDR	BDLIVWDRO	NRLLVWRGO
ECLLYWBCR	FGLIYWORO	HOLIVWPBR	BQLLVWRDU	BOLIYWGPR
EDLIYWRGP	KPLIYWDQR	NQLLYWBRU	BRLLYWGBB	NRLLVVOBP
NDILTWCCO	RGLIYWBOU	NRLLYVQDB	FBILYWGDP	RCLIVWGCO
NCLLTWCQR	BDLLVVDQO	FRLLVWQPD	KBLLYWQBU	NDILVWRBP
RGLLVVDDB	NRLLYWRGU	BGLLYVDRP	NRILVWDDR	NOILYVRRD
BOLIVVDOR	RDLLVVGQP	RDLIYWRDU	HPILYWGBO	RBLLVWGGO
RDLIYWPDR	NPLIVWDGO	FBLLYWDBB	RRLLYWDBU	RGLLYWRQR
EDLLVVDDO	BGILYWDGU	NCILVVCDP	EDLIYWDRO	BDLLTWQGO
NDILYWGPU	BGLLYWRPO	RGLLVWRDR	EOILYVDDO	NCLLTWCBO
HBILVVROD	NQLLVWBPO	BGLIYWDBB	NPLIVWDDR	KBLLVWRBP
FQLLVWBOR	RRLLYWRDR	RGLLVWBGO	EDILYWPRD	RPLIYWDBP
FDLLYWRGR	NDILYVQDO	EBLLVWDPP	FOLLVVGOR	NBLLYWGRU
RDLIYWQGR	BBLLYWBGO	NQLIVWROU	BDILVVODO	BRLLYWBRP
NGILYWRPP	RDLLVWGRU	KGLLVWBGR	KBLLTWGBD	RQLIYWDGR
EGLIVWGDP	NPLLVVDBD	FGLLVWBDB	NPLLTWROB	FDLLYWRGU
EPLLYWOBU	NDLIVWDRU	NPLLYWGBR	NBLLVWGRB	RRLLVWPDB
EQLLYWRDR	EGLLVWBRD	NPILYWDGU	FDLLVWRQO	RRLLYWGGR

```
RPLLVWBDU    NPLIVWCCU    NQLLVVCCU    RDLLVVQGO    RBLLTVDDD

BQLLYWPBR    BPLLVWPDR    BPLIYWDBB    BDLIYWGBR    BRLLVWBDU

BQLIYWBBP    BBLLVVDDB    NBLLVWGBU    RDILYWBBO    NBLLVVCCO

FDLLTWDGU    NBLLVVDDU    FDILVWQQO    RDILVWDBU    BCLLTWCRO

HQILYVGBP    RDLLTWQBO    HDLLTWBBR    NGLIYWQBR    FBLLYWRBP

NBILVWQGB    BBLIYWDGR    BDILVWPQO    HBLITWGOR    BPLLVWDPU

NBLIVWDPO    NRLLVWBBO    HPLLVVDBB    ECILVWCGO    FBILYWDGO

BBLLVVBDD    BQILTWDBD    NPLLVWBGR    FDILYWQPO    EGILVWDDR

FDLLTWRGP    NBLLVVOPO    BQLLYWBBU    NGILVWPDU    BRLLYWGPR

BQLLYVCCU    HDILTWPDR    EGILYWDBO    BBILYWQRP    RDLLTVDDO

NDLIVVCCO    ERLLYWBGB    RDLLYVDBU    NDILYVDPO    NQILYWBRO

RDLLTWBPP    HRLIVWGBO    NBLIYWPQB    BDLIVWBGB    FRLIVWDDB

KDILVWBBO    RDLIVWGPO    EBLLTWGQD    NRILVWDGU    NPLLVVCCO

NDILYVGGO    BBLLYWBGR    KBILYWRDU    BGLLTWPQP    BQLLYVDBO

EDLIYWGQR    NBLIYWGBB    NRLIYWDBR    KPLIVWRRD    NRLIYWPQO

EBILYWDDR    BGLIVWOBR    RDLIYWPGR    KPLITWOPP    BQLLTWOBR

BPLTYWCCR    BBLLVWBQB    NQLLVWPRR    EDLLVWQPU    BBILVWDBP

NGILYWRPO    EDLLYVGPP    BGILYWGBO    BBLLTWDDR    BCLLYVCBO

EQLLVWGGU    NPLLYWBBU    BDILYWGRR    BBLLYWGRR    FOLLTWDBR

RGLLTWQDR    NBLLVVRDP    NDILYVPDO    RQLLTWQDR    RDLITWCCB
```

ECLLVVCCU	BCLIYWRBU	BCLIVVROB	EQLIVWGCR	BCLLYVBBO
ERLIVWGCB	RRLITWRCD	FCILVWRDU	KCLLYVRPR	RCILVWBGU
NPLLVVGCR	FBLLYVCDU	HDILVVRCU	KDLLTVBCO	BPLIVWCQU
RBILTWBCD	RPLIVWCBB	EDLLYVBCR	EBLLVWBCR	BQILVWCRR
ECILVWPDU	FBILYWCBO	RDLIVVPCP	BBLLTWGCU	RCLLYVBBB
BCLLVVGBO	RQLLYVBCU	FCLIYWBRB	BCLIVVDDR	ECLLYVGPB
RGLLVVBCO	BCLLYVQBR	RDLLVVCBU	BPILTWRCD	NCILYVRDR
FDLITWCBP	FRLLYVPCP	RQILYVDCU	RRLLTWGCU	BRILYWCPR
BRLLTWGCU	RPLLVWRRR	FBLIVWRQP	BPLIVWDBU	BPLLVVDBB
RPLLYVPDR	EPILYWDPR	EBILYWRDR	EQLLVWBBR	BBILVWPDR
RBLIYVOPB	ERLLVWRBO	EBLLVWRQR	EDLIVVQRD	BCLLTVGCO
RBLLTWBDR	RGLITWRDO	EBLLVVOQR	FPLLYWPBR	BQLIVWGRR
RPLIVVOBP	BQLIYVRDB	FCILVWCPR	NBLIYWBBU	NRLLYVQBR
KPLIVWBGR	RDLLVVBRB	FDLLVVQQR	NPLITWPDB	FBLIVWCCR
ERLLVWPBO	FBILYWDPR	NBILVVGQP	BBLLVVGDU	BDLLVVBQR
RDLIVWPRR	FDILTWDRO	BPLLTWBPP	NBILVWGBR	KBLIVWGBR
NPLLYVRQR	KRLLYVBGU	RQILVWBGU	FRLLTWORR	EBLLVWBRO
BQLLTWRRO	EBLIYWRDR	NGLLTWBRU	NRLLTWQBU	BRLLTWDRU
RBILVWPDR	RGILTWDRO	EDLLTWBGU	FDLITWPDO	FQLLVWRRR
BRILYVDPP	NBLITWDBO	EBILVWBQP	RCLLVVPCU	BQLLTWQPR

```
RPLLVUUQB    RRLIVVRDQ    FQILVVBGO    RQLIVVGBQ    EBILYWUBR

RDIITWPUP    UBLLVVRQO    UCLLTUCDU    KBLLTVBBB    UPILVVCCB

BDIIYVUPP    UQIIVWBPP    NULLVVGUU    NRIIVWBUB    BDIIVVDRU

BBILYUQRB    NULLVUUDR    EQLIVUDQU    FRLIVVDGQ    KGILIWPBU

NPILTWRPR    BQLLVUUBB    UGLLTUDDQ    NULLTWUBR    URLLYUBRP

NBLLIUUDP    FQLIIWGQU    KULIVVGGR    UBLLYVRGU    KQLLIUQUP

UUIIVWBDP    BULLVUBBP    RCIIVUCDQ    FGIIVVDBP    EUIIVWGDQ

KRLLIVPDU    NBLITWGUR    NRLITWGUU    BRIIVWBQR    KULIVWUPQ

FGIIVWDUR    FPILYVCBR    EBLIYVPCR    EBIIVWCBQ    UGLLVVUCR

UCIIVWBUP    KBIIVVCBB    UPILIWBCP    UCILYVPQR    EULLVUCGU

ECILVVBUP    NGILTVBCU    FDLIVUBCU    FCLIVUBDU    NBILYUCUR

FDILIVCDR    UGIITWQCO    RCIIVWUBR    UUIIVWCQO    ECILIUCCP

BCLIVUBGU    BCILYUBUB    FULLVVBCR    UBLIVVCDR    KDIIIVCDU

FUIIVWCUP    EBLIYUBCO    UCILVUDGU    KCILVUUBO    KQIIVVPCU

NDLIIVBCU    EGILVVCRU    FRILTWRCQ    NBLIVUBCR    RRILVVRCQ

FPLLVUBCU    NGIIIVDCB    BCLIVURQU    UBLITWPCO    RRIITWCPO

BUTTVWCBU    UBIIVWCGR    UPLIVUCGP    RUILTWBCR    FQILIVCDO

EUILVVCPP    NDIIVUCBR    ECLLIVQPO    NDILTUCBU    KQIIVVCRU

NPLIVUCUB    EQIIVWCUQ    EUIIVVDCP    BCLLVUPUR    KULIVVBCU

FCILTVQDU    ERIIYVCDU    UCLIVVQQR    NBLIYUUCQ    KULIVUCGR
```

EBLIVWUUU   RPLIVVUGQ   BBLLVVUUR   NBLITVUBP   ECILVUCUU

BQLIYVUUR   URILTWBBO   ERIIVUCCO   UCILVUCPR   BULITWUQU

UBILIWPDU   RBLIIWBBR   KRLIVUPBU   FCIIYUPCR   FULLTVUDO

BBILVVRBU   UCILVVCUU   FULIIWGGQ   KPLIYUURU   FQLLIVDUQ

UQILVVUQO   RDLITUPQU   FBLLTUDPR   NULIVUGUB   EDLIIUCCU

UCLIIVDCU   FRIITWQBB   KUIIYURBP   EQLIVUDUR   NDLIIVQUU

FCLLIVCUU   FULLVVBBQ   NGIIVVBUB   BBILVUUDQ   EDLLTURPU

BQILVURPU   FULIVVGGU   UULLTWBUO   UBILVWBUR   EUILVUBQP

UULIVWRRR   RUILVUBDU   UULITWDBQ   KPILTVUPP   RBLIVUGRU

KGIITUGQO   FDILTUBDR   EQILIUGDO   EDILTVPUP   NRLIVUPBU

FPLIYVUBO   FDIIVVQUO   BDIIVVUUP   BBLIIVDGU   UGILTWBBQ

RBLLTVBUB   ERLLIUDQR   KRLIVVBUR   UUILTWBDQ   EDIIVURDU

KULIYUBUB   FBLIVUGQU   NBILIVGQU   UGLIVUQUP   FGLIVUQBR

FQILVUQPU   EQLIVVQUR   BBILVVBBU   BBILIVCCU   FULLTVPRP

KRIIIWPBO   FUILVVQGU   BUIIIWDQQ   EBLITVCCU   NQIITVGQU

FQIIVVBGB   EQLIVUUDU   NGIIIWPRU   UQILVUGBO   FBILTWBRU

FRLLTVUDR   RUIIVUQDB   BGIIVVGPR   NUIIVUQBP   RPILVUQUB

UPILYVUQB   EBIITWPDU   RUILTVGQB   RGIIVUQUP   BUILYVBRR

UUIIVWQGU   EBLIVVUQB   FGLLTVBBQ   EDIIVVGBR   RGILVVBUR

URIIVUQDP   NPIIVUDPU   UPLLIVDBO   UQIITWDUB   NUILVVRRQ

```
EBIIYUDUB    RUIIVVDUB    FUILTWQUU    URIIVWUPO    RBLLIUUDU

FPLIVUPQU    EUIIYWURU    FPIIVURDB    KQIIIVUDB    KUILTUBGB

RBILVVUBR    UULLVUQRR    UPLITVCCQ    UPILVVQRU    RDIITVUCQ

FULIVVCUQ    BUIIIWUCB    ERIIVUBCB    UCLIIURDB    FCLIIVQUO

UUILTUCDB    BPILTUCBR    BBIIIVQCB    BBIITUCGB    UUILVURCO

UCILTVDUU    EULLIUQCU    EPLITVCBR    UUILVVCBR    KCILIVBUU

UQLIIURCP    RGIIIUCDU    ECIIIUDQO    EGLITUCRQ    UULLTVCRU

UULIIWCRU    KULITUUCB    ECILVVUUR    EQLITUPCU    ECILTUDUU

BCIIYURUU    BPIIVUUCO    UDILTUCRU    FPLLIUBCU    FCIIVUUQO

ECLIVUURU    EPILIVCPB    FUIIVUDCU    FCIIIVPDO    RUILTVCRU

EULIVVCUU    FBIIVVCRU    EUILTVCBB    FBIIVVCBU    FUILVUPCR

EUILTVGCR    UBIIYVCUB    FBLITUBCO    FDIITUBCO    FBLLIUCRR

UDILIVUCB    RPIITVGCU    FBILVUCUQ    EGLITUCUO    UBIITWCUB

ERLIIUCDU    FCIIVVUQR    KUILIUGCU    UBIIYUCGU    BRIIVUUCB

RPIITVQCU    UGIIVUGCU    URILVUCRU    BULLIUCBQ    FULIVUCBR

ECIIVVUUP    UCIIVUDBU    BBIITVPCB    UPIIVVCQQ    UPILTVCUP

EPTTYVCUR    ERLIVUUCQ    RCIIIWBUU    UBIIIWQCQ    ECIIIWUBB

ECILTVPPR    UPIIYUBCO    RBIIVUUCO    UBLIVVCUQ    ERLITVCBR

ECILTUUDR    RUIIYUCRU    FBILVUUCU    RCIIYUBUR    BCIIYUPUR

UCIIIWBPO    EULIIWUCQ    KCILIUQUU    UUIIVWUCQ    FBIIVUGCQ
```

```
UPIIVVBQU    RUIIVUGUB    UDIITVPBO    URILTUDBQ    UULITWRUU

UPLLIUBQU    UQIITVBPP    FULIVVURQ    FBIIYUBUB    RGLITUUUO

ERILVUPUR    UQLITVPRU    RBIITVBGQ    NBIITURQR    EULITVBUP

BUILYUUUU    EUIIYUBQU    EULLTUURO    RUILTUQBU    UBIIVVBQR

UDIIIVDPR    ERILIUDBU    NGIITVUPU    FRIIVVRPR    UQILVUUPR

ECIIIUBCB    URIIVVBQQ    UDLIVUUUQ    FUILVUQUR    FDIIIVRRO

UUILVUPBO    UPIIVWUUU    UBLLTVPUU    FPILIUBBP    EBLITUQRU

FPIIVVBUB    UBLIVVUPU    BUIIVUPQQ    UQLLTURUU    NULIIUGBU

FUIIVVRGR    FBIITUPQP    NQIIIUQPR    EUILIUQDQ    EUILVUPUO

UQIITUDQR    EBILVVUUQ    RGIIIUQGR    UBIITWUGU    RPIITWUUU

KBIITVRBR    NBIIVUUUB    BPLIIVUGR    UBLIVVBUR    URIIVVGUO

FQIITUPGO    UBLITVUQB    UQIITUCCU    ERLIIVRGR    NRIITVBBU

UDILTVRUU    FBLLIVRUR    BGIITVQUR    RUIIVVGUQ    FUILVUUBO

UPIIYVPUB    UPIITVCCU    FUILIVBDU    EULLTUPBR    BUIIYVUUB

UULIYVUPR    EUILVUBRQ    UGIIIWBRR    FBIIIWUBO    UULIVVQUR

BBLITURRU    URLIVVURR    KULIIUGRU    KPIIVUPUU    UPIIIWBQU

EUILVUPRR    UCIIIUBCP    FULIIVRDQ    ERLITUUDU    KRIITVRBR

EBILTVRBU    RGIITVRRU    UGIIVVPUO    NPIIIVDUU    FGILVUUUU

UUILVUQBU    URILVVUBQ    UGILTVRPU    FULLIUBRO    FULIIVBQO

KUIIVUBUO    NBILIUPRU    NUIITVRGU    EUILIVPGB    RBILTUBUO
```

EUIIVPUQ UUIITVPQU FRLITUUUQ EUIIVUBUQ FUIIIUBQO

FBLIIUUUO FPIIIVGUU UCIIIUUCU UPLIIVBUU UUIITVQUB

RUIIIUUDR EUILIVUBU UPLITVUUR BUIITUBRU UDIIIUUGU

FUILTUUPU UULLIVUUQ UUIIVVURU URIITVGUU FUILTVUUU

BULIIUUBR NUIIIVPUR UPILIVUPU UBILIUBBQ EBIIIUBUP

UBLIIUUUP FULIIVRUQ EPIIIVQUU UUIITUDBU KRIIIVUUU

FQIIIURBR UBLITUUPU FUILTUPUQ KUIITVUUU UBIITUUDR

UPIIVUBUR EUIIVUUPQ FUILIVPUR UPIIIVPQU ERILIUBUU

ERILIUUBU URIITUCUU UCIITUUBU UUIITUPCU UCILIUUUR

UCIIIUBPR URIIIVUCU UCIITUUUB UUIIIVCPU UPIITUUPO

FUILIUBUU RQIIIUUUU BBIIIUUPU FUIITUGUU UULIIVBUQ

FRIIIUGUU UBIIIVGUU FPIITUBUQ URILIVUUU FBIIIUUQQ

UULITUUBR FBIITUUUO UUIIIVBPB UBLITUUUU FGIIIUUBU

FULIIUURR UBIIIVBBU EUIITVUPU NUIITUUUR ERIIIUQUU

FUIIIUDUU EUIIIURRO UPIIIVUQU UBIIIVUQU UQILIUUUU

EUIIIVPBU FPIIIUUGR UBIIIUCUU UUIIIVUCR UUIIIUQPR

UQTTIUUUQ UUIITUBPU URIIIVUPQ EBIIIURUU EUIITUPUU

UUIITUPUO UULIIUUBQ EUIITURUR FRIIIUUBR UUIIIVRRQ

UUIIIVBRQ UPIIIUPBU EUIIIVUBU UUILIUBUU FUIIIVUUU

FUIIIVUUR UBIIIUUBU UUIIIUBBU UUIITUUUU UUIIIUUUU

Climates: Architecture and
the Planetary Imaginary

Columbia Books on
Architecture and the City
An imprint of the Graduate
School of Architecture,
Planning and Preservation
Columbia University
407 Avery Hall
1172 Amsterdam Avenue
New York, NY 10027
arch.columbia.edu/books

The Avery Review
A digital periodical of critical
essays on architecture
www.averyreview.com

Lars Müller Publishers
Zurich, Switzerland
www.lars-mueller-publishers.com

© 2016 Lars Müller Publishers
and the Trustees of Columbia
University in the City of New York
Essays © the authors
All rights reserved

ISBN 978-3-03778-494-5

Printed in Germany

This book has been produced
through the Office of the Dean,
Amale Andraos and the Office of
Publications at Columbia University
GSAPP.

Director of Publications
James Graham

Managing Editor
Caitlin Blanchfield

Associate Editor
Alissa Anderson

Copyeditor
Ellen Tarlin

Designed by
Neil Donnelly
Sean Yendrys

Printing and Binding
Kösel, Altusried-Krugzell, Germany

Paper
Munken Polar
170 g/m^2, 150 g/m^2, 130 g/m^2,
120 g/m^2, 100 g/m^2, 80 g/m^2

Avery Review website and
identity designed by Eric Hu,
Nothing in Common

Library of Congress Cataloging-
in-Publication Data
Title: Climates : architecture and
the planetary imaginary / edited
by James Graham with Caitlin
Blanchfield, Alissa Anderson, Jordan
Carver, and Jacob Moore.
Description: New York : Columbia
Books on Architecture and the City,
2016. |
    Includes bibliographical references
and index.
Identifiers: LCCN 2016015773 |
ISBN 9783037784945 (alk. paper)
Subjects: LCSH: Architecture and
climate.
Classification: LCC NA2541 .C544
2016 | DDC 720/.47—dc23
LC record available at https://lccn.
loc.gov/2016015773